34th Winter Workshop on Nuclear Dynamics (WWND 2018)

Journal of Physics: Conference Series Volume 1070

Deshaies, Guadeloupe
25 – 31 March 2018

ISBN: 978-1-5108-7199-1
ISSN: 1742-6588

Printed from e-media with permission by:

Curran Associates, Inc.
57 Morehouse Lane
Red Hook, NY 12571

Some format issues inherent in the e-media version may also appear in this print version.

Printed by Curran Associates, Inc. (218)

For permission requests, please contact the Institute of Physics at the address below.

Institute of Physics
Dirac House, Temple Back
Bristol BS1 6BE UK

Phone: 44 1 17 929 7481
Fax: 44 1 17 920 0979

techtracking@iop.org

Additional copies of this publication are available from:

Curran Associates, Inc.
57 Morehouse Lane
Red Hook, NY 12571 USA
Phone: 845-758-0400
Fax: 845-758-2633
Email: curran@proceedings.com
Web: www.proceedings.com

Table of contents

Volume 1070

2018

34th Winter Workshop on Nuclear Dynamics
25–31 March 2018, Deshaies, Guadeloupe

Accepted papers received: 24 July 2018
Published online: 18 September 2018

Preface

Papers

IOP Conf. Series: Journal of Physics: Conf. Series **1070** (2018) 011001 doi:10.1088/1742-6596/1070/1/011001

PREFACE

These are the proceedings of the 34th edition of the Winter Workshop on Nuclear Dynamics, which was held on the Caribbean island of Guadeloupe. As in previous years the unique character of this conference series has allowed us to bring scientists from various fields of nuclear physics together to discuss their scientific achievements.

At the high-energy frontier very exciting results not only from the heavy-ion collisions but also smaller systems at the LHC were shown. At the lower energies the RHIC beam energy scan is underway to span the gap between the SPS and RHIC and search for critical phenomena in the nuclear matter phase diagram. This program is complemented by spin and small systems measurements at the highest RHIC energies. The future programs of sPHENIX and the planned electron-ion collider (EIC) were represented as well.

Theory progress has been made across all these different energy and system size regimes, and the new data, in particular from the LHC and RHIC, are motivating more detailed modeling and a deeper understanding of the underlying physics.

These proceedings of the 34th Winter Workshop on Nuclear Dynamics again provide a snapshot of the status of the field. The articles, many of which are written by some of the most promising young scientists in the field, are conveying the present excitement that permeates all subfields of nuclear science.

Joerg Aichelin (SUBATECH, Universite de Nantes, France)
Rene Bellwied (University of Houston, USA)
Elena Bratkovskaya (GSI, Darmstadt, Germany)
Anthony Timmins (University of Houston, USA)

IOP Conf. Series: Journal of Physics: Conf. Series **1070** (2018) 011002 doi:10.1088/1742-6596/1070/1/011002

Peer review statement

All papers published in this volume of *Journal of Physics: Conference Series* have been peer reviewed through processes administered by the proceedings Editors. Reviews were conducted by expert referees to the professional and scientific standards expected of a proceedings journal published by IOP Publishing.

IOP Conf. Series: Journal of Physics: Conf. Series **1070** (2018) 012001 doi:10.1088/1742-6596/1070/1/012001

Quantum entanglement in the initial and final state of relativistic heavy ion collisions

Rene Bellwied

Physics Department, University of Houston, 617 SR1 Building, Houston, TX 77204, USA

E-mail: bellwied@uh.edu

Abstract. The possibility of quantum entanglement leading to a seemingly thermal distribution of the initial partonic state that maps to the final hadronic state in the evolution of the deconfined phase generated in relativistic heavy ion collisions is discussed in the context of early thermalization and final state particle distributions obtained in experiments at RHIC and the LHC.

1. Introduction

The creation of the deconfined phase of QCD, the Quark-Gluon Plasma (QGP), in heavy-ion collision experiments at RHIC and the LHC has generated increasing theoretical activity, aimed at understanding the quantum properties of matter under extreme conditions. It is believed that the deconfined system initially is in a gluon condensed state that evolves to a quark-gluon phase before transitioning to a confined hadronic phase. At high temperatures and low chemical potential, μ_B, this transition is an analytic crossover [1], which is followed by a hadronic phase that is bracketed by the chemical freeze-out on the high temperature side and the kinetic freeze-out on the low temperature side, assuming thermal equilibrium has been reached at the time of hadronization. The notion of common freeze-out surfaces for all produced particles requires the system to be in, at least, local equilibrium at the onset of the hadronic phase, but furthermore the commonly accepted hydrodynamical description of the preceding partonic phase requires the quantum system to be thermalized at very early times (< 1 fm/c). Since such early thermalization is likely impossible to reach through interactions between the parton constituents, several interesting ideas have been proposed, ranging from plasma instabilities leading to thermalization to Hawking-Unruh radiation giving the impression of a thermalized system [2, 3]. Here we investigate the idea that seemingly local thermalization in the initial state could be attributed to quantum entanglement and we pose the question whether the entanglement could survive the subsequent system evolution and ultimately lead to the final state particle multiplicity distributions, in particular for composite light nuclei and multi-quark states, which were measured at RHIC and LHC.

The underlying theory is based on recent measurements in an isolated quantum many body system made of ultracold atoms [4]. Here it was found that a local disturbance can lead to apparent thermalization through quantum entanglement for an isolated subset of the system. A series of recent theory papers aimed at applying the same theory to the emission spectrum obtained in, either deep inleastic scattering or from an expanding quantum string [6, 5], in order to explain the apparent thermal emission obtained in elementary relativistic e^+e^- and

IOP Conf. Series: Journal of Physics: Conf. Series **1070** (2018) 012001 doi:10.1088/1742-6596/1070/1/012001

pp collisions [7]. The goal of this manuscript is to use a similar argument in order to obtain a microscopic understanding of the high yields of composite objects, such as light or hypernuclei, in heavy ion collisions at RHIC and LHC. The novel idea here is that inital quantum entanglement in the QCD system does not only lead to thermal particle spectra, but it also defines the partonic composition and the abundance of each final state hadron, up to the most composite quark structure. The particles in question, namely light or hyper-nuclei, feature binding energies in the hundreds of keV range, but their yield is in agreement with calculations based on a thermally equilibrated system of a temperature around the QGP transition temperature, i.e. around 150-160 MeV [8]. These original results at RHIC energies were recently confirmed through detailed measurements with ALICE at the LHC [9]. The production mechanism is not clear since purely hadronic coalescence models, which only require a wave function overlap between baryons near the kinetic freeze-out, i.e. at much lower temperatures when the hadronic interactions cease, yield similar cross sections for light nuclei than the thermal models. In a thermal many body system near the QCD transition it is rather inconceivable that this loosely bound objects would survive the surrounding heat bath. On a macroscopic level it was found, though, that the thermodynamic entropy per baryon in a fireball, is fixed when the system is in its most compressed state [10, 11], although no microscopic explanation for this phenomenon exists. Here we try to argue that the thermodynamic entropy of the final state can be related to the entanglement entropy in the initial state and thus quantum entanglement drives the final state particle multiplicities up to the heaviest measured particles, namely the light nuclei at RHIC and the LHC, even in heavy ion collisions.

2. The initial state

The idea of applying quantum entanglement to the evolution of a parton collision inside a proton-proton system seems natural since any parton in the color neutral proton should be entangled transversely with the other valence quarks through a density matrix that includes mixed states. In particular, the sum of all hadronic final states have to contain mixed states since the entanglement of the valence quarks leads to several components of the proton's wave function [6]. This type of quantum entanglement can be viewed as transverse entanglement and should be complemented with a longitudinal component that builds up when determining the particle production from a dynamically evolving QCD string produced in the collision. For example in PYTHIA, the different regions in a string are entangled in rapidity space and any given region is described by a mixed state reduced density matrix [5]. Overall the combined longitudinal and transverse quantum entanglement makes the entanglement entropy volume dependent and thus an extensive quantity. The coherent state vacuum at the onset of the collision can therefore be described by entangled pairs of quasi-particles with opposite wave numbers. Since at these times in the evolution the system is strongly coupled we can apply a conformal field theory, which yields a relationship between entropy and time-dependent temperature that was first deduced by Calabrese and Cardy [12], The following equations should be valid locally, where the local extent is determined by the volume over which the entanglement holds, i.e.

$$L = \tau \Delta \eta \tag{1}$$

in this case the time-dependent temperature becomes

$$T = \frac{1}{2\pi\tau} \tag{2}$$

where the entropy is defined as

$$S(\tau, \Delta\eta) = \frac{c}{3} ln(\frac{2\tau}{\epsilon} sinh(\Delta\eta/2)) + const. \tag{3}$$

see also [5]. As can be seen from the equations above, the temperature falls when the time τ, and thus the entanglement volume, increases.

3. Dynamical evolution from the initial state to the phase transition

The question is whether this type of quantum entanglement could determine the thermal-like yields and spectra of the final state hadrons. In the calculations the quantum entanglement is spatially motivated, meaning the density matrices are given by the parton distributions which relate simply to the entanglement entropy. If the density is high, the system consists of many states all with equal probability and thus maximally entangled. This is easiest visualized for low-x where the system can be considered a pure maximally dense gluon state. The picture becomes more complicated if we allow quarks to contribute to the number of states, and thus one first propagates a pure glue system through the partonic phase of the QGP. This can be done through a quenched lattice QCD calculation. In the end the quark degrees of freedom need to contribute to the number of entangled states, also in order to push the volume of the entangled system and the hadronization temperature to realistic values. For a maximally entangled state the final volume reflects the fireball at hadronization, and thus it is no surprise that the resulting temperature is close to the Hagedorn temperature. If one assume parton-hadron duality then this system could potentially freeze-out into hadrons and the entanglement entropy would simply translate into a thermodynamic entropy obtained from the Boltzmann distribution of the final state. This step requires, though, that the entanglement during the dynamic evolution of the partonic system is not destroyed, in other words the system does not decohere through partonic interactions. Any hadronic interactions after freeze-out do not impact the initial particle yield since the emission from the quantum entangled system determines the confined state. If this principle goes as far as preserving a multi-quark bag which forms composite objects such as light nuclei at freeze-out, then quantum entanglement can serve as an explanation for the seemingly thermal yields of light nuclei that have binding energies that are several orders of magnitude lower than the thermalization temperature at freeze-out.

4. Experimental evidence in the final state

Historically parton-hadron duality was used to explain the perturbative part of the spectrum in elementary collisions [13], but the apparent similarity between the gluon densities required to explain the nuclear suppression factors measured in relativistic heavy-ion collisions at RHIC and the number of charged hadrons measured in the detector can also lead to the application of the model in larger multi-body systems [14]. In this picture the gluon saturation scale will determine the final state hadron multiplicities. and lead to a geometrical scaling as a function of an energy dependent saturation scale [15, 16]. Although this scaling intially seemed to be broken in heavy-ion collisions at LHC energies, new studies point out that, when taking into account the effect of DGLAP evolution on the scale, an agreement with experimental data can be reached which preserves the idea of LHPD [17].

Any LPHD scenario neglects all collective effects in AA collisions, and therefore should be viewed as the opposite of a model describing a system that expands in the transverse and longitudinal direction according to ideal hydrodynamical equations of motion. Interestingly, though, in the ideal hydrodynamical case and in LHPD the entropy, and thus the multiplicity of the relevant degrees of freedom stays constant during the evolution. This leads in both cases to the fact that the final particle multiplicity is effectively equal to the initial gluonic multiplicity. Although appealing as a concept, for any hydrodynamical model there is still no quantitative theoretical understanding of the thermalization process. Here the quantum entanglement argument, which leads to an entanglement entropy that maps onto the thermodynamic entropy, offers an intuitive picture.

4.1. Violation of KNO scaling in elementary collisions

Experimental evidence for the potential equivalence between entanglement entropy of the initial system and the von Neumann (thermodynamic) entropy of the final system can be verified through rather well established experimental effects in particle distributions. Without explicitly determining the gluon density one can measure a breakdown of the scaling relation between the final hadron multiplicity and the number of scatterings for each parton (KNO scaling) [18]. If KNO scaling holds, then the scaling of particle production would be independent of the process. Recent LHC results have confirmed KNO scaling even at the highest energies as long as the pseudo-rapidity window is small. A larger coverage, though, leads to the breakdown of the process independent scaling, which is well documented by plotting the energy dependence of the higher cumulants of the multiplicity distribution [19]. There are large multiplicity, and thus cumulant, fluctuations in the soft sector which lead to non-Poissonian cumulants. Any entanglement in the soft gluon sector leads thus to a violation of the KNO scaling, and the predictions based on gluon saturation and enhanced soft multi-parton scattering are in good agreement with the measured cumulants.

4.2. Thermal emission of composite objects at the QCD phase transition

The mapping between the inital gluon density and the final hadron density is also realized in heavy ion systems in the context of the aforementioned semi-classical gluon saturation model to describe the initial state, i.e. the so-called Color Glass Condensate (CGC) [15]. It is important to realize that the scaling violations determined in elementary pp collisions can also be applied to understand parton-hadron duality in the heavy ion system, although also in the heavy ion system the duality needs to be extended from the perturbative regime to the highly non-perturbative part of the spectrum, i.e. soft gluon contributions will dominate the final particle spectrum. A more detailed analysis of the higher moments of the multiplicity distribution as a function of system size and collision energy can certainly shed light on the link between saturation level and final charged particle multiplicity.

The advantage that a heavy ion particle distribution has to offer is that the abundance of rare particle species enables the application of a thermal fit over many more orders of magnitude.The rather tenuous link to thermal behavior in elementary collisions [20] thus gets replaced with solid fits based on statistical equilibrium models [21]. If the yield of rare light nuclei and hypernuclei indeed continues to follow the trend of reflecting a particular freeze-out temperature, which is identical to the chemical freeze-out of all basic particle species, as was shown recently by the ALICE collaboration [9], then the measured temperature could be connected to the entanglement entropy as shown in Eqs.1-3.The 'temperature' then reflects a degree of quantum entanglement, which is described by the entanglement entropy specific for a particular time and volume near the hadronization transition. An alternative explicit link between the entanglement 'temperature' and the thermal temperature measured from the spectra of particles was derived in [22], where the gluonic system is frozen in early in the evolution through a rapid quench that generates a highly excited multi-particle state.The thermal spectrum originates from the event horizon formed by the acceleration of the color field. Any mapping between the calculated temperature and the temperature obtained from the final hadron spectra then relies on the clustering of color sources during hadronization. This theory tries to link the kinetic freeze-out temperatures to the entanglement 'temperature' whereas here we link the particle multiplicities at the chemical freeze-out temperature, i.e. the thermodynamic entropy per baryon, to the entanglement entropy.

The evolving picture is that the dynamical evolution of the initial saturation state through the strong coupling regime, during the QCD crossover, is likely not changing the entropy of the system very much, in which case a one to one correspondence between the initial entanglement entropy and the final thermodynamic entropy could be postulated. This gives a microscopic explanation to the early findings by Siemens and Kapusta [10] that showed that, at lower energies near the nuclear liquid-gas phase transition, the entropy per baryon of the final system is apparently fixed in the compressed and hot state of the produced fireball, which led later on to the 'snowball in hell' interpretation by Braun-Munzinger and Stachel [23].

Unfortunately the interpretation of the measured light nuclei abundances at the LHC is not unique. The thermal model emission yields are very close to coalescence predictions. Coalescence is a process that solely requires sufficient wave function overlap among hadrons in the final state, and thus no equilibrated system, in order to yield composite objects, whereas in the thermal model the production yield depends on the final chemical freeze-out parameters (T and μ_b). The entanglement scenario would be closer to the thermal production in that it yields a common parameter, T, for all particle species, although it gives this parameter a different meaning. Here the apparent temperature from a thermal equilibrium fit will only reflect the initial maximum level of entanglement, which would stay constant as a function of collision energy as soon as the saturation scale is reached. Therefore it is not surprising that the chemical freeze-out temperature follows the Hagedorn temperature and becomes energy independent at RHIC and LHC energies. The formation of light nuclei will then reflect the longitudinal and transverse entanglement in the initial QCD phase but the yields can still be predicted by a model assuming thermalization at a fixed temperature.In this context it is experimentally relevant that recent calculations, based on a common emission temperature, show that multi-strange multi-quark objects or hypernuclei are within range of the high statistics runs at the LHC scheduled from 2021 on [24].

5. Outlook

The model of quantum entanglement of the initial state produced in a relativistic heavy ion collision might give rise to complex particle production in the final hadronic state. The entanglement entropy dominated by the soft gluon sector in a gluon saturated system could be matched one to one to a thermodynamic entropy of the final state hadron system. Since the entropy per baryon would then be driven by the entropy per gluon any complex baryonic structure could be traced back to an initial QCD string that carries its own quantum entanglement properties based on the mixed state density matrix of the initial state.

One of the more far-reaching consequences of such an approach to particle formation in the QCD crossover is the fact that nuclei production up to a certain complexity of the final state is determined by the initial state saturation scale. At which point the initial state entanglement is exhausted needs to be calculated in a more detailed approach.

6. Acknowledgments

This work is supported by the U.S. Department of Energy under Contract No. DEFG02-07ER41521

References

[1] Y. Aoki, G. Endrodi, Z. Fodor, S. D. Katz and K. K. Szabo, Nature **443**, 675 (2006)
[2] M. Strickland, J. Phys. G **34**, S429 (2007)
[3] P. Castorina, D. Kharzeev and H. Satz, Eur. Phys. J. C **52**, 187 (2007)
[4] A.M. Kaufman,M.E. Tai, A.Lukin, M. Rispoli, R.Schittke, P.M. Presiss, M. Greiner Science 353, 794 (2016)
[5] J. Berges, S. Floerchinger and R. Venugopalan, JHEP **1804**, 145 (2018)
[6] D. E. Kharzeev and E. M. Levin, Phys. Rev. D **95**, no. 11, 114008 (2017)
[7] O. K. Baker and D. E. Kharzeev, arXiv:1712.04558 [hep-ph].
[8] A. Andronic, P. Braun-Munzinger, J. Stachel and H. Stocker, Phys. Lett. B **697**, 203 (2011)
[9] J. Adam *et al.* [ALICE Collaboration], Phys. Rev. C **93**, no. 2, 024917 (2016)
[10] P. J. Siemens and J. I. Kapusta, Phys. Rev. Lett. **43**, 1486 (1979).

IOP Conf. Series: Journal of Physics: Conf. Series **1070** (2018) 012001 doi:10.1088/1742-6596/1070/1/012001

[11] L. P. Csernai and J. I. Kapusta, Phys. Rept. **131**, 223 (1986).
[12] P. Calabrese and J. L. Cardy, J. Stat. Mech. **0406**, P06002 (2004)
[13] M. A. Shifman, hep-ph/0009131.
[14] Y. L. Dokshitzer, V. A. Khoze and S. I. Troian, J. Phys. G **17**, 1585 (1991).
[15] E. Iancu and R. Venugopalan, In *Hwa, R.C. (ed.) et al.: Quark gluon plasma* 249-3363 [hep-ph/0303204]
[16] D. Kharzeev, E. Levin and M. Nardi, Phys. Rev. C **71**, 054903 (2005)
[17] T. Lappi, Eur. Phys. J. C **71**, 1699 (2011)
[18] Z. Koba, H. B. Nielsen and P. Olesen, Nucl. Phys. B **40**, 317 (1972)
[19] V. Khachatryan *et al.* [CMS Collaboration], JHEP **1101**, 079 (2011)
[20] F. Becattini, P. Castorina, A. Milov and H. Satz, Eur. Phys. J. C **66**, 377 (2010)
[21] J. Stachel, A. Andronic, P. Braun-Munzinger and K. Redlich, J. Phys. Conf. Ser. **509**, 012019 (2014)
[22] X. Feal, C. Pajares and R. A. Vazquez, arXiv:1805.12444 [hep-ph].
[23] P. Braun-Munzinger and J. Stachel, J. Phys. G **21**, L17 (1995)
[24] A. Dainese *et al.*, Frascati Phys. Ser. **62** (2016) [arXiv:1602.04120 [nucl-ex]].

IOP Conf. Series: Journal of Physics: Conf. Series **1070** (2018) 012002 doi:10.1088/1742-6596/1070/1/012002

Lattice thermodynamics at finite chemical potential from analytical Continuation

Jana N. Guenther[1,2], Szabolcs Borsanyi[1], Zoltan Fodor[1,3,4], Sandor K. Katz[4], K. K. Szabó[1,3], Attila Pasztor[1],Israel Portillo[5] and Claudia Ratti[5]

1 University of Wuppertal, Department of Physics, Wuppertal D-42097, Germany
2 University of Regensburg,Department of Physics, Regensburg D-93053, Germany
3 Jülich Supercomputing Centre, Jülich D-52425, Germany
4 Eötvös University, Budapest 1117, Hungary
5 University of Houston, Department of Physics, Houston, TX 77204, USA

E-mail: `Jana.Guenther@t-online.de`

Abstract. An efficient way to study the QCD phase diagram at small finite density is to extrapolate thermodynamical observables from imaginary chemical potential. We present results for fluctuations of baryon number to order $(\mu_B/T)^6$. The results at real chemical potentials are obtained through analytical continuation of simulations at imaginary chemical potentials. We also calculate higher order fluctuations of strangeness and electric charge to obtain the Taylor coefficients of the baryon number fluctuation with vanishing strangeness. We compare to the STAR proton fluctuation data to characterize the chemical freeze-out in view of the lattice results.

1. Introduction

When investigating Quantum Chromodynamics (QCD) an important but challenging goal is the study of the phase diagram. At there chemical potential lattice QCD predicts a smooth crossover between hadrons and the quark gluon plasma [1, 2, 3, 4, 5], taking place in the temperature range $T \simeq 145 - 165$ MeV. Due to the sign problem lattice QCD is unable to study the region with final chemical potential.

With the advent of the second Beam Energy Scan (BESII) at the Relativistic Heavy Ion Collider (RHIC), scheduled for 2019-2020, there is a renewed interest in the heavy ion community towards the phases of QCD at moderate-to-large densities. A rich theoretical effort is being developed in support of the experimental program; several observables are being calculated, in order to constrain the existence and location of the QCD critical point and to observe it experimentally.

Fluctuations of conserved charges (electric charge Q, baryon number B and strangeness S) are important observables for the finite-density investigations. One possible way to extend lattice results to finite density is to perform Taylor expansions of the thermodynamic observables around chemical potential $\mu_B = 0$ [6, 7, 8, 9, 10]: fluctuations of conserved charges are directly related to the Taylor expansion coefficients of such observables. They allow for a comparison between theoretical and experimental results to extract the chemical freeze-out temperature T_f and chemical potential μ_{Bf} as functions of the collision energy [11, 12, 13, 14]. The higher order

IOP Conf. Series: Journal of Physics: Conf. Series **1070** (2018) 012002 doi:10.1088/1742-6596/1070/1/012002

fluctuations are also an important signature for the critical endpoint, as they give access to the correlation length [8, 15, 16].

We present results for χ_2^B to χ_6^Bin the temperature range 140 MeV $\leq T \leq$ 220 MeV. Several diagonal and non-diagonal fluctuations of conserved charges up to sixth-order are available in our resent paper [17].

We simulate the lower-order fluctuations at imaginary chemical potential and extract the higher order fluctuations as derivatives of the lower order ones at $\mu_B = 0$. This method has been successfully used in the past and proved to lead to a more precise determination of the higher order fluctuations, compared to their direct calculation [18, 19, 20]. The latter requires the evaluation of several terms and is affected by a signal-to-noise ratio which is decreasing as a power law of the spatial volume V, with an exponent that grows with the order of the susceptibility. Reliable results would therefore require extremely large volumes.

To connect to experimental results, we calculate the ratio of the cumulants of the net-baryon number distribution as functions of temperature and chemical potential by means of their Taylor expansion in powers of μ_B/T. This is possible by combining different diagonal and non-diagonal fluctuations to obtain a result at the strangeness neutral point and with $\langle n_Q \rangle = 0.4 \langle n_B \rangle$.

2. Lattice details

We use a tree-level Symanzik improved gauge action, with four times stout smeared staggered fermions. We simulate $2+1+1$ dynamical quarks where the light flavors are tuned in a way to reproduce the physical pion and kaon mass and we set $\frac{m_c}{m_s} = 11.85$ [21]. For the zero temperature runs we use large volumes which full fill $Lm_\pi > 4$. The scale is determined via f_π. More Details can be found in [22].

The maximal useful value of μ_B is $\mu_B = i\pi T$ because of the Roberge-Weiss transition [23]. We simulate at eight different values of μ_B given as: $\mu_B^{(j)} = iT\frac{j\pi}{8}$ for $j \in \{0,1,2,3,4,5,6,7\}$. The analysis is done purely on a $48^3 \times 12$ lattice without continuum extrapolation. All simulations are done at $\mu_Q = \mu_S = 0$.

3. Analysis for χ^B

We start with the analysis for $\chi_2^B(T)$, $\chi_4^B(T)$ and $\chi_6^B(T)$. Our goal is to calculate these quantities at zero chemical potential, using the imaginary chemical potential data up to $\chi_B^4(T, \hat{\mu}_B)$. In this work we extract these derivatives at a fixed temperature. Therefor the results for different temperatures are completely independent.

We consider the ensembles at a fixed temperature T. For each value of imaginary $\mu_B \neq 0$ we determine χ_1^B, χ_2^B, χ_3^B and χ_4^B from simulation, while for $\mu_B = 0$ only χ_2^B and χ_4^B can be used, since χ_1^B and χ_3^B are odd functions of μ_B and therefore equal to zero.

We make the ansatz for the pressure:

$$\chi_0^B(\hat{\mu}_B) = c_0 + c_2\hat{\mu}_B^2 + c_4\hat{\mu}_B^4 + c_6\hat{\mu}_B^6 + c_8\hat{\mu}_B^8 + c_{10}\hat{\mu}_B^{10}, \tag{1}$$

where the Taylor expansion coefficients c_n are related to the baryon number fluctuations χ_n^B by: $n!c_n = \chi_n^B$. Our data do not allow for an independent determination of c_8 and c_{10}. Nevertheless, in order to have some control over these terms we make the assumption

$$|\chi_8^B| \lesssim \chi_4^B \tag{2}$$

$$|\chi_{10}^B| \lesssim \chi_4^B \tag{3}$$

or in terms of the c_n coefficients

$$8!c_8 \lesssim 4!c_4$$
$$10!c_{10} \lesssim 4!c_4.$$

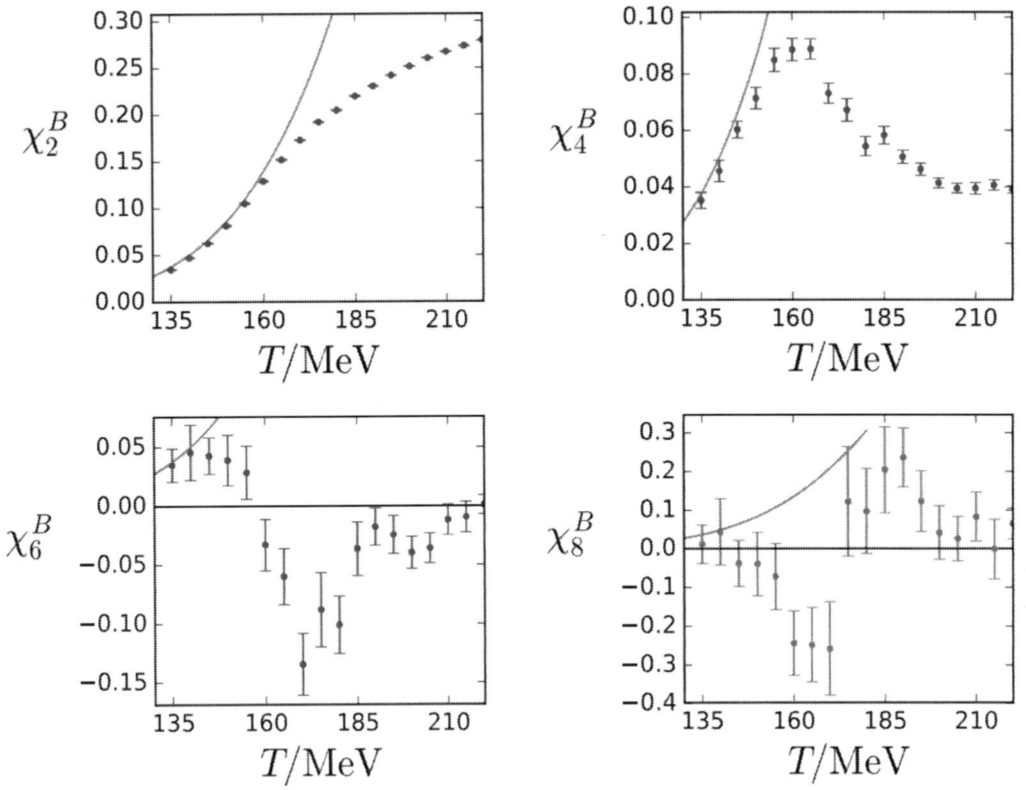

Figure 1. Results for χ_2^B, χ_4^B, χ_6^B and an estimate for χ_8^B as functions of the temperature, obtained from the single-temperature analysis. We plot χ_8^B in green to point out that its determination is guided by a prior, which is linked to the χ_4^B observable by Eq. (4). The red curve in each panel corresponds to the Hadron Resonance Gas (HRG) model result.

We can then rewrite our ansatz as

$$\chi_0^B(\hat{\mu}_B) = c_0 + c_2\hat{\mu}_B^2 + c_4\hat{\mu}_B^4 + c_6\hat{\mu}_B^6 + \frac{4!}{8!}c_4\epsilon_1\hat{\mu}_B^8 + \frac{4!}{10!}c_4\epsilon_2\hat{\mu}_B^{10}. \tag{4}$$

where ϵ_1 and ϵ_2 are drawn randomly from a normal distribution with mean -1.25 and variance 2.75. We use the same distribution for all temperatures. For a detailed discussion of the choice of parameters see [17].

For this ansatz we calculate the following derivatives, which are the actually simulated lattice observables:

$$\chi_1^B(\hat{\mu}_B) = 2c_2\hat{\mu}_B + 4c_4\hat{\mu}_B^3 + 6c_6\hat{\mu}_B^5 + \frac{4!}{7!}c_4\epsilon_1\hat{\mu}_B^7 + \frac{4!}{9!}c_4\epsilon_2\hat{\mu}_B^9 \tag{5}$$

$$\chi_2^B(\hat{\mu}_B) = 2c_2 + 12c_4\hat{\mu}_B^2 + 30c_6\hat{\mu}_B^4 + \frac{4!}{6!}c_4\epsilon_1\hat{\mu}_B^6 + \frac{4!}{8!}c_4\epsilon_2\hat{\mu}_B^8 \tag{6}$$

$$\chi_3^B(\hat{\mu}_B) = 24c_4\hat{\mu}_B + 120c_6\hat{\mu}_B^3 + \frac{4!}{5!}c_4\epsilon_1\hat{\mu}_B^5 + \frac{4!}{7!}c_4\epsilon_2\hat{\mu}_B^7 \tag{7}$$

$$\chi_4^B(\hat{\mu}_B) = 24c_4 + 360c_6\hat{\mu}_B^2 + c_4\epsilon_1\hat{\mu}_B^4 + \frac{4!}{6!}c_4\epsilon_2\hat{\mu}_B^6. \tag{8}$$

We perform a correlated fit for the four measured observables, thus obtaining the values of c_2, c_4 and c_6 for each temperature, and the corresponding χ_2^B, χ_4^B and χ_6^B. We repeat the fit for 1000 random draws for ϵ_1 and ϵ_2. The result is weighted using the Akaike Information Criterion [24]. Through these weights we get a posterior distribution from the prior distribution. Our final estimate for χ_8^B represents this posterior distribution. We do not show the posterior for χ_{10}^B, which is mostly noise.

These results are shown in Fig. 1, together with an estimate of χ_8^B, related to χ_4^B by Eq. (4).

4. Cumulants

For a comparison with heavy ion collision experiments the cumulants of the net baryon distribution are a useful tool. The first four cumulants are the mean M_B, the variance σ_B^2, the skewness S_B and the kurtosis κ_B. By forming appropriate ratios, we can cancel out explicit volume factors. However the measured distributions themselves may still depend on the volume, which one should take into account, when comparing to experiments.

Heavy ion collisions with lead or gold take place with at $\mu_B > 0$, $\langle n_S \rangle = 0$ and $\langle n_Q \rangle = 0.4\langle n_B \rangle$. Since our simulations are done at $\mu_S = \mu_Q = 0$ and $\mu_B \neq 0$ we have to do some calculations to arrive at the same observables that are measured in experiments (see for example [25]). We investigate three different rations of cumulants and write each as a Taylor expansion:

$$\frac{M_B}{\sigma_B^2} = \frac{\chi_1^B(T,\hat{\mu}_B)}{\chi_2^B(T,\hat{\mu}_B)} = \hat{\mu}_B r_{12}^{B,1} + \hat{\mu}_B^3 r_{12}^{B,3} + \dots \tag{9}$$

$$\frac{S_B\sigma_B^3}{M_B} = \frac{\chi_3^B(T,\hat{\mu}_B)}{\chi_1^B(T,\hat{\mu}_B)} = r_{31}^{B,0} + \hat{\mu}_B^2 r_{31}^{B,2} + \dots \tag{10}$$

$$\kappa_B\sigma_B^2 = \frac{\chi_4^B(T,\hat{\mu}_B)}{\chi_2^B(T,\hat{\mu}_B)} = r_{42}^{B,0} + \hat{\mu}_B^2 r_{42}^{B,2} + \dots \tag{11}$$

The μ_B dependence of the $\chi_i^B(T,\hat{\mu}_B)$ can be again written in terms of the Taylor expansion:

$$\chi_{i,j,k}^{BQS}(\hat{\mu}_B) = \chi_{i,j,k}^{BQS}(0) + \hat{\mu}_B\left[\chi_{i+1,j,k}^{BQS}(0) + q_1\chi_{i,j+1,k}^{BQS}(0) + s_1\chi_{i,j,k+1}^{BQS}(0)\right] \tag{12}$$

$$+\frac{1}{2}\hat{\mu}_B^2\left[\chi_{i+2,j,k}^{BQS}(0) + s_1^2\chi_{i,j+2,k}^{BQS}(0) + q_1^2\chi_{i,j,k+2}^{BQS}(0)\right. \tag{13}$$

$$\left.+2q_1s_1\chi_{i,j+1,k+1}^{BQS}(0) + 2s_1\chi_{i+1,j+1,k}^{BQS}(0) + 2q_1\chi_{i+1,j,k+1}^{BQS}(0)\right] + \dots \tag{14}$$

$$\tag{15}$$

with

$$q_j = \frac{1}{j!}\frac{d^j\hat{\mu}_Q}{(d\hat{\mu}_B)^j}(0) \tag{16}$$

$$s_j = \frac{1}{j!}\frac{d^j\hat{\mu}_S}{(d\hat{\mu}_B)^j}(0) \tag{17}$$

We can now use the constrains $\langle n_S \rangle = 0$ and $\langle n_Q \rangle = 0.4\langle n_B \rangle$ which can be rewritten as

$$\chi_1^Q = 0.4\chi_1^B, \qquad \chi_1^S = 0 \tag{18}$$

to determine $r_{ij}^{B,k}$ coefficients form the equations 9, 10 and 11. However we now need to know not only the behaviour of the χ_i^B but also of derivatives with respect to μ_S and μ_Q. For now our

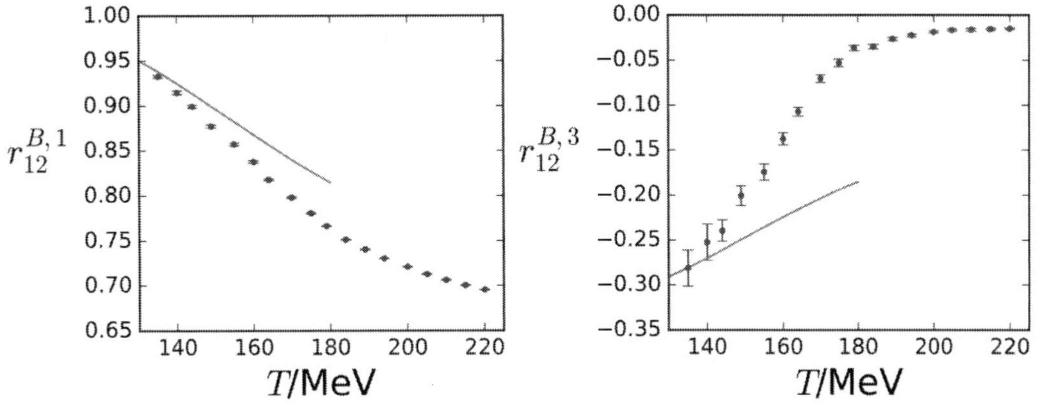

Figure 2. Taylor expansion coefficients for $\frac{M_B}{\sigma_B^2} = \frac{\chi_1^B(T,\hat{\mu}_B)}{\chi_2^B(T,\hat{\mu}_B)}$ as functions of the temperature: $r_{12}^{B,1}$ (left panel) and $r_{12}^{B,3}$ (right panel).

simulations are restricted to ensembles with finite μ_B. Therefore the μ_S and μ_Q derivatives have to be calculated directly and without the support from the fit that we used in the μ_B direction. We calculate various $\chi_{i,j,k}^{B,Q,S}$ with the appropriate values of j and k and all possible values for i so that

$$i + j + k \leq 4. \tag{19}$$

For each group of fluctuations with the same j and k we perform a fit analogous to the procedure described in section 3. This is sufficient to determine the first to $r_{ij}^{B,k}$ coefficients for all three observables. The results are shown in figure 2, 3 and 4. For higher order coefficients, higher order derivatives in μ_S and μ_Q are needed. The direct measurements have a rapidly increasing error with each derivative and very large statistics would be needed to improve our calculations in that manner. Another possibility would be add ensembles with finite μ_S and μ_Q and do a similar fit as for the μ_B direction. This approach has been used in [19].

5. Error Analysis

For a reliable comparison between experimental measurements and theoretical calculations, the error estimate is an important ingredient. Our statistical error is estimated through the jackknife method. For our systematic error there are several sources. We determine our systematic error by the histogram method described in [26], where each analysis is weighted with the Akaike information criteria. We include the influence of the number of points in the μ_B direction, by either including or ignoring the data from our highest value of μ_B. A very important source for our systematic error is the influence of the higher order contributions in μ_B. This effect was estimated by adding the higher order terms with pre-factors ϵ_1 and ϵ_2 as described in Section 3. We consider 1000 different ϵ pairs and add the different analyses to our histogram. The width of the histogram using Akaike weights corresponding to the fit quality gives the systematic errors for the fit coefficients, and from the same histogram we obtain the posterior distributions for ϵ_1. The physical quantities that are constrained only by the posterior distribution are plotted with green symbols.

These histograms are built independently for each number (j and k) of μ_S and μ_Q derivatives. When calculating the systematic for the cumulants (see section 4) we need to calculate different combinations from the available analyses. We will always take the fluctuations from the same

IOP Conf. Series: Journal of Physics: Conf. Series **1070** (2018) 012002 doi:10.1088/1742-6596/1070/1/012002

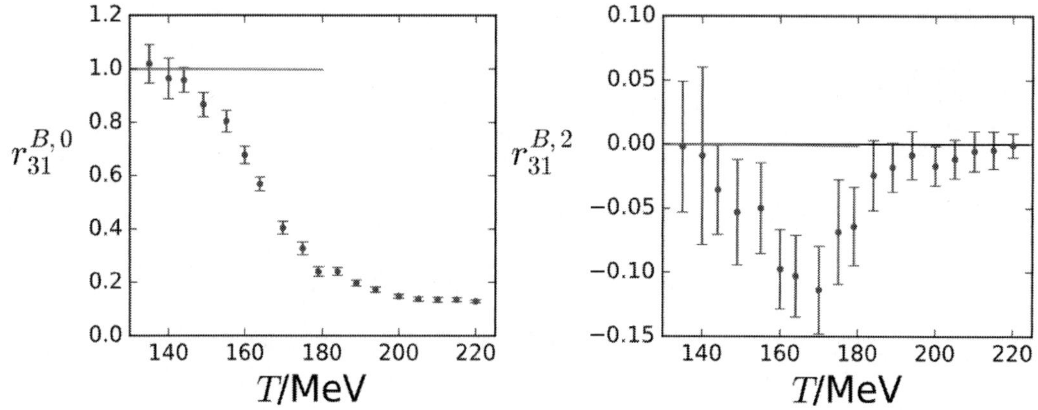

Figure 3. Taylor expansion coefficients for $\frac{S_B\sigma_B^3}{M_B} = \frac{\chi_3^B(T,\hat{\mu}_B)}{\chi_1^B(T,\hat{\mu}_B)}$ as functions of the temperature: $r_{31}^{B,0}$ (left panel) and $r_{31}^{B,2}$ (right panel).

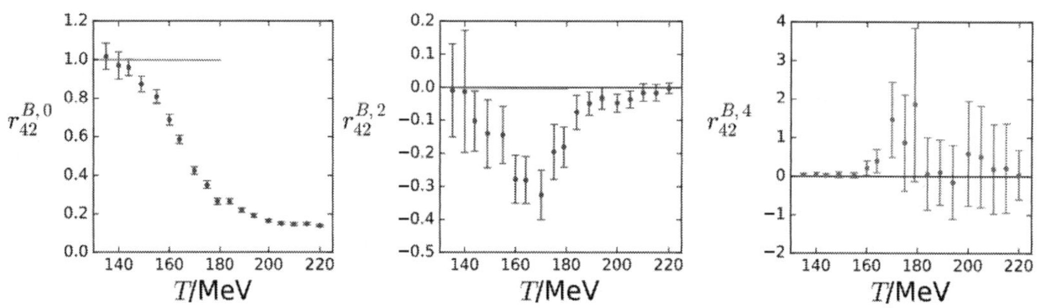

Figure 4. Taylor expansion coefficients for $\kappa_B\sigma_B^2 = \frac{\chi_4^B(T,\hat{\mu}_B)}{\chi_2^B(T,\hat{\mu}_B)}$ as functions of the temperature: $r_{42}^{B,0}$ (left panel) $r_{42}^{B,2}$ (middle panel), $r_{42}^{B,4}$ (right panel). The latter is not obtained independently, but by means of the prior ansatz (see text): for this reason, we plot it in green.

channel from the same analysis. However we have 15 channels and 2000 analyses in each resulting in 2000^{15} combinations, which would be a very large computational effort. Instead we determine 1000 different combinations via important sampling:

(i) For each channel we draw a random number r from a uniform distribution between zero and one

(ii) We add the normalized weights from our 2000 different analyses to the sum $s \geq r$ with the nth analysis

(iii) We take for this combination the nth analysis and proceed to the next channel

In this way, $\mathcal{O}(100)$ random combinations of χ_{ijk}^{BSQ} results already give convergence for each discussed quantity and its error bar. For the results in this paper we used 1000 such random combinations. This procedure assumes that between different j,k pairs the prior distribution is uncorrelated.

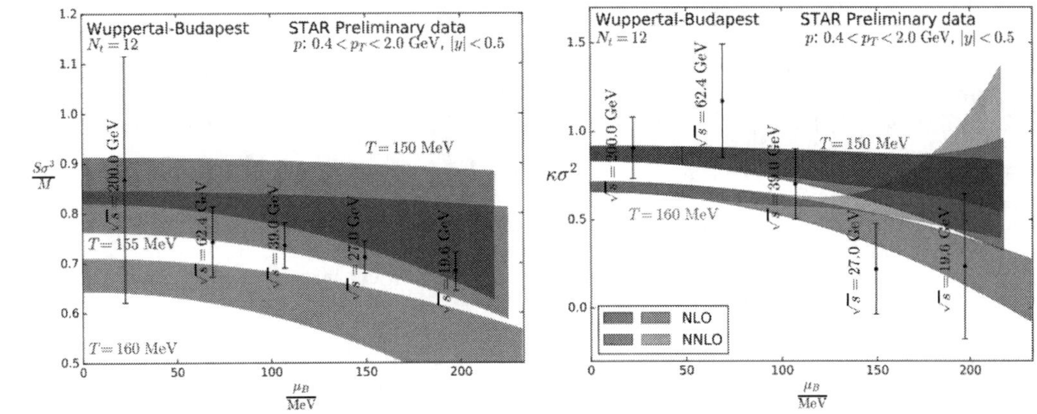

Figure 5. $S_B\sigma_B^3/M_B$ (left panel) and $\kappa_B\sigma_B^2$ (right panel) extrapolated to finite chemical potential. The left panel is extrapolated up to $\mathcal{O}(\hat{\mu}_B^2)$. In the right panel, the darker bands correspond to the extrapolation up to $\mathcal{O}(\hat{\mu}_B^2)$, whereas the lighter bands also include the $\mathcal{O}(\hat{\mu}_B^4)$ term.

6. Conclusion

We have calculated the baryon number fluctuation as well as several diagonal and non-diagonal fluctuations of electric charge, baryon number and strangeness up to sixth-order on a $48^3 \times 12$ lattice. The analysis has been performed simulating the lower order fluctuations at eight different values of zero and imaginary chemical potential μ_B, and extracting the higher order fluctuations as derivatives of the lower order ones at $\mu_B = 0$. The chemical potentials for electric charge and strangeness have both been set to zero in the simulations. From these fluctuations, we have constructed ratios of baryon number cumulants as a functions of T and μ_B, by means of a Taylor series which takes into account the experimental constraints $\langle n_S \rangle = 0$ and $\langle n_Q \rangle = 0.4\langle n_B \rangle$. These ratios qualitatively explain the behavior observed in the experimental measurements by the STAR collaboration as functions of the collision energy.

Acknowledgments

This project was funded by the DFG grant SFB/TR55. This work was supported by the Hungarian National Research, Development and Innovation Office, NKFIH grants KKP126769 and K113034. An award of computer time was provided by the INCITE program. This research used resources of the Argonne Leadership Computing Facility, which is a DOE Office of Science User Facility supported under Contract DE-AC02-06CH11357. The authors gratefully acknowledge the Gauss Centre for Supercomputing e.V. (www.gauss-centre.eu) for funding this project by providing computing time on the GCS Supercomputer JUQUEEN[27] at Jülich Supercomputing Centre (JSC) as well as on HAZELHEN at HLRS Stuttgart, Germany. This material is based upon work supported by the National Science Foundation under grants no. PHY-1654219 and OAC-1531814 and by the U.S. Department of Energy, Office of Science, Office of Nuclear Physics, within the framework of the Beam Energy Scan Theory (BEST) Topical Collaboration. C.R. also acknowledges the support from the Center of Advanced Computing and Data Systems at the University of Houston.

IOP Conf. Series: Journal of Physics: Conf. Series **1070** (2018) 012002 doi:10.1088/1742-6596/1070/1/012002

References

[1] Aoki Y, Endrodi G, Fodor Z, Katz S D and Szabo K K 2006 *Nature* **443** 675–678 (*Preprint* hep-lat/0611014)

[2] Aoki Y, Borsanyi S, Durr S, Fodor Z, Katz S D, Krieg S and Szabo K K 2009 *JHEP* **06** 088 (*Preprint* 0903.4155)

[3] Borsanyi S, Fodor Z, Hoelbling C, Katz S D, Krieg S, Ratti C and Szabo K K (Wuppertal-Budapest) 2010 *JHEP* **09** 073 (*Preprint* 1005.3508)

[4] Bhattacharya T *et al.* 2014 *Phys. Rev. Lett.* **113** 082001 (*Preprint* 1402.5175)

[5] Bazavov A *et al.* 2012 *Phys. Rev.* **D85** 054503 (*Preprint* 1111.1710)

[6] Allton C R, Ejiri S, Hands S J, Kaczmarek O, Karsch F, Laermann E, Schmidt C and Scorzato L 2002 *Phys. Rev.* **D66** 074507 (*Preprint* hep-lat/0204010)

[7] Allton C R, Doring M, Ejiri S, Hands S J, Kaczmarek O, Karsch F, Laermann E and Redlich K 2005 *Phys. Rev.* **D71** 054508 (*Preprint* hep-lat/0501030)

[8] Gavai R V and Gupta S 2008 *Phys. Rev.* **D78** 114503 (*Preprint* 0806.2233)

[9] Basak S *et al.* (MILC) 2008 *PoS* **LATTICE2008** 171 (*Preprint* 0910.0276)

[10] Kaczmarek O, Karsch F, Laermann E, Miao C, Mukherjee S, Petreczky P, Schmidt C, Soeldner W and Unger W 2011 *Phys. Rev.* **D83** 014504 (*Preprint* 1011.3130)

[11] Karsch F 2012 *Central Eur. J. Phys.* **10** 1234–1237 (*Preprint* 1202.4173)

[12] Bazavov A *et al.* 2012 *Phys. Rev. Lett.* **109** 192302 (*Preprint* 1208.1220)

[13] Borsanyi S, Fodor Z, Katz S D, Krieg S, Ratti C and Szabo K K 2013 *Phys. Rev. Lett.* **111** 062005 (*Preprint* 1305.5161)

[14] Borsanyi S, Fodor Z, Katz S D, Krieg S, Ratti C and Szabo K K 2014 *Phys. Rev. Lett.* **113** 052301 (*Preprint* 1403.4576)

[15] Stephanov M A, Rajagopal K and Shuryak E V 1999 *Phys. Rev.* **D60** 114028 (*Preprint* hep-ph/9903292)

[16] Cheng M *et al.* 2008 *Phys. Rev.* **D77** 014511 (*Preprint* 0710.0354)

[17] Borsanyi S, Fodor Z, Guenther J N, Katz S K, Szabo K K, Pasztor A, Portillo I and Ratti C 2018 (*Preprint* 1805.04445)

[18] Guenther J N, Bellwied R, Borsanyi S, Fodor Z, Katz S D, Pasztor A, Ratti C and Szab K K 2017 *Nucl. Phys.* **A967** 720–723 (*Preprint* 1607.02493)

[19] D'Elia M, Gagliardi G and Sanfilippo F 2017 *Phys. Rev.* **D95** 094503 (*Preprint* 1611.08285)

[20] Bonati C, D'Elia M, Negro F, Sanfilippo F and Zambello K 2018 (*Preprint* 1805.02960)

[21] McNeile C, Davies C T H, Follana E, Hornbostel K and Lepage G P 2010 *Phys. Rev.* **D82** 034512 (*Preprint* 1004.4285)

[22] Bellwied R, Borsanyi S, Fodor Z, Guenther J, Katz S D, Ratti C and Szabo K K 2015 *Phys. Lett.* **B751** 559–564 (*Preprint* 1507.07510)

[23] Roberge A and Weiss N 1986 *Nucl. Phys.* **B275** 734–745

[24] Akaike H 1992 *Information theory and an extension of the maximum likelihood principle* (Springer)

[25] Adamczyk L *et al.* (STAR) 2014 *Phys. Rev. Lett.* **112** 032302 (*Preprint* 1309.5681)

[26] Durr S *et al.* 2008 *Science* **322** 1224–1227 (*Preprint* 0906.3599)

[27] Jülich Supercomputing Centre 2015 *Journal of large-scale research facilities* **A1** 1

IOP Conf. Series: Journal of Physics: Conf. Series **1070** (2018) 012003 doi:10.1088/1742-6596/1070/1/012003

Freeze-out properties from net-Kaon fluctuations at RHIC

Claudia Ratti[a], Rene Bellwied[a], Jacquelyn Noronha-Hostler[b], Paolo Parotto[a], Israel Portillo Vazquez[a], Jamie M Stafford[a]

[a] *Department of Physics, University of Houston, Houston, TX, USA 77204*
[b] *Department of Physics and Astronomy, Rutgers University, Piscataway, NJ USA 08854*

E-mail: cratti@uh.edu

Abstract. We calculate the net-kaon mean-over-variance in the hadron resonance gas model and compare it to the experimental data by the STAR collaboration. We show that it is not possible to match the experimental values using the freeze-out parameters obtained from a combined fit of net-proton and net-electric charge mean-over-variance. At the highest collision energies, kaons need about 10-15 MeV higher freeze-out temperatures than the light hadrons. We also predict the variance-over-mean and skewness-times-variance for net-Λ particles at the light and strange chemical freeze-out parameters. It turns out that these Λ fluctuations are sensitive to the difference in the freeze-out temperatures observed in our analysis.

1. Introduction

Relativistic heavy ion collisions, currently taking place at the Relativistic Heavy Ion Collider (RHIC) and the Large Hadron Collider (LHC), are routinely creating a new phase of matter, the Quark-Gluon Plasma (QGP), which permeated the universe just a few microseconds after the Big Bang. First principle simulations of the theory of strong interactions (QCD) on a discretized lattice show that the transition from hadronic degrees of freedom to the QGP is an analytical crossover at low baryonic density [1, 2, 3].

After hadronization, it is possible to define two important points in the evolution of a heavy-ion collision: the chemical freeze-out, at which all inelastic collisions between the hadrons cease (the chemical composition of the system is fixed at this point), and the thermal freeze-out, at which also elastic collisions cease (the particle spectra are fixed at this point). The measured particle yields and fluctuations carry information about the chemical freeze-out. Thermal fits of particle yields and ratios allow us to extract the freeze-out temperature and chemical potential, $\{T_f, \mu_{Bf}\}$ [4, 5, 6, 7, 8]. It was pointed out that, at small baryon densities, there is a tension between the yields of light particles versus strange particles [9]: it appears that light hadrons prefer a lower T_f compared to strange hadrons. A similar effect was observed at RHIC [10].

Additional information can be gained by studying fluctuations of conserved charges: fluctuations are more sensitive than yields to the freeze-out parameters [11], and they can be calculated from first-principle lattice QCD simulations [12, 13, 14, 15, 16, 17, 18], thus providing a valuable tool to investigate this issue further (for a recent review, see e.g. [19]). The STAR collaboration recently published experimental measurements for the energy dependence

IOP Conf. Series: Journal of Physics: Conf. Series **1070** (2018) 012003 doi:10.1088/1742-6596/1070/1/012003

of the fluctuations of net-protons [20], net-charge [21], and net-kaons [22]. Experimentally, one can only measure charged particles, so that K^0's, π^0's, and neutrons are not included in these measurements. A previous study in the Hadron Resonance Gas (HRG) model with all experimental effects, such as acceptance cuts in p_T and rapidity and isospin randomization [23], found that the net-proton and net-charge fluctuations indicate a lower chemical freeze-out temperature than the one quoted in the thermal fits [24].

In our analysis [25], we calculate the net-kaon fluctuations in the HRG model, taking into account resonance decays and with the same experimental cuts in rapidity and momentum, and compare them to the recent STAR data from the Beam Energy Scan [22]. We find that, even when using the most up-to-date particle data list as an input for the model, the kaons need larger freeze-out temperatures, compared to the light hadrons. We also predict the values of the Λ fluctuations, calculated in the HRG model at the freeze-out parameters of the kaons and of the light hadrons. The results show a clear separation, which can hopefully be resolved by the forthcoming experimental results.

2. Results and conclusions

We calculate the net-kaon fluctuations using the following formula:

$$\tilde{\chi}_n^{K^\pm} = \sum_i^{N_{HRG}} (Pr_{i\to K^\pm} S_i)^n \frac{d_i}{4\pi^2} \frac{\partial^n}{\partial \mu_S^n} \left\{ \int_{-0.5}^{0.5} \mathrm{d}y \int_{0.2}^{1.6} \mathrm{d}p_T \times \right.$$
$$\left. \times \frac{p_T \sqrt{p_T^2 + m_k^2} \mathrm{Cosh}[y]}{(-1)^{B_k+1} + \exp((\mathrm{Cosh}[y]\sqrt{p_T^2+m_k^2} - (B_i\mu_b + S_i\mu_S + Q_i\mu_Q))/T)} \right\}. \quad (1)$$

Here $Pr_{i\to K^\pm} = Br_{i\to K^\pm} n_i(K^\pm)$ is the probability for a resonance i to decay into a charged kaon where $Br_{i\to K^\pm}$ is the branching for the resonance i to decay into $K^\pm$ and $n_i(K^\pm)$ is the number of times particle i appears in the channel $K^\pm$. We use the same acceptance cuts as described in [22]. Notice that, for $n \geq 2$, there are cross-terms appearing in the above equation.

Since the experimental error-bar is smaller for lower order fluctuations, and the possible sources of non-thermal fluctuations should have a negligible effect in this case, we consider only the ratio of net-kaon mean over variance χ_1^K/χ_2^K in our analysis. Since we need two quantities to independently fit T_f and μ_{Bf}, we calculate χ_1^K/χ_2^K along the isentropic trajectories from Ref. [26], obtained using Lattice QCD results for the Taylor reconstructed QCD phase diagram at finite μ_B. These isentropes were determined by starting from the chemical freeze-out points for light hadrons from Ref. [23], calculating S/N_B at those points, and imposing that the ratio is conserved on the corresponding trajectory. In this way we take into account the possibility that kaons can freeze-out at a different moment in the evolution of the system at a given collision energy, related to the light particle freeze-out point by the conservation of S/N_B. The strangeness and electric charge chemical potentials are fixed by imposing the following conditions to match the experimental situation

$$\begin{aligned} \sum_{i\in S} n_i(T,\mu_B,\mu_Q,\mu_S) &= 0 \\ \sum_{i\in Q} n_i(T,\mu_B,\mu_Q,\mu_S) &= 0.4 \sum_{i\in B} n_i(T,\mu_B,\mu_Q,\mu_S). \end{aligned} \quad (2)$$

This procedure allows us to determine both $\{T_f, \mu_{Bf}\}$ for kaons. The curves for χ_1^K/χ_2^K as functions of the temperature, calculated along the isentropes, are shown in Fig. 1. They are compared to the experimental values for the same quantity, from which the chemical freeze-out temperature can be determined. In a recent paper [27], Bluhm and Nahrgang performed a fit of

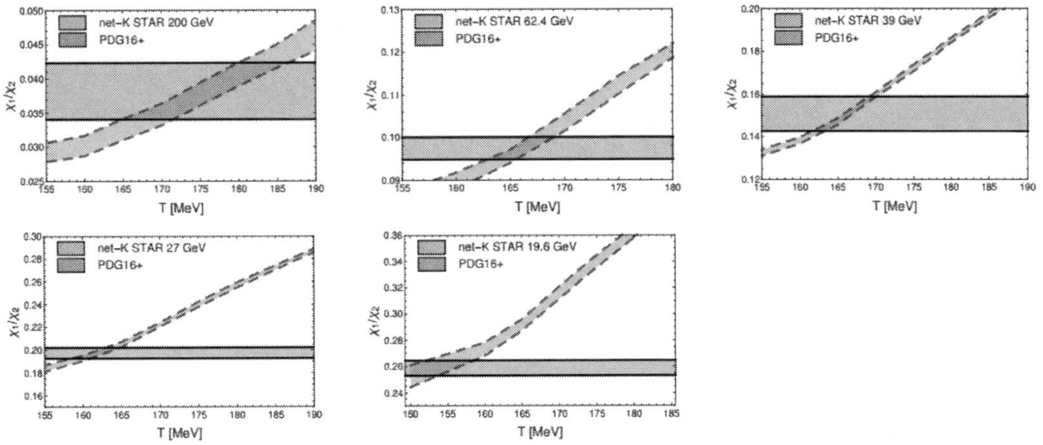

Figure 1. χ_1^K/χ_2^K calculated along the Lattice QCD isentropic trajectories (pink band) in the HRG model, compared to $(M/\sigma^2)_K$ data from [22] (gray band) at different collision energies.

χ_1^K/χ_2^K and the strange anti-baryon over baryon yield ratios, in order to independently fit T_f and μ_{Bf}.

In Fig. 2 we show the freeze-out temperature and chemical potentials that we obtain from our analysis, compared to T_f and μ_{Bf} obtained previously [23] through a combined fit of net-proton and net-electric charge fluctuations at RHIC. Also shown are two points corresponding to thermal fits from the STAR collaboration at $\sqrt{s} = 39$ GeV [10], obtained fitting all measured ground-state yields (orange point) and only protons, pions and kaons (blue point). As it is clear from the figure, the light freeze-out points from fluctuations agree with the thermal fits of light particles, while the freeze-out points obtained from net-kaon fluctuations agree with the thermal fits including multi-strange hadrons.

In Fig. 3 we show our predictions for net-Λ $\chi_2^\Lambda/\chi_1^\Lambda$ (left panel) and $\chi_3^\Lambda/\chi_2^\Lambda$ (right panel) as functions of the collision energy, calculated at the values of T_f and μ_{Bf} extracted from the fit of χ_1^K/χ_2^K (red points), and from the combined fit of χ_1^p/χ_2^p and χ_1^Q/χ_2^Q (blue points). Both observables show a clear separation between the two scenarios, that the future experimental results will hopefully be able to resolve.

In conclusion, our analysis points out a tension between the freeze-out parameters obtained, within the same HRG model, from a combined fit of χ_1^p/χ_2^p and χ_1^Q/χ_2^Q vs. χ_1^K/χ_2^K. This tension, already pointed out in thermal fits at the LHC and RHIC, could be due to different effects. From lattice QCD there is an indication that strange particles might hadronize at a higher temperature [15], naturally leading to a higher chemical freeze-out temperature as well. Interactions in the hadronic phase might be partially responsible for this observation [28, 29, 30, 31, 32, 33]. Hopefully, future experimental data on net-Λ fluctuations will help to clarify this issue.

Acknowledgements
This material is based upon work supported by the National Science Foundation under grants no. PHY-1654219 and OAC-1531814 and by the U.S. Department of Energy, Office of Science, Office of Nuclear Physics, within the framework of the Beam Energy Scan Theory (BEST) Topical Collaboration. We also acknowledge the support from the Center of Advanced Computing and Data Systems at the University of Houston. The work of R. B. is supported through DOE grant

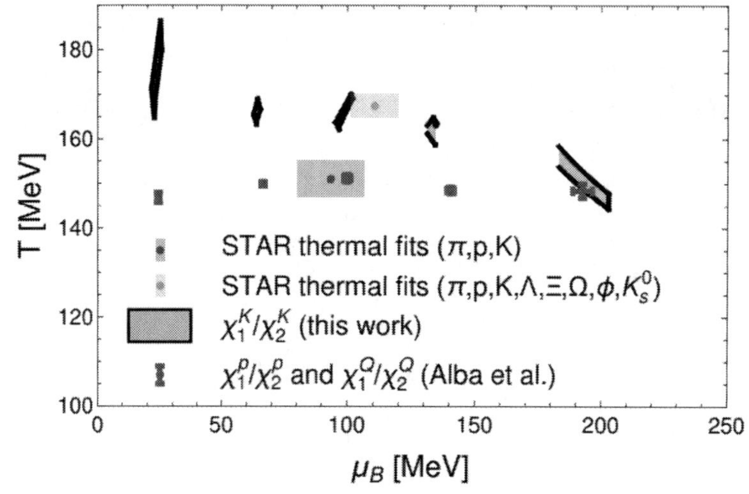

Figure 2. Freeze-out parameters at the highest five energies from the Beam Energy Scan at RHIC. The red points were obtained from the combined fit of χ_1^p/χ_2^p and χ_1^Q/χ_2^Q [23], the gray bands are obtained from the fit of χ_1^K/χ_2^K in this work. Also shown are the freeze-out parameters obtained by the STAR collaboration at $\sqrt{s}$ = 39 GeV [10] from thermal fits to all measured ground-state yields (orange point) and only to protons, pions and kaons (blue point).

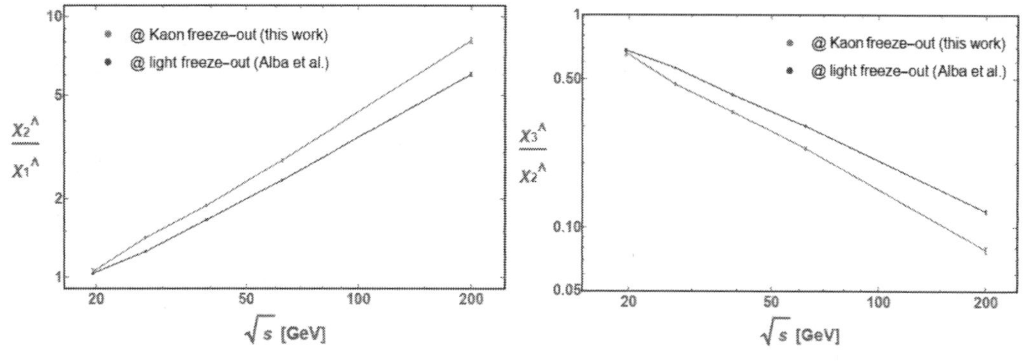

Figure 3. Left panel: $\chi_2^\Lambda/\chi_1^\Lambda$ as a function of $\sqrt{s}$. Right panel: $\chi_3^\Lambda/\chi_2^\Lambda$ as a function of $\sqrt{s}$. In both panels, the red points are calculated at the values of T_f and μ_{Bf} extracted from the fit of χ_1^K/χ_2^K, while the blue points are calculated at the values of T_f and μ_{Bf} extracted from the combined fit of χ_1^p/χ_2^p and χ_1^Q/χ_2^Q in Ref. [23].

DEFG02-07ER41521.

References

[1] Aoki Y, Endrodi G, Fodor Z, Katz S D and Szabo K K 2006 *Nature* **443** 675–678 (*Preprint* hep-lat/0611014)
[2] Borsanyi S, Fodor Z, Hoelbling C, Katz S D, Krieg S, Ratti C and Szabo K K (Wuppertal-Budapest) 2010 *JHEP* **09** 073 (*Preprint* 1005.3508)
[3] Bazavov A *et al.* 2012 *Phys. Rev.* **D85** 054503 (*Preprint* 1111.1710)
[4] Cleymans J, Oeschler H and Redlich K 1999 *Phys. Rev.* **C59** 1663 (*Preprint* nucl-th/9809027)
[5] Andronic A, Braun-Munzinger P and Stachel J 2009 *Phys. Lett.* **B673** 142–145 [Erratum: Phys. Lett.B678,516(2009)] (*Preprint* 0812.1186)

[6] Manninen J and Becattini F 2008 *Phys. Rev.* **C78** 054901 (*Preprint* 0806.4100)
[7] Abelev B I *et al.* (STAR) 2010 *Phys. Rev.* **C81** 024911 (*Preprint* 0909.4131)
[8] Aggarwal M M *et al.* (STAR) 2011 *Phys. Rev.* **C83** 034910 (*Preprint* 1008.3133)
[9] Floris M 2014 *Nucl. Phys.* **A931** 103–112
[10] Adamczyk L *et al.* (STAR) 2017 *Phys. Rev.* **C96** 044904 (*Preprint* 1701.07065)
[11] Alba P, Bellwied R, Bluhm M, Mantovani Sarti V, Nahrgang M and Ratti C 2015 *Phys. Rev.* **C92** 064910
[12] Karsch F 2012 *Central Eur. J. Phys.* **10** 1234–1237
[13] Bazavov A *et al.* 2012 *Phys. Rev. Lett.* **109** 192302
[14] Borsanyi S, Fodor Z, Katz S D, Krieg S, Ratti C and Szabo K K 2013 *Phys. Rev. Lett.* **111** 062005
[15] Bellwied R, Borsanyi S, Fodor Z, Katz S D and Ratti C 2013 *Phys. Rev. Lett.* **111** 202302
[16] Borsanyi S, Fodor Z, Katz S D, Krieg S, Ratti C and Szabo K K 2014 *Phys. Rev. Lett.* **113** 052301
[17] Bazavov A *et al.* 2016 *Phys. Rev.* **D93** 014512
[18] Noronha-Hostler J, Bellwied R, Gunther J, Parotto P, Pasztor A, Vazquez I P and Ratti C 2016 (*Preprint* 1607.02527)
[19] Ratti C 2018 (*Preprint* 1804.07810)
[20] Adamczyk L *et al.* (STAR) 2014 *Phys. Rev. Lett.* **112** 032302 (*Preprint* 1309.5681)
[21] Adamczyk L *et al.* (STAR) 2014 *Phys. Rev. Lett.* **113** 092301 (*Preprint* 1402.1558)
[22] Adamczyk L *et al.* (STAR) 2017 (*Preprint* 1709.00773)
[23] Alba P, Alberico W, Bellwied R, Bluhm M, Mantovani Sarti V, Nahrgang M and Ratti C 2014 *Phys. Lett.* **B738** 305–310
[24] Andronic A, Braun-Munzinger P, Redlich K and Stachel J 2017 (*Preprint* 1710.09425)
[25] Bellwied R, Noronha-Hostler J, Parotto P, Portillo Vazquez I, Ratti C and Stafford J M 2018 (*Preprint* 1805.00088)
[26] Gunther J, Bellwied R, Borsanyi S, Fodor Z, Katz S D, Pasztor A and Ratti C 2017 *EPJ Web Conf.* **137** 07008 (*Preprint* 1607.02493)
[27] Bluhm M and Nahrgang M 2018 (*Preprint* 1806.04499)
[28] Becattini F, Steinheimer J, Stock R and Bleicher M 2017 *Phys. Lett.* **B764** 241–246 (*Preprint* 1605.09694)
[29] Rapp R and Shuryak E V 2001 *Phys. Rev. Lett.* **86** 2980–2983 (*Preprint* hep-ph/0008326)
[30] Rapp R and Shuryak E V 2002 *Nucl. Phys.* **A698** 587–590 (*Preprint* hep-ph/0104006)
[31] Rapp R 2002 *Phys. Rev.* **C66** 017901 (*Preprint* hep-ph/0204131)
[32] Steinheimer J, Aichelin J and Bleicher M 2013 *Phys. Rev. Lett.* **110** 042501 (*Preprint* 1203.5302)
[33] Becattini F, Bleicher M, Kollegger T, Schuster T, Steinheimer J and Stock R 2013 *Phys. Rev. Lett.* **111** 082302 (*Preprint* 1212.2431)

IOP Conf. Series: Journal of Physics: Conf. Series **1070** (2018) 012004 doi:10.1088/1742-6596/1070/1/012004

Hadron resonance gas with repulsive interactions

P. Huovinen
Institute of Theoretical Physics, University of Wroclaw, 50204 Wrocław, Poland
E-mail: pasi.huovinen@ift.uni.wroc.pl

P. Petreczky
Physics Department, Brookhaven National Laboratory, Upton, NY 11973, USA
E-mail: petreczk@bnl.gov

Abstract. We study QCD equation of state and fluctuations of baryon number using the hadron resonance gas model with repulsive mean field. We show that the inclusion of repulsive mean field into the resonance gas model leads to better description of the lattice results. The repulsive mean field is particularly important for the higher order baryon number fluctuations.

1. Introduction

QCD thermodynamics at net zero baryon density below the chiral crossover can be described by hadron resonance gas (HRG) model. The idea behind this model is that interactions between hadrons can be accounted for by including resonances as additional free particles. Typically resonances are treated as zero width states. Equation of State obtained in lattice QCD agree well with HRG predictions (see e.g. Refs. [1, 2, 3, 4] for some recent works). The fluctuations and correlations of conserved charges, defined as derivatives of the pressure with respect to chemical potentials can also be studied in HRG model (see e.g. Ref. [5]). The HRG model can be justified using the relativistic virial expansion based on the S-matrix approach [6]. In most cases after summation over spin and isospin channels, the second virial coefficient is dominated by the resonance contribution, and thus the interacting gas of hadrons can be approximated as a gas of non-interacting hadrons and hadronic resonances [7]. The cancellation of non-resonant contributions is somewhat accidental [8] and does not happen for all thermodynamic quantities [9]. Furthermore, not all hadronic interactions are dominated by resonances. One example is the nucleon-nucleon interactions, which are non-resonant. Finally, close to the chiral transition one may worry about in-medium modifications of hadrons [10]. Lattice calculation indicate that negative parity baryon masses experience shift compared to their zero temperature values [1]. Therefore, it is important to check the HRG model by comparing it to recent lattice QCD calculations of different thermodynamic quantities.

For higher order fluctuations of conserved charges the agreement between the lattice and HRG is no longer good for temperatures close to the chiral transition. It has been argued that this is due to repulsive baryon-baryon interactions [14, 15, 16]. As the baryon density increases

[1] Here we note that extraction of hadron properties at non-zero temperature is a very difficult task, see e.g. discussions in Refs. [11, 12, 13].

the role of repulsive interactions will increase. Therefore in order to extend HRG to non-zero baryon density repulsive baryon-baryon interactions have to be included in the description. As a step toward this goal we will study in this contribution the trace anomaly and fluctuations of baryon number using HRG with repulsive mean field.

2. Nucleon gas with repulsive mean field and virial expansion

Using the virial expansion the pressure of the nucleon gas can be written as [17]

$$p(T,\mu) = p_0(T)\cosh(\beta\mu) + 2b_2(T)T\cosh(2\beta\mu),\ \beta = 1/T. \tag{1}$$

Here

$$p_0(T) = \frac{4M^2T^2}{\pi^2}K_2(\beta M) \tag{2}$$

is the pressure of free nucleon gas at zero chemical potential, and the second virial coefficient can be written as

$$b_2(T) = \frac{2T}{\pi^3}\int_0^\infty dE(\frac{ME}{2}+M^2)K_2\left(2\beta\sqrt{\frac{ME}{2}+M^2}\right)\frac{1}{4i}\mathrm{Tr}\left[S^\dagger\frac{dS}{dE}-\frac{dS^\dagger}{dE}S\right], \tag{3}$$

with S being the scattering S-matrix and E is the kinetic energy in the lab frame. Furthermore, M is the nucleon mass and $K_2(x)$ is the Bessel function of second kind. Using the parameterization of the S-matrix in terms of the phase shifts and their numerical values the virial coefficient $b_2(T)$ can be calculated, and it turns out that $b_2(T)$ is negative [17]. For the comparison with the mean-field approach discussed below it is convenient to write the above expression for the pressure in the following form

$$p(T,\mu) = p_0(T)(\cosh(\beta\mu) + \bar{b}_2(T)K_2(\beta M)\cosh(2\beta\mu)), \tag{4}$$

with

$$\bar{b}_2(T) = \frac{2Tb_2(T)}{p_0(T)K_2(\beta M)}. \tag{5}$$

being the reduced virial coefficient.

In the mean-field approach the pressure can be written as [17]

$$p(T,\mu) = T(n_b+\bar{n}_b) + \frac{K}{2}(n_b^2+\bar{n}_b^2), \tag{6}$$

where n_b and $\bar{n}_b$ are the densities of nucleons and anti-nucleons, respectively, defined by the following self-consistent relations:

$$n_b = 4\int\frac{d^3p}{(2\pi)^3}e^{-\beta(E_p-\mu+U)},\ \ \bar{n}_b = 4\int\frac{d^3p}{(2\pi)^3}e^{-\beta(E_p+\mu+\bar{U})},\ E_p^2=p^2+M^2. \tag{7}$$

Here $U = Kn_b$ and $\bar{U} = K\bar{n}_b$ are the mean-field potentials for nucleons and anti-nucleons. The chemical potential corresponding to the net nucleon density is denoted by μ. The above form of the pressure ensures thermodynamic consistency, i.e. $\partial p/\partial\mu = n_b - \bar{n}_b$. Since we are mostly interested in the region of not too high baryon densities one can expand the above expressions in $\beta U = \beta K n_b$ and $\bar{U} = K\bar{n}_b$ and keep only the leading order terms in K. Then we get

$$p(T,\mu) = T(n_b^0+\bar{n}_b^0) - \frac{K}{2}\left(\left(n_b^0\right)^2+\left(\bar{n}_b^0\right)^2\right), \tag{8}$$

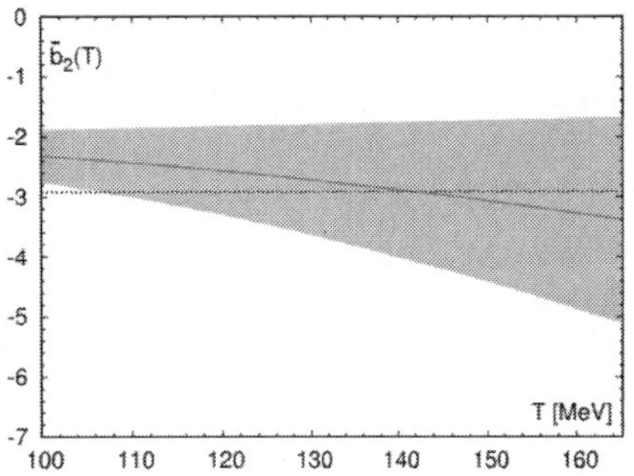

Figure 1. The reduced second virial coefficient $\bar{b}_2$ as function of temperature, and the mean-field parameter K.

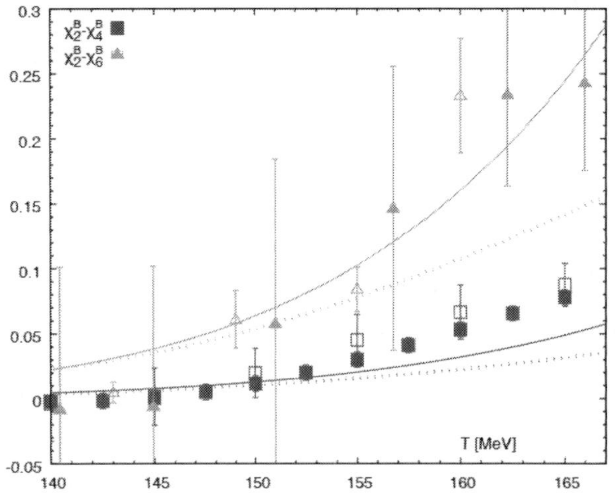

Figure 2. The differences between fourth order and second order baryon number fluctuations (squares) as well as the the sixth and second order baryon number fluctuations (triangles). The dashed lines correspond to the exact mean-field calculations, while the solid line to the expanded mean field (see text). The filled symbols correspond to lattice results of Ref. [18]. The open symbols are the lattice data for χ_4^B from Ref. [19] and χ_6^B from Ref. [20], respectively.

with n_b^0 ($\bar{n}_b^0$) being the free nucleon(anti-nucleon) densities. The above equation can also be written as

$$p(T, \mu) = \frac{4T^2M^2}{\pi^2} K_2(\beta M) \cosh(\beta\mu) - 4K \frac{T^2M^4}{\pi^4} K_2^2(\beta M) \cosh(2\beta\mu). \quad (9)$$

Now we see that the virial expansion and the expanded mean-field approach give very similar results. From the comparison of these two results one can determine the value of K in a limited temperature range. Since $\bar{b}_2(T)$ turns out to be negative, $K > 0$, i.e. the mean field is repulsive. The temperature dependence of $\bar{b}_2$ is shown in Fig. 1 and it turns out to be relatively mild except maybe at the highest temperature. The value of $\bar{b}_2$ is consistent with $K = 250\,\mathrm{MeV/fm}^3$. The largest value allowed for K by the virial expansion is around $K = 450\,\mathrm{MeV/fm}^3$. It is straightforward to generalize the mean-field approach described above to a multicomponent system if one assumes that the repulsive mean-field is the same for all ground state baryons [17] and the baryon resonances are not affected by the mean field. Within this approach we calculate baryon number fluctuations defined as derivatives of the pressure with respect to baryon chemical potential [17]

$$\chi_n^B = T^n \frac{\partial^n p/T^4}{\partial \mu_B^n}. \tag{10}$$

In Fig. 2 we show the differences of the baryon number fluctuations, $\chi_4^B - \chi_2^B$ and $\chi_6^B - \chi_2^B$ obtained in HRG model with the mean field and the value of $K = 450\,\mathrm{MeV/fm}^3$. The model can qualitatively explain these differences as one can see from Fig. 2, however, at quantitative level it underpredicts the lattice data even for the largest allowed value of K. We note that at temperatures above 150 MeV the expanded mean-field expression (8) is no longer accurate and one should use the exact mean-field expressions with particle densities determined self-consistently [17]. The exact mean-field results are shown as dashed lines in the figure.

3. Comparison with lattice QCD

In this section we compare the calculations of HRG with repulsive mean field with lattice QCD results on the trace anomaly and fluctuations of baryon number up to sixth order. First we note that there are many baryon resonances not listed in Particle Data Group (PDG) but are predicted by quark models and lattice QCD. These are the so-called missing states. It was found that it is important to include these missing states in HRG description in order to describe the strangeness fluctuations and baryon-strangeness correlation [21]. Therefore, in our analysis we will include the missing baryons from the quark model calculations of Ref. [22, 23] and label the corresponding results as QM-HRG, whereas the HRG based on the PDG resonance list [24] is labeled PDG-HRG. In the previous analysis it was assumed that baryon resonances are not affected by the mean field [17]. This is not realistic because if the density of ground state baryons is reduced there will be also fewer baryon resonances in the system. Therefore, in the present analysis we included the effect of the mean field on the baryon resonances as well. Furthermore, we use a more realistic value for the mean-field parameter, $K = 250\,\mathrm{MeV/fm}^3$. In Fig. 3 we show our calculations for the trace anomaly, $\epsilon - 3P$ compared to the lattice results [2, 3] at zero baryon chemical potential, $\mu_B = 0$. We clearly see the effects of the missing states on the trace anomaly. These states lead to significant increase of the trace anomaly around temperatures of 150 MeV and higher compared to the HRG calculations with only PDG states (PDG-HRG). The HRG calculations with repulsive mean field are shown as dashed lines. The effects of the repulsive interactions turns out relatively small, because the contributions of baryons is small too. The situation will be different at sufficiently large μ_B, when the contribution of mesons and baryons to the equation of state is comparable.

In Fig. 4 we show our results for second order baryon number fluctuations. Again we show results for QM-HRG and PDG-HRG with and without the effect of the repulsive mean field. First, we see the the inclusion of missing baryons is important also here. If all the missing baryons are included the HRG overshoots the lattice results. The repulsive mean field has an opposite effect and is comparable in size. Therefore, it restores the agreement with the lattice data. Higher order baryon number fluctuations are more sensitive to the effects of the repulsive interactions as shown in Fig. 5. The HRG-QM model again overpredicts the lattice results and lies significantly above the HRG-PDG result. The repulsive interactions reduce the HRG prediction and lead to reasonably good agreement with the lattice data. Note that the effects of the repulsive mean field are now larger than in the previous analysis shown in Fig. 2 even though we use smaller value of K. This is because the resonances are also affected by the repulsive mean field. We not that at temperatures above 155 MeV the HRG model does not do a good job in describing the lattice results even when repulsiv meanfield is included.

IOP Conf. Series: Journal of Physics: Conf. Series **1070** (2018) 012004 doi:10.1088/1742-6596/1070/1/012004

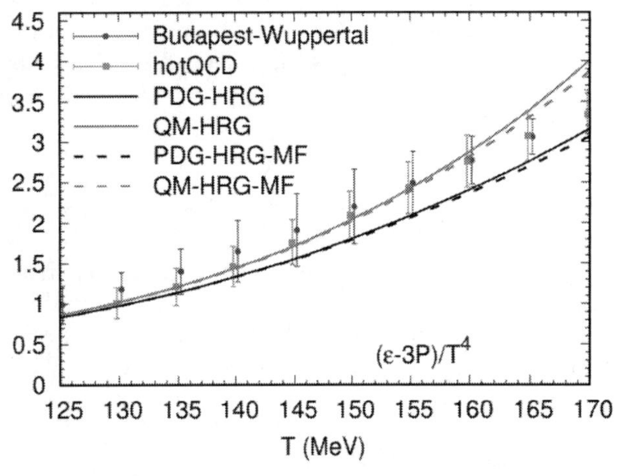

Figure 3. The trace anomaly in HRG model with the Particle Data Group (PDG-HRG) and Quark Model (QM-HRG) lists of resonances compared to the lattice results [2, 3]. The dashed lines correspond to HRG model with repulsive mean field.

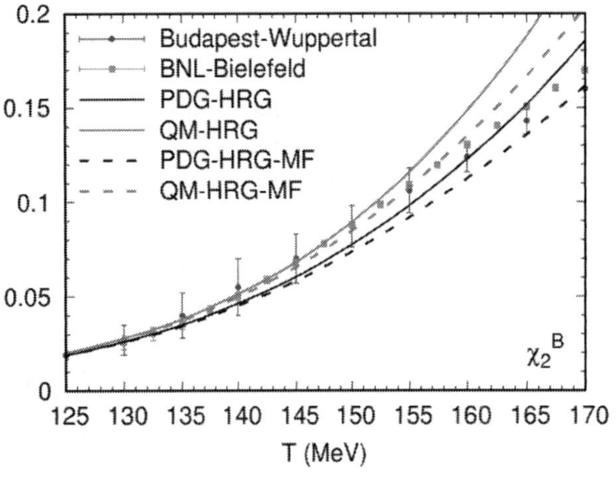

Figure 4. The second order baryon number fluctuations in HRG model with the Particle Data Group (PDG-HRG) and Quark Model (QM-HRG) lists of resonances compared to the lattice results [18, 19]. The dashed lines correspond to HRG model with repulsive mean field.

4. Conclusion

We studied the equation of state and fluctuations of baryon number within the HRG framework, where the effect of the repulsive baryon-baryon interactions are included using the mean-field approach. We also studied the effects of missing baryon resonances on these quantities. We have found that the missing states lead to significant increase of thermodynamic quantities. In the case of baryon number fluctuations missing states cause the HRG model to overshoot the lattice results for $T \geq 150$ MeV. The repulsive mean field has the opposite effect. The effects of repulsive interactions are small for the trace anomaly but are significant for baryon number fluctuations, where they are needed to bring the HRG calculations in agreement with the lattice

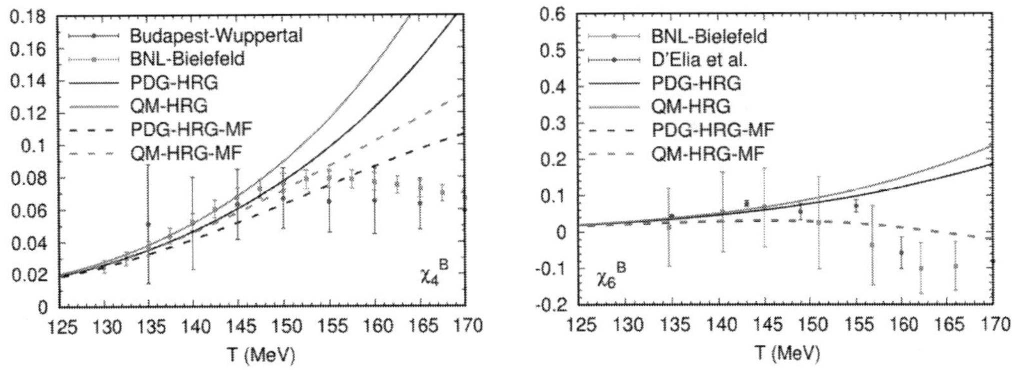

Figure 5. The fourth order (left) and sixth order (right) baryon number fluctuations in HRG-QM and HRG-PDG models. The lattice results for χ_4^B are from Refs. [18, 19], while the lattice results for χ_6^B are from [18, 20]. The dashed lines correspond to HRG model with repulsive mean field.

results for temperatures $T < 155$ MeV. This implies that when extending the HRG model to calculate equation of state at non-zero baryon density the repulsive interactions have to be taken into account.

Acknowledgment

This work was supported by U.S.Department of Energy under Contract No. DE-SC0012704, and by National Science Center, Poland, under grant Polonez DEC-2015/19/P/ST2/03333 receiving funding from the European Union's Horizon 2020 research and innovation program under the Marie Skłodowska-Curie grant agreement No 665778.

[1] Wuppertal-Budapest, S. Borsanyi *et al.*, JHEP **09**, 073 (2010), 1005.3508.
[2] S. Borsanyi *et al.*, Phys. Lett. **B730**, 99 (2014), 1309.5258.
[3] HotQCD, A. Bazavov *et al.*, Phys. Rev. **D90**, 094503 (2014), 1407.6387.
[4] A. Bazavov, P. Petreczky, and J. H. Weber, Phys. Rev. **D97**, 014510 (2018), 1710.05024.
[5] P. Huovinen and P. Petreczky, Nucl. Phys. **A837**, 26 (2010), 0912.2541.
[6] R. Dashen, S.-K. Ma, and H. J. Bernstein, Phys. Rev. **187**, 345 (1969).
[7] R. Venugopalan and M. Prakash, Nucl. Phys. **A546**, 718 (1992).
[8] C. Fernández-Ramírez, P. M. Lo, and P. Petreczky, (2018), 1806.02177.
[9] P. M. Lo, B. Friman, K. Redlich, and C. Sasaki, Phys. Lett. **B778**, 454 (2018), 1710.02711.
[10] G. Aarts *et al.*, JHEP **06**, 034 (2017), 1703.09246.
[11] I. Wetzorke, F. Karsch, E. Laermann, P. Petreczky, and S. Stickan, Nucl. Phys. Proc. Suppl. **106**, 510 (2002), hep-lat/0110132.
[12] F. Karsch *et al.*, Nucl. Phys. **A715**, 701 (2003), hep-ph/0209028.
[13] S. Datta, F. Karsch, P. Petreczky, and I. Wetzorke, Phys. Rev. **D69**, 094507 (2004), hep-lat/0312037.
[14] M. Albright, J. Kapusta, and C. Young, Phys. Rev. **C92**, 044904 (2015), 1506.03408.
[15] V. Vovchenko, M. I. Gorenstein, and H. Stoecker, Phys. Rev. Lett. **118**, 182301 (2017), 1609.03975.
[16] V. Vovchenko *et al.*, (2017), 1707.09215.
[17] P. Huovinen and P. Petreczky, Phys. Lett. **B777**, 125 (2018), 1708.00879.
[18] A. Bazavov *et al.*, Phys. Rev. **D95**, 054504 (2017), 1701.04325.
[19] S. Borsanyi *et al.*, Phys. Rev. Lett. **113**, 052301 (2014), 1403.4576.
[20] M. D'Elia, G. Gagliardi, and F. Sanfilippo, Phys. Rev. **D95**, 094503 (2017).
[21] A. Bazavov *et al.*, Phys. Rev. Lett. **113**, 072001 (2014), 1404.6511.
[22] U. Loring, B. C. Metsch, and H. R. Petry, Eur.Phys.J. **A10**, 447 (2001), hep-ph/0103290.
[23] D. Ebert, R. N. Faustov, and V. O. Galkin, Phys. Rev. **D79**, 114029 (2009), 0903.5183.
[24] Particle Data Group, C. Patrignani *et al.*, Chin. Phys. **C40**, 100001 (2016).

IOP Conf. Series: Journal of Physics: Conf. Series **1070** (2018) 012005 doi:10.1088/1742-6596/1070/1/012005

Preliminary Results from the Parton-Hadron-Quantum-Molecular Dynamics (PHQMD) Transport Approach

J Aichelin[1,2], E Bratkovskaya[3,4], A Le Fèvre[3], V Kireyev[5] and Y Leifels[3]

[1] SUBATECH, Université de Nantes, EMN, IN2P3/CNRS 4 rue Alfred Kastler, 44307 Nantes cedex 3, France

[2] Frankfurt Institute for Advanced Studies, Ruth Moufang Str. 1, 60438 Frankfurt, Germany

[3] GSI Helmholtzzentrum für Schwerionenforschung GmbH, Planckstr. 1, 64291 Darmstadt, Germany

[4] Institut fr Theoretische Physik, Johann Wolfgang Goethe-Universität, Max-von-Laue-Str. 1, 60438 Frankfurt am Main, Germany

[5] Joint Institute for Nuclear Research, Joliot-Curie 6, 141980 Dubna, Moscow region, Russia

Abstract. We present first results of the novel microscopic n-body dynamical transport approach PHQMD (Parton-Hadron-Quantum-Molecular-Dynamics) for the description of particle production and cluster formation (including hypernuclei) in heavy-ion reactions at relativistic energies. The PHQMD extends the established PHSD (Parton-Hadron-String-Dynamics) transport approach by introducing n-body quantum molecular dynamic type propagation of hadrons. This allows for a dynamical description of cluster formation based on the FRIGA (Fragment Recognition In General Application) model.

1. Introduction

After the discovery of a new state of matter, a plasma of quarks and gluons (QGP), in the experiments at the Relativistic Heavy Ion Collider (RHIC) at Brookhaven and at the Large Hadron Collider (LHC) at CERN, the next goal is to study this new state at finite baryon density where some of the theories predict a first order phase transition [1, 2]. Two accelerators are presently under construction, the Facility for Antiproton and Ion Research (FAIR) in Darmstadt and the Nuclotron-based Ion Collider fAcility (NICA) in Dubna which will become operational in the next years. Moreover, presently running the Beam Energy Scan (BES) program at RHIC provides the experimental data at this energy regime, too. There scientific goal is to study those observables which may carry information on the existence of the plasma and especially of its phase transition in the hadronic world.

Fragment production at midrapidity is in general a very promising tool to study the phase space distribution of hadrons during the expansion. The multiplicity of the produced clusters is directed related to the phase space distribution of baryons at there creation point and therefore a change of the fluctuations like in the neighborhood of a first order phase transition will be directly reflected in the fragment multiplicity. In addition, without identifying fragments single particle observables cannot be correctly predicted. In 1.5 AGeV Au+Au collisions, the highest energy where data on clusters formation are available presently, only 65% of the total baryon charge is observed as protons [3]. The rest is predominantly bound in small clusters which show a different rapidity distribution, a different flow and a different p_T spectrum. Therefore, even if one aims only at the investigation of single baryon spectra one has to identify fragments, otherwise a precise prediction is not possible.

The hypernuclei are one of the most promising probes to study the interaction between non-strange and strange baryons as well as the reaction dynamics of the heavy-ion collisions. The production of hypernuclei may shed light on our theoretical understanding of the dynamics of heavy-ion reactions which cannot be addressed by other probes. In particular, the formation of heavy projectile/target like hypernuclei elucidates the physics at the interface between spectator and projectile matter. Since the strange baryon comes necessarily from the overlap region the multiplicity as well as the rapidity distribution of the hypernuclei are related to the probability that strange baryons of a given rapidity are absorbed by the spectator matter by means of their potential interaction with the cold spectator matter. On the other hand, mid-rapidity hypernuclei test the phase space distribution of baryons in the expanding participant matter, especially whether strange and non strange baryons are in thermal equilibrium and whether their space time distribution is similar.

Here present preliminary resuls of the novel Parton-Hadron-Quantum-Molecular Dynamics (PHQMD) approach which allows to identify fragments is based on the collision integrals of the Parton-Hadron-String Dynamics (PHSD) approach [4, 5, 6, 7, 8] and the n-body potential interactions of QMD type models [11, 9, 10]. The original PHSD propagation in mean-field way (realized within the parallel ensemble method) is kept as an option, too, which will allow to investigate the differences in the both approaches.

The PHQMD includes the following features for the dynamical fragment identification: the final fragment yield can either be determined with a minimum spanning tree (MST) procedure [11] or with cluster finding algorithms based on the Simulated Annealing Clusterization Algorithm (SACA) [12].

i) The MST algorithm is based on spatial correlations and it is effective in finding the clusters at the end of the reaction. However, it is a rather simplified approach, which ignores the momentum correlations and energy balance in the found clusters.

ii) In order to identify the cluster formation at the early time of reactions, when the collisions between the nucleons are still going on and the nuclear density is high, the SACA method has to be used. The SACA finds this most bound configuration at any moment during the heavy-ion reaction with the help of the simulated annealing technique [12]. Combined with the QMD approach [11], SACA has been successfully used in the past to study spectra of fragments of a given charge by enclosing quantum shell effects, pairing as well as symmetry energy and hyperon nucleon interactions [14]. This possibility offers not only to study the physical origin of fragment formation but also clarifies experimental observations like bimodality [14, 15] or the large experimental inverse slope of the energy spectra which is about twice as large as the binding energy per nucleon of clusters [16].

2. The Parton-Hadron Quantum Molecular Dynamics (PHQMD) approach

The PHQMD approach unites the collision integrals of the PHS approach with the n-body interactions similar as in the Quantum Molecular Dynamics (QMD) approach.

2.1. Initialization

In the QMD approach the single-particle Wigner density of the the nucleon i ,f_i, is given by (we use $\hbar = c = 1$)

$$f_i(\mathbf{r}, \mathbf{p}, t) = \frac{1}{\pi^3 \hbar^3} e^{-\frac{2}{L}(\mathbf{r}-\mathbf{r}_i(t))^2} e^{-\frac{L}{2\hbar^2}(\mathbf{p}-\mathbf{p}_i(t))^2} \quad . \tag{1}$$

The total one-body Wigner density is the sum of the Wigner densities of all nucleons. To initialize the nucleons we choose randomly the position $\mathbf{r_i}(t=0)$ in a sphere with the radius $R = 1.14\ A^{1/3}$ We take care that the distribution is smooth by requiring a minimal phase space distance between the nucleons.

Having placed the nucleons in coordinate space we calculate the expectation value of the potential energy, $< U(\mathbf{r_i}, t) >$, of each nucleon i.

$$\begin{aligned} < U(\mathbf{r_i}) > &= \sum_j \int d^3r d^3r' d^3p d^3p' \\ & U_{ij}(\mathbf{r}, \mathbf{r}') f_i(\mathbf{r}, \mathbf{p}, t) f_j(\mathbf{r}', \mathbf{p}', t) \end{aligned} \tag{2}$$

with

$$\begin{aligned} U_{ij}(\mathbf{r}, \mathbf{r}') &= U_{\mathrm{Skyrme}} + U_{\mathrm{Coul}} \\ &= \frac{1}{2} t_1 \delta(\mathbf{r} - \mathbf{r}') + \frac{1}{\gamma+1} t_2 \delta(\mathbf{r} - \mathbf{r}') \rho^{\gamma-1}(\mathbf{r}) \\ & + \frac{1}{2} \frac{Z_i Z_j e^2}{|\mathbf{r} - \mathbf{r}'|}. \end{aligned} \tag{3}$$

Initially the expectation value of the potential energy
$< U(\mathbf{r_i}) >$ is negative. As the next step we chose then randomly the momenta $\mathbf{p_i}(t=0)$ with the requirement that the nucleons are bound

$$0 < \sqrt{m^2 + \mathbf{p_i}^2(t=0)} - m < - < U(\mathbf{r_i}) > . \tag{4}$$

where m is the nucleon mass. This procedure gives a lower momentum to those nucleons which are located close to the surface because there the density is lower. Finally we take care that $\sum_i \mathbf{p_i}(t=0) = 0$ by adding a common momentum to all nucleons.

With so determined momenta and positions we calculate the average binding energy of the nucleons and compare the result with the Weizscker mass formula. It turned out the we underestimate slightly the average binding energy independent of the size of the nucleus. To obtain the right binding energy we multiply finally all momenta by a common factor which is close to one and the same for all nuclei. It depends on the value of L.

Before the nuclei collide target and projectile are boosted into the nucleus nucleus center of mass frame and get Lorentz contracted.

2.2. Propagation

The propagation of the Wigner density is determined by a generalized Ritz variational principle [11], which has been developed for the Time Dependent Hartree Fock approach. Assuming that the width of the wave function is time independent and that the n-body Wigner density is the direct product of the single particle Wigner densities we obtain for the centroids of the Gaussian single particle wave functions :

$$\dot{r_i} = \frac{\partial \langle H \rangle}{\partial p_i} \qquad \dot{p_i} = -\frac{\partial \langle H \rangle}{\partial r_i} \quad , \tag{5}$$

where the expectation value of the total Hamiltonian is

$$\begin{aligned} \langle H \rangle &= \langle T \rangle + \langle V \rangle \\ &= \sum_i (\sqrt{p_i^2 + m^2} - m) + \sum_i \sum_{j>i} \int \mathrm{d}\mathbf{r}\, \mathrm{d}\mathbf{r}' \mathrm{d}\mathbf{p}\, \mathrm{d}\mathbf{p}' \\ & \cdot \quad f_i(\mathbf{r}, \mathbf{p}, t)\, V_{ij}(\mathbf{r}, \mathbf{r}', \mathbf{p}, \mathbf{p}') f_j(\mathbf{r}', \mathbf{p}', t) \quad . \end{aligned} \tag{6}$$

The potential consists of several terms:

$$V_{ij}(\mathbf{r},\mathbf{r}') = V_{\text{Skyrme}} + V_{\text{Coul}} = t_1\delta(\mathbf{r}-\mathbf{r}') + t_2\delta(\mathbf{r}-\mathbf{r}')\rho^{\gamma-1}(\mathbf{r}) + \frac{Z_iZ_je^2}{|\mathbf{r}-\mathbf{r}'|}. \quad (7)$$

The single-particle potential resulting from the convolution of the distribution functions f_i and f_j with the V_{Skyrme} interaction can be expressed as

$$< U_i^{Skyrme}(\mathbf{r_i},t) >= \alpha\left(\frac{\rho_{int}(\mathbf{r_i},t)}{\rho_0}\right) + \beta\left(\frac{\rho_{int}(\mathbf{r_i},t)}{\rho_0}\right)^\gamma, \quad (8)$$

where ρ_{int} is the interaction density obtained by convoluting the distribution function of a particle with the distribution functions of all other particles of the surrounding medium.

$$\rho_{int}(\mathbf{r_i},t) = C\sum_{j\neq i}(\frac{4}{\pi L})^{3/2}e^{-\frac{4}{L}(\mathbf{r_i}(t)-\mathbf{r_j}(t))^2} \quad (9)$$

C is a constant which has to be adjusted to reproduce the density profile of the nuclei.

The nuclear equation of state (EoS) describes the variation of the energy $E(T=0,\rho/\rho_0)$ when changing the nuclear density in infinite matter to values different from the saturation density ρ_0 for zero temperature. In infinite matter the density is position independent and we can use eq.(8) to connect our Hamiltonian with nuclear matter properties because for a given value of γ the parameters t_1, t_2 in eq. (7) are uniquely related to the coefficients α, β of the EoS, eq. (8). Values of these parameters for the different model choices can be found in Tab. 1.

	α (MeV)	β (MeV)	γ
S	-390	320	1.14
H	-130	59	2.09

Table 1. Parameter sets for the nuclear equation of state used in the QMD model

3. Results

QMD type approaches have been very successfully in the past to describe many details of the fragment formation at energies below $E_{kin} = 200\ AMeV$ [17, 14, 16, 17, 18]. They could reproduce charge yields, fragment multiplicities, fragment spectra and complex phenomena like bimodality. At these energies the fragmentation of spectator matter is the dominate mechanism and cluster identifications methods like the minimum spanning tree method or the SACA method [19] could identify the produced cluster.

The MST method identifies cluster at the end of the reaction when nucleons and fragments are well separated in phase space. In MST two nucleons are considered as part of a cluster if their distance is smaller than $r_{min} < 3fm$. Additional momentum cuts do not change the result because unbound nucleons have separated at this late time to a distance larger than r_{min}. The drawback of the MST method is that it can identify fragments only at the end of the reaction and therefore it can not be used to study how fragments are formed during the reaction.

To overcome this limitation we have developed the Simulated Annealing Cluster Algorithm (SACA) approach [19]. It is based on the idea of Dorso and Randrup [13] that the most bound configuration of nucleons and fragments during the reaction has a large overlap with the final distribution of fragments and free nucleons. Dorso and Randrup could demonstrate this for small systems and Puri and al. [19] found out that it is also true for large systems. To obtain the most bound configuration during the reaction one applies a simulated annealing algorithm. In this algorithm in each iteration step one identifies one nucleon. If this nucleon is bound it can become unbound or can become part of another cluster. If it is unbound it can become part of a fragment. After each iteration step one calculates the total binding energy of the system neglecting the interaction among nucleons which belong to different fragments. If the new configuration is more bound it is kept, if it is less bound it is kept with a given probability to avoid that the procedure ends in a local and not in a global minimum. The binding energy is calculated with the same interaction which is also used for the time evolution of the nucleons (eq. (3)). For details we refer to [19].

The PHQMD proton rapidity distribution for central AuAu collisions at beam energies at 1.5 $AGeV$ is shown in Fif. 1 and compared with the experimental result of the FOPI collaboration [20]. We display as wll the distribution of all charges. First of all we see that the stopping in this reaction is quit nicely reproduced giving the right rapidity distribution. Seecond we see that quite a number of all charges are bound in clusters. PHQMD overpredict even the number of bound nucleons at midrapidity resulting in an underestimation of the height of the experimental proton rapiduty distribution. Next we investigate the results of

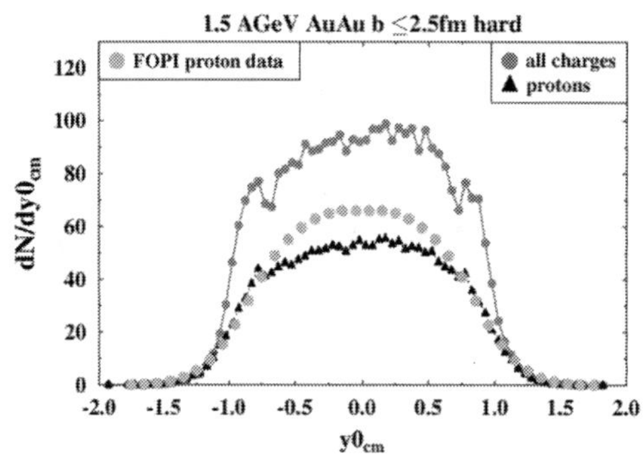

Figure 1. Scaled experimental rapidity distribution, $y_0 = y/y_{cm}$, of free protons observed in central AuAu reactions at 1.5 $AGeV$ [20] in comparison with the PHQMD results. The red line shows the rapidity distribution of all bound and unbound protons whereas the black line indicates those protons which are considered as free after the fragments have been determined by MST.

the ALADIN collaboration which has measured the fragment formation at beam energies of around 800 AMeV [21]. This is presently the highest energy for which data on heavy fragments are available. For this investigation we use a hard equation of state. One of the key result of the Aladin collaboration was the "rise and fall" of the multiplicity of intermediate mass fragments $3 \leq Z \leq 30$ emitted in forward direction. Plotting this multiplicity as a function of the sum of all forward emitted bound charges, $\sum_i Z_i \, \Theta(Z_i - (1+\epsilon))$ one obtains a curve which is for Au projectiles independent of the beam energy in the intervall 600 $AMeV \leq E_{beam} \leq 1000$ $AMeV$ and also independent of the target size. Up to now this "rise and fall" has not been described by transport approaches. In Fig. 2 we display our results in comparison with the minimum bias ALADIN data. We see clearly the rise of the multiplicity of intermediate mass fragments for small values of Z_{bound} independent of the identification of fragments which we use. The multiplicity is also correctly reproduced for large Z_{bound} if we apply the SACA algorithm but

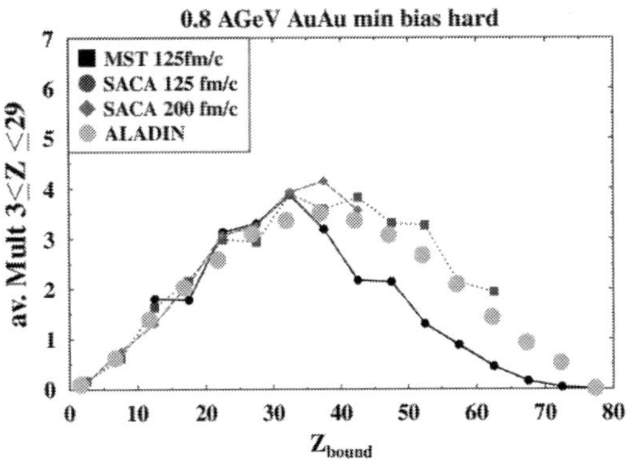

Figure 2. Rise and all of the multiplicity of intermediate mass fragments as a function of the total bound charge. Both quantities are measured for forward emitted fragments. We compare the results of PHQMD for different approaches to identify fragments with the experimental data [21].

MST underestimates there the multiplicity. This is due to the fact that in MST in the average one of the intermediate mass fragments which is identified by SACA is considered in MST as part of the largest fragment. We can conclude that PHQMD reproduces one of the most complex fragment data in this energy domain.

It is interesting to see what our present fragment algorithm gives if we increase the beam energy. In Fig.3 we display the rapidity distribution of fragments for central AuAu collisions at $E_{beam} = 3.5\ AGeV$. The fragments are identified with the MST algorithm. We display this distribution for two times 30fm/c and 90 fm/c and for Z=2 fragments as well as for hyper-nuclei, i.e. Λs and Σ^0s which form a cluster with at least one other nucleon. Besides in the target and projectile rapidity region where fragment formation and interactions still continue because the nucleons are close not only in momentum but also in coordinate space the fragment distribution changes little between 30 and 90 fm/c. This confirms that - in contradistinction to the coalescence model - in PHQMD the fragment yield is time independent and that the fragments are formed dynamically. We see as well an appreciable number of Z=2 fragments and hyper-nuclei. Even if the MST method does not allow for a quantitative description of the yield our calculation shows the potential of the future experiments of the BM@N detector at Dubna to study the formation of light hyper-nuclei.

4. Conclusion

We have presented preliminary results of a new transport model, PHQMD, to study the dynamics of heavy ion collisions at beam energies between 500 $AMeV$ and the highest LHC energies. It extends the studies with the QMD model at lower beam energies. This model includes the collision dynamics as well as the partonic interaction from the PHSD approach. It extends the PHSD approach by replacing the one body mean field dynamics by a n-body quantum molecular dynamics which propagates all correlations and fluctuations and which is therefore a basis to study the fragment formation in heavy ion reactions.

Acknowledgments: We acknowledge support from GSI and IN2P3/CNRS and the COST Action THOR, CA15213.

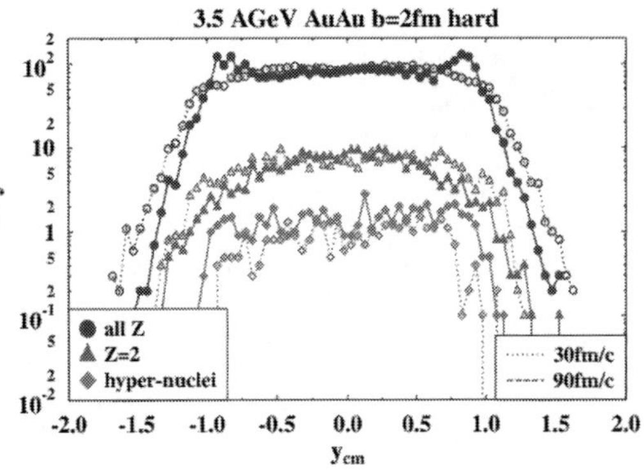

Figure 3. Rapidity distribution of bound and unbound protons, of Z=2 fragments and of hyper-nuclei for central AuAu collisions at $\sqrt{s} = 200\ AGeV$.

References

[1] M. Asakawa and K. Yazaki, Nucl. Phys. A **504**, 668 (1989).
[2] M.A. Stephanov, Prog. Theor. Phys. Suppl. **153**, 139 (2004).
[3] W. Reisdorf *et al.* [FOPI Collaboration], Nucl. Phys. A **848** (2010) 366.
[4] W. Cassing and E. L. Bratkovskaya, Phys. Rev. C **78**, 034919 (2008).
[5] W. Cassing, Eur. Phys. J. ST **168**, 3 (2009).
[6] W. Cassing and E. L. Bratkovskaya, Nucl. Phys. A **831**, 215 (2009).
[7] E. L. Bratkovskaya, W. Cassing, V. P. Konchakovski and O. Linnyk, Nucl. Phys. A **856**, 162 (2011).
[8] O. Linnyk, E. L. Bratkovskaya, W. Cassing, Prog. Part. Nucl. Phys. **87**, 50 (2016).
[9] R. Marty and J. Aichelin, Phys. Rev. **C 87**, 034912 (2013).
[10] [40] R. Marty, E. Bratkovskaya, W. Cassing and J. Aichelin, Phys. Rev. **C 92**, 015201 (2015).
[11] [22] J. Aichelin, Phys. Rept. **202** (1991) 233.
[12] R. K. Puri and J. Aichelin, J. Comput. Phys. **162**, 245 (2000).
[13] C. Dorso, J. Randrup, Phys. Lett. **B 301**, 328 (1993).
[14] A. Le Fevre *et al.*, Phys. Rev. C **80** (2009) 044615.
[15] P. B. Gossiaux, R. Puri, C. Hartnack and J. Aichelin, Nucl. Phys. **A 619**, 379 (1997).
[16] A. L. Fevre and J. Aichelin, Phys. Rev. Lett. **100**, 042701 (2008).
[17] K. Zbiri *et al.*, Phys. Rev. C **75** (2007) 034612.
[18] R. Nebauer *et al.* [INDRA Collaboration], Nucl. Phys. A **658** (1999) 67.
[19] R. K. Puri and J. Aichelin, J. Comput. Phys. **162** (2000) 245.
[20] W. Reisdorf *et al.* [FOPI Collaboration], Nucl. Phys. A **781** (2007) 459
[21] A. Schuttauf *et al.*, Nucl. Phys. A **607** (1996) 457.

IOP Conf. Series: Journal of Physics: Conf. Series **1070** (2018) 012006 doi:10.1088/1742-6596/1070/1/012006

Electromagnetic emissivity of hot and dense matter

T Song[1], E L Bratkovskaya[2,3 1], W Cassing[1], P Moreau[2]

[1] Institute for Theoretical Physics, University of Giessen, Giessen, Germany
[2] Institute for Theoretical Physics, University of Frankfurt, Frankfurt, Germany
[3] GSI Helmholtzzentrum für Schwerionenforschung GmbH, Darmstadt, Germany

E-mail: E.Bratkovskaya@gsi.de

Abstract. In this study we investigate dilepton production in heavy-ion collisions from $\sqrt{s_{\rm NN}} = 8$ GeV to 5 TeV with a focus on the competition between the thermal QGP radiation and the semi-leptonic decays from correlated $D-$meson pairs. As a 'tool' we employ the parton-hadron-string dynamics (PHSD) transport approach incorporating for the first time a fully microscopic treatment of the charm dynamics and their semi-leptonic decays. We find that the dileptons from correlated $D-$meson decays dominate the 'thermal' radiation from the QGP in central Pb+Pb collisions at the intermediate masses (1.2 GeV $< M <$ 3 GeV) for $\sqrt{s_{NN}} > 40$ GeV, while for $\sqrt{s_{\rm NN}} = 5$ to 20 GeV the contribution from $D, \bar{D}$ decays to the intermediate mass dilepton spectra is subleading such that one should observe a rather clear signal from the QGP radiation. We, furthermore, study the p_T-spectra and the $R_{AA}(p_T)$ of single electrons at different energies.

1. Introduction

Relativistic heavy-ion collisions are well suited to generate hot and dense matter in the laboratory although within small space-time regimes. Whereas in low energy collisions one produces dense nuclear matter with moderate temperature T and large baryon chemical potential μ_B, ultra-relativistic collisions at Relativistic Heavy Ion Collider (RHIC) or Large Hadron Collider (LHC) energies produce extremely hot matter at small baryon chemical potential. In order to explore the phase diagram of strongly interacting matter as a function of T and μ_B both type of collisions are mandatory. According to lattice calculations of quantum chromodynamics (lQCD) [1, 2, 3, 4], the phase transition from hadronic to partonic degrees of freedom (at vanishing baryon chemical potential μ_B=0) is a crossover. This phase transition is expected to turn into a first order transition at a critical point (T_r, μ_r) in the phase diagram with increasing baryon chemical potential μ_B. Since this critical point cannot be determined theoretically in a reliable way the beam energy scan (BES) program at RHIC aims to find the critical point and the phase boundary by gradually decreasing the collision energy [5, 6]. Furthermore, new facilities such as FAIR (Facility for Antiproton and Ion Research) and NICA (Nuclotron-based Ion Collider fAcility) are under construction to explore in particular the intermediate energy range where one might study also the competition between chiral symmetry restoration and deconfinement as suggested in Refs. [7, 8].

Since the partonic phase in relativistic heavy-ion collisions appears only for a couple of fm/c, it is quite a challenge for experiment to investigate its properties. In particular the electromagnetic

[1] speaker

emissivity of strongly interacting matter is a subject of longstanding interest [9, 10, 11, 12] and is explored also in relativistic nucleus-nucleus collisions, where the photons (and dileptons) measured experimentally provide a time-integrated picture of the collision dynamics. After decades of experimental and theoretical studies it has become clear that dileptons with invariant masses below about 1.2 GeV preferentially stem from hadronic decays providing some glance at the modification of hadron properties in the dense and hot hadronic medium (cf. [12, 13] and references therein) while the intermediate mass regime 1.3 GeV $< M <$ 3 GeV should provide information about 'thermal' dileptons from the QGP ($q+\bar{q} \rightarrow e^+e^-$, $q+\bar{q} \rightarrow g+\gamma^*, g+q(\bar{q}) \rightarrow$, $q(\bar{q})+e^+e^-$) as well as the amount of correlated open charm (semi-leptonic) decays from early production of $c\bar{c}$ pairs [14, 15, 16, 17, 18, 19, 20]. Whereas at RHIC and LHC energies the background from $D\bar{D}$ pairs overshines the contribution from the QGP in the intermediate mass regime [13], one might expect to find some window in bombarding energy where the partonic sources dominate since the charm production drops rapidly with decreasing bombarding energy. In this contribution we intend to quantify this expectation and to identify optimal systems for future measurements at FAIR/NICA or at the RHIC Beam-Energy-Scan (BES) as well as at the Super Proton Synchrotron (SPS) by the NA61 collaboration.

We here employ the microscopic parton-hadron-string dynamics (PHSD) approach, which differs from the conventional Boltzmann-type models (with or without mean-fields [21, 22]) in the aspect [23] that the degrees-of-freedom for the QGP phase are off-shell massive strongly-interacting quasi-particles that generate their own mean-field potential. The masses of the dynamical quarks and gluons in the QGP are distributed according to spectral functions whose pole positions and widths, respectively, are defined by the real and imaginary parts of their self-energies [13]. The partonic propagators and self-energies, furthermore, are defined in the dynamical quasiparticle model (DQPM) in which the strong coupling and the self-energies are fitted to lattice QCD results [24]. We recall that the PHSD approach has successfully described numerous experimental data in relativistic heavy-ion collisions from the Alternating Gradient Synchrotron (AGS), SPS, RHIC to LHC energies [13, 23, 25, 26, 27, 28, 29, 30, 31, 32]. More recently, the charm production and propagation has been explicitly implemented in the PHSD and detailed studies on the charm dynamics and hadronization/fragmention have been performed at top RHIC and LHC energies in comparison to the available data [33, 34, 35].

2. Charm pairs from p+p collisions

As pointed out in the Introduction the charm quark ($c\bar{c}$) pairs are produced through initial hard nucleon-nucleon scattering in relativistic heavy-ion collisions. We employ the PYTHIA event generator to produce the heavy-quark pairs and modify their transverse momentum and rapidity such that they are similar to those from the FONLL calculations at RHIC and LHC energies (cf. Ref. [35]). At SPS and lower energies we do not employ any modification of the PYTHIA results. Fig. 1 a) shows the charm production cross section for p+p collisions (as implemented in PHSD) as a function of the invariant energy $\sqrt{s_{NN}}$ which is fitted to a wide range of experimental data. We can see a rather fast drop of the $c\bar{c}$ cross section with decreasing energy especially close to threshold. Note, however, that the data show an uncertainty of about a factor of two which implies a corresponding uncertainty in the following PHSD calculations.

2.1. Multiplicities for $c\bar{c}$ pairs in central Pb+Pb reactions

We recall that in heavy-ion reactions the number of $c\bar{c}$ pairs produced is approximately given by the number of binary nucleon-nucleon collisions $N_{bin}(b)$ (at given impact parameter b) times the probability to produce a $c\bar{c}$ pair in an inelastic nucleon-nucleon collision at given $\sqrt{s_{NN}}$ which is given by the ratio of the $c\bar{c}$ cross section to the inelastic $N+N$ cross section [36, 37]. The scaling of the $c\bar{c}$ multiplicity with the number of binary $N+N$ collisions is rather well reproduced in actual PHSD calculations where additionally the smearing of $\sqrt{s_{NN}}$ by Fermi motion is taken

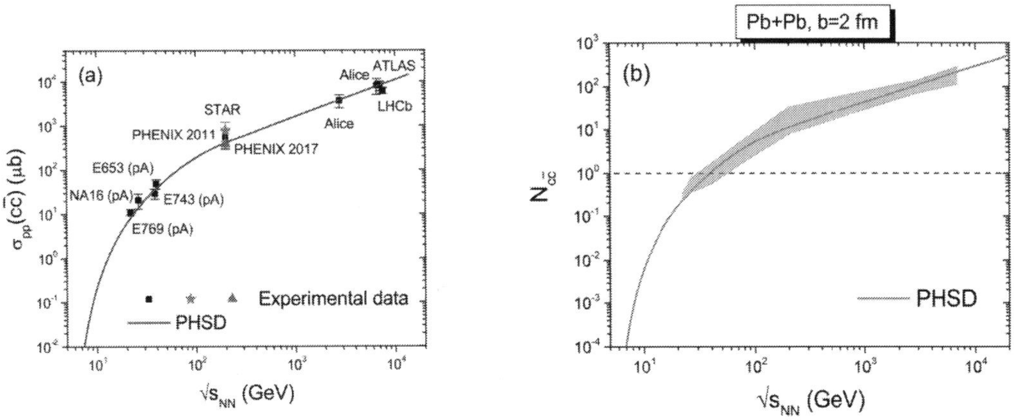

Figure 1. a) The $c\bar{c}$ pair cross section in p+p reactions as a function of the invariant energy $\sqrt{s}$ as implemented in PHSD. The symbols denote experimental data from Refs. [38, 39, 40]. b) The number of primary $c\bar{c}$ pairs in Pb+Pb collisions at b=2 fm as a function of $\sqrt{s_{NN}}$. The shaded area in (b) shows the uncertainty in the number of $c\bar{c}$ pairs due to the uncertainty in the charm production cross section in p+p collisions.

into account as well as fluctuations in the number of binary nucleon-nucleon collisions $N_{bin}(b)$ on an event by event basis. The corresponding PHSD results for Pb+Pb collisions (at b=2 fm) are displayed in Fig. 1 b) as a function of $\sqrt{s_{NN}}$ and demonstrate that the average $c\bar{c}$ pair multiplicity in central collisions is far below unity at SPS and FAIR/NICA energies. In this case we may gate in the PHSD calculations on events with a single $c\bar{c}$ pair - selected by Monte-Carlo from the number of possible binary $N + N$ reactions - and follow the dynamics of the charm quarks throughout the time evolution in PHSD, i.e. partonic scattering, hadronization by coalescence or fragmentation, and final hadronic rescattering of charmed mesons and baryons (see below). At the end all observables have to be multiplied by the probability for the charm event as illustrated in Fig. 1 b). The shaded area in Fig. 1 b) shows the uncertainty in the number of $c\bar{c}$ pairs due to the uncertainty of the charm cross section in p+p collisions (cf. Fig. 1 a)). Note that for $\sqrt{s_{NN}} < 20$ GeV no data are available and the number of $c\bar{c}$ pairs entirely stem from a parameterized function which takes into account the phase space of the final states.

The heavy quarks and antiquarks produced in early hard collisions - as described above - in case of heavy-ion collisions interact with the dressed lighter off-shell partons in the QGP. The cross sections for the heavy-quark scattering with massive off-shell partons have been calculated by considering explicitly the mass spectra of the final-state particles in Refs. [41, 42]. The elastic scattering of heavy quarks in the QGP is treated by including the non-perturbative effects of the strongly interacting quark-gluon plasma (sQGP) constituents, i.e. the temperature-dependent coupling $g(T/T_c)$ which rises close to T_c, the multiple scattering etc. [43]. The multiple strong interactions of quarks and gluons in the sQGP are encoded in their effective propagators with broad spectral functions (imaginary parts). As pointed out above, the effective propagators, which can be interpreted as resummed propagators in a hot and dense QCD environment, have been extracted from lattice data in the scope of the DQPM [24, 43].

The produced charm and bottom quarks from p+p collisions are hadronized by emitting soft gluons, which is denoted by 'fragmentation'. As in Refs. [33, 34] we use the fragmentation function of Peterson [44]. In heavy-ion collisions the formation of $D-$ and B-mesons also occurs via coalescence - preferentially at lower p_T - and is described in detail in Refs. [33, 34]. In the hadronic medium the heavy mesons interact with the lighter mesons and baryons according to cross sections that have been calculated in an effective lagrangian approach with heavy-quark spin symmetry [45, 46, 47]. Finally, after freeze-out of the $D-$mesons they produce single electrons through semi-leptonic decays [48] with the branching ratios given by the PYTHIA event generator.

3. Dilepton production channels

We recall that in the hadronic sector PHSD is equivalent to the Hadron-String-Dynamics (HSD) transport approach [49] that has been used for the description of pA and AA collisions from SIS to SPS energies and has lead to a fair reproduction of hadron abundances, rapidity distributions and transverse momentum spectra as well as dilepton spectra. In particular, HSD incorporates off-shell dynamics for vector mesons and a set of vector-meson spectral functions [50] that covers possible scenarios for their in-medium modification, i.e. in particular a collisional broadening of the vector resonances. Note that in the off-shell transport description, the hadron spectral functions change dynamically during the propagation through the medium and evolve towards the on-shell spectral function in the vacuum. The dilepton production by a (baryonic or mesonic) resonance R decay can be schematically presented in the following way: e.g.. in a first step a resonance R might be produced in baryon-baryon (BB) or meson-baryon (mB) collisions. Then this resonance can couple to dileptons directly (e.g., Dalitz decay of the Δ resonance: $\Delta \to e^+e^-N$) or decays to a meson m (+ baryon) , which produces dileptons via direct decays (ρ, ω, ϕ) or Dalitz decays (π^0, η, ω). The resonance R might also decay into another resonance R' which later produces dileptons via Dalitz decay. Note, that in the combined model the final particles – which couple to dileptons – can be produced also via non-resonant mechanisms, i.e. 'background' channels at low and intermediate energies or string decay at high energies. In addition to the hadronic channels above we account for the '4π' channels, i.e. the dilepton production in the two-body reactions $\pi + \rho$, $\pi + \omega$, $\rho + \rho$, $\pi + a_1$ as described in detail in Ref. [51]. The latter provide the background from hadronic channels in the intermediate mass regime 1.2 GeV $< M <$ 3 GeV [51], which is not shown explicitly in this study since the contribution of '4π' channels is much smaller than the contribution from open charm decays and the QGP radiation.

We recall that the influence of in-medium effects on the vector mesons (ρ, ω, ϕ) has been extensively studied within the PHSD approach in the past (cf. Refs. [13, 50, 51]) and it has been shown that the collisional broadening scenario for the in-medium vector-meson spectral functions is consistent with experimental dilepton data from SPS to LHC energies in line with the findings by other groups [12]. Accordingly, in the present study we will adopt the collisional broadening scenario for the vector-meson spectral functions as the 'default' scenario.

In order to address the electromagnetic radiation of the partonic phase, off-shell cross sections of $q\bar{q} \to \gamma^*$, $q\bar{q} \to \gamma^* g$ and $qg \to \gamma^* q$ $(\bar{q}g \to \gamma^* \bar{q})$ reactions - taking into account the effective propagators for quarks and gluons from the DQPM - have been calculated in Ref. [52]. Here γ^* stands for the e^+e^- or $\mu^+\mu^-$ pair. Dilepton production in the QGP - as created in early stages of heavy-ion collisions - is calculated by implementing these off-shell processes into the PHSD transport approach on the basis of the same partonic propagators as used for the time-evolution of the partonic system. For a review on electromagnetic production channels within PHSD we refer the reader to Ref. [13] and for the details of the dilepton cross sections from off-shell partonic channels to the Appendix of Ref. [53].

4. Results for heavy-ion reactions

Since the matter produced in heavy-ion collisions is extremely dense, the interactions with the bulk matter suppresses heavy flavors at high-p_T. On the other hand, the partonic or nuclear matter is accelerated outward (exploding), and a strong flow is generated via the interactions of the bulk particles and the repulsive scalar interaction for partons. Since the heavy flavor

IOP Conf. Series: Journal of Physics: Conf. Series **1070** (2018) 012006 doi:10.1088/1742-6596/1070/1/012006

strongly interacts with the expanding matter, it is also accelerated outwards. Such effects of the medium on the heavy-flavor dynamics are expressed in terms of the nuclear modification factor defined as

$$R_{\mathrm{AA}}(\mathrm{p_T}) \equiv \frac{\mathrm{dN_{AA}/dp_T}}{\mathrm{N_{binary}^{AA}} \times \mathrm{dN_{pp}/dp_T}}, \tag{1}$$

where N_{AA} and N_{pp} are, respectively, the number of particles produced in heavy-ion collisions and that in p+p collisions, and $N_{\mathrm{binary}}^{\mathrm{AA}}$ is the number of binary nucleon-nucleon collisions in the heavy-ion collision for the centrality class considered. Note that if the heavy flavor does not interact with the medium in heavy-ion collisions, the numerator of Eq. (1) will be similar to the denominator. For the same reason, a $\mathrm{R_{AA}}$ smaller (larger) than one in a specific $\mathrm{p_T}$ region implies that the nuclear matter suppresses (enhances) the production of heavy flavors in that transverse momentum region.

4.1. Nuclear modification of dielectrons from heavy flavor

In this subsection we focus on the $c\bar{c}$ dynamics and the dielectrons produced from heavy flavor pairs and their modification in relativistic heavy-ion collisions.

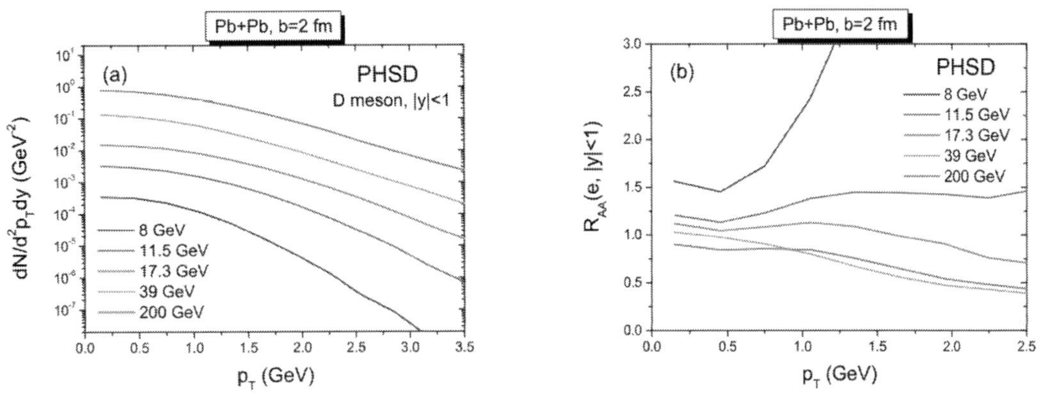

Figure 2. The transverse momentum spectra of D mesons (a) and the $R_{AA}(p_T)$ of single electrons from semi-leptonic decay of D mesons (b) as a function of the transverse momentum p_T in central Pb+Pb collisions from PHSD at $\sqrt{s_{NN}}$ = 8, 11.5, 17.3, 39 and 200 GeV at midrapidity.

Fig. 2 (a) shows the transverse momentum spectra of D mesons in central Pb+Pb collisions at $\sqrt{s_{NN}} = 8$, 11.5, 17.3, 39, and 200 GeV for $|y| < 1$. Since the cross section for charm production increases with collision energy as shown in Fig. 1 (a), the transverse momentum spectrum of D meson enhances strongly with increasing collision energy and also becomes harder. Fig. 2 (b) displays the nuclear modification factor of single electrons from D meson semi-leptonic decays at mid-rapidity ($|y| < 1$) for the same set of central Pb+Pb collisions. We mention that for the semi-leptonic decays of heavy flavors we use the subroutine 'pydecay' of the PYTHIA event generator [54]. Contrary to the R_{AA} at RHIC and LHC energies we find ratios well above unity at $\sqrt{s_{NN}} = 8$ and 11.5 GeV which implies an enhancement of the yield (at higher momenta) rather than the familiar suppression at RHIC and LHC. The enhanced R_{AA} at low energies (8 and 11.5 GeV) may be dominantly attributed to the Fermi motion of nucleons in the colliding nuclei, which does not exist in p+p collisions and slightly increases the collision energy in binary nucleon-nucleon scattering. Since the collision energies are close to the threshold energy for charm production, where the production cross section increases rapidity as shown in Fig. 1 (a), a small enhancement of the collision energy gives a sizeable increase of the charm production and subsequently the decay products.

We note in passing that the invariant mass spectrum of dielectrons changes little for $\sqrt{s_{NN}}$ = 17.3 GeV when including the charm interactions in the hot and dense medium, while it is considerably suppressed at large invariant mass at $\sqrt{s_{NN}}$ = 200 GeV. For further details we refer the reader to Ref. [53].

4.2. Excitation function of dielectron production in Pb+Pb collisions from $\sqrt{s_{\mathrm{NN}}}$ =8 to 200 GeV

The dileptons produced in relativistic heavy-ion collisions can be classified into three parts: i) dileptons from heavy flavor pairs, ii) from partonic scatterings in the QGP phase, and iii) from hadronic interactions in the hadronic (HG) phase. In this subsection we compare the separate contributions in central Pb+Pb collisions at various energies from 8 to 200 GeV.

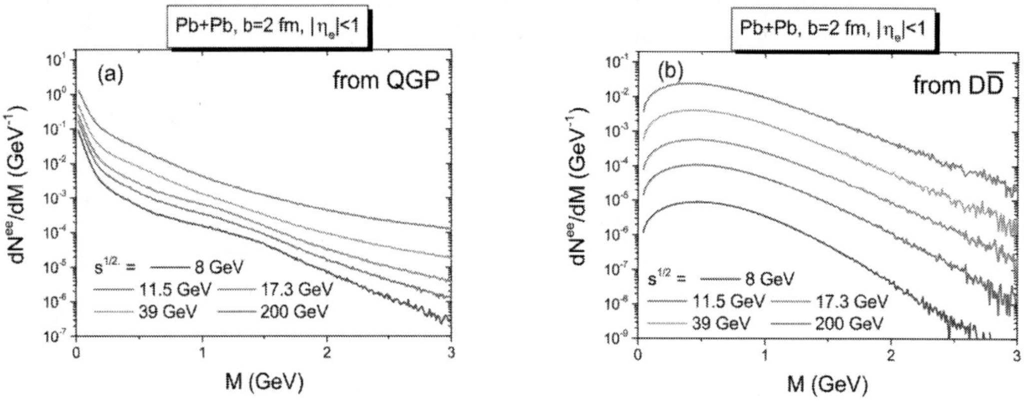

Figure 3. The invariant mass spectra of dileptons from the QGP (a), and $D\bar{D}$ pairs (b) in central Pb+Pb collisions at $\sqrt{s_{NN}}$ = 8, 11.5, 17.3, 39 and 200 GeV from the PHSD.

Fig. 3 shows the dielectron mass spectra from partonic interactions in the QGP (a) and from the semi-leptonic decays of $D\bar{D}$ pairs (b) in central Pb+Pb collisions at $\sqrt{s_{NN}}$ = 8, 11.5, 17.3, 39, and 200 GeV at mid-pseudorapidity $|\eta_e| < 1$ for the leptons. We mention that the contribution from the hadronic channels increases only moderately with collision energy (in line with the hadron abundances) [53], while the contribution from the QGP raises more steeply (in line with the enhanced space-time volume of the QGP phase) and that from $D\bar{D}$ pairs is most dramatically increasing (in line with the number of $c\bar{c}$ pairs, cf. Fig. 1b)). Accordingly, the contribution from heavy flavor is small at low-energy collisions, but becomes more and more important with increasing collision energy in competition with the production from the QGP channels.

In order to show the separate contributions explicitly, we compare in Fig. 4 the contributions from the QGP (red lines) and from $D\bar{D}$ pairs (green lines) with the total dielectron spectrum (blue lines) at different collision energies for central Pb+Pb collisions. In low-energy collisions the dielectrons from hadronic channels dominate in the low-mass region and those from partonic interactions dominate in the intermediate-mass range while the contribution from $D\bar{D}$ pairs is negligible. With increasing collision energy the contribution from $D\bar{D}$ pairs becomes more and more significant and comparable to that from partonic interactions at $\sqrt{s_{NN}} \approx$ 39 GeV in the

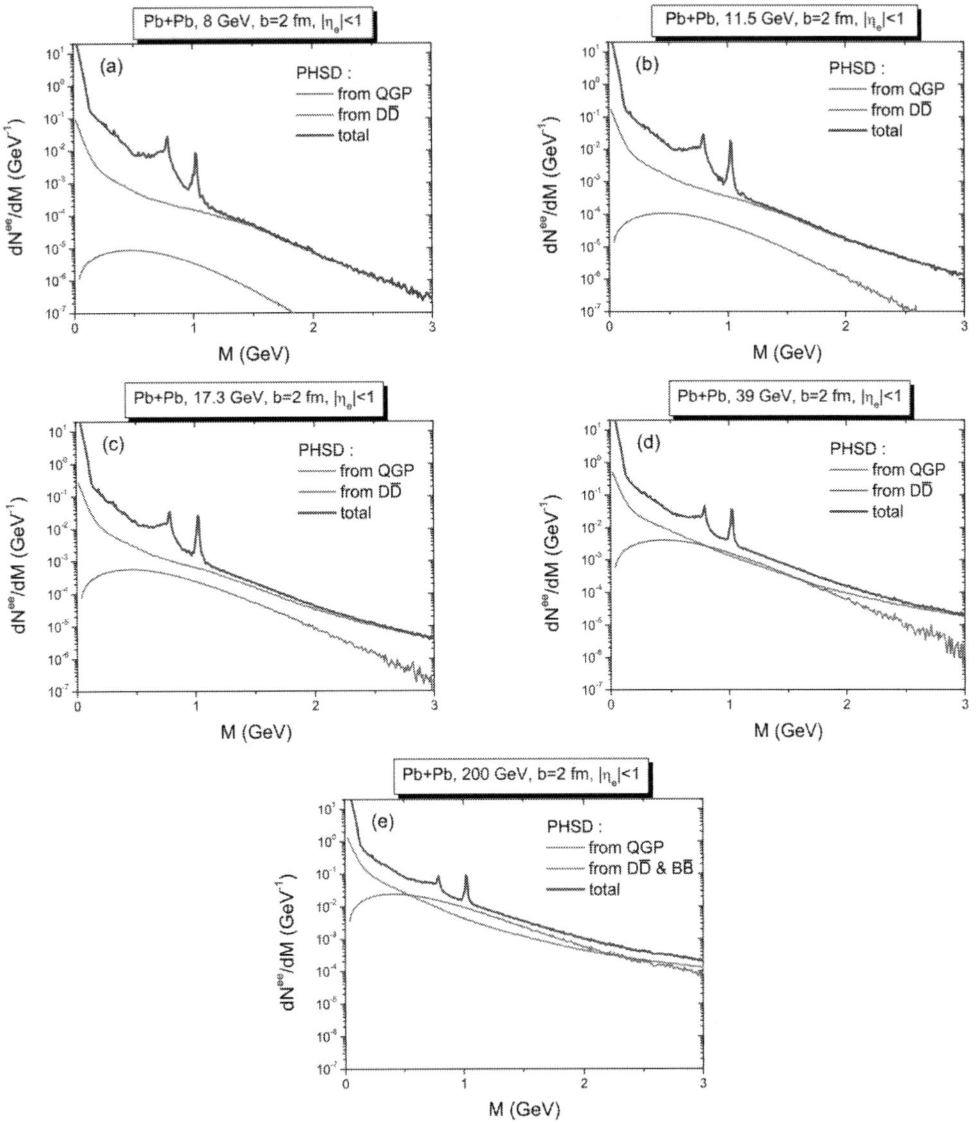

Figure 4. The invariant mass spectra of dileptons from partonic interactions (red lines) and from $D\bar{D}$ pairs (green lines) together with the total dielectron spectrum (blue lines) in central Pb+Pb collisions at $\sqrt{s_{NN}}$ = 8, 11.5, 17.3, 39 and 200 GeV from the PHSD at mid-pseudorapidity for the leptons.

intermediate-mass range. Finally, it overshines the partonic contribution at $\sqrt{s_{NN}} = 200$ GeV (and above).

Fig. 5 compares the contributions from $D\bar{D}$ pairs (green lines) to three partonic channels, i.e. $q+\bar{q} \to e^+ + e^-$, $q+\bar{q} \to g + e^+ + e^-$, and $q(\bar{q}) + g \to q(\bar{q}) + e^+ + e^-$, for intermediate mass dileptons (1.2 GeV $< M <$ 3 GeV) as a function of collision energy $\sqrt{s_{NN}}$ for Pb+Pb collisions at b=2 fm. The figure clearly shows that the contribution from partonic interactions, especially from $q+\bar{q} \to e^+ + e^-$, dominates the intermediate-mass range in low-energy collisions. However,

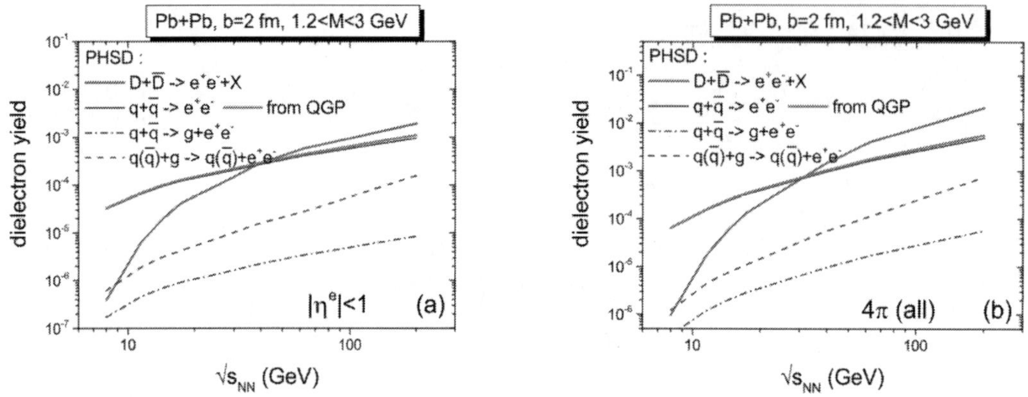

Figure 5. The contributions to intermediate-mass dielectrons (1.2 GeV $< M <$ 3 GeV) from $D\bar{D}$ pairs (green lines), different channels of partonic interactions, $q + \bar{q} \to e^+ + e^-$, $q + \bar{q} \to g + e^+ + e^-$, $q(\bar{q}) + g \to q(\bar{q}) + e^+ + e^-$ (see legend) as a function of $\sqrt{s_{NN}}$ for Pb+Pb collisions at b=2 fm for midrapidity leptons (a) and in full 4π (b). The red solid line displays the sum of the partonic contributions.

the contribution from $D\bar{D}$ pairs rapidly increases with increasing collision energy, because the scattering cross section for charm production grows fast above the threshold energy as shown in Fig. 1 (a). It overshines the contribution from partonic interactions around $\sqrt{s_{NN}} \approx 40$ GeV and dominates at higher energies. Since the detectors of different collaborations have a different acceptance, we show in Fig. 5 (b) the results without any acceptance cuts, while Fig. 5 (a) shows the results for a mid-pseudorapidity cut on leptons of $|\eta^e| < 1$. However, the contributions from the partonic interactions and from $D\bar{D}$ pairs show a similar behavior in both cases.

One of most important issues in heavy-ion physics is to find and study the properties of partonic nuclear matter which is created in a small space-time volume in relativistic heavy-ion collisions. To this end one needs observables that are not blurred by hadronic interactions. Our results in Figs. 4 and 5 clearly demonstrate that the window to study partonic matter by dielectrons at intermediate masses without substantial background from heavy flavor decays opens for collision energies $\sqrt{s_{NN}} < 40$ GeV.

5. PHSD versus experimental data

In this section we compare the invariant mass spectra of dielectrons from the PHSD to the experimental data in Au+Au collisions from $\sqrt{s_{NN}} = 19.6$ to 200 GeV from the STAR collaboration and those in Pb+Pb collisions from the ALICE collaboration at $\sqrt{s_{NN}} = 2.76$ TeV. We note that the experimental data from the STAR collaboration and those from the ALICE collaboration have different centralites and different acceptance cuts. The STAR data are obtained for minimum-bias Au+Au collisions and electrons and positrons with transverse momenta $p_T \geq 0.2$ GeV and pseudo-rapidities $|\eta^e| < 1.0$. On other hand, the ALICE data are obtained for 0-10 % central Pb+Pb collisions and the electrons and positrons with transverse momenta $p_T \geq 0.4$ GeV and pseudo-rapidities $|\eta^e| < 0.8$. The sensitivity of the invariant mass spectra of dielectrons to the cross section for charm production and cuts in p_T and pseudo-rapidity η^e is discussed in more detail in the appendix of Ref. [53].

The first five panels of Fig. 6 show the invariant mass spectra of dielectrons from the Beam-Energy-Scan (BES) at $\sqrt{s_{NN}} = 19.6$, 27, 39, and 62.4 GeV and from the top RHIC energy. As discussed in the previous subsection, the contribution from hadrons is dominant in the low-

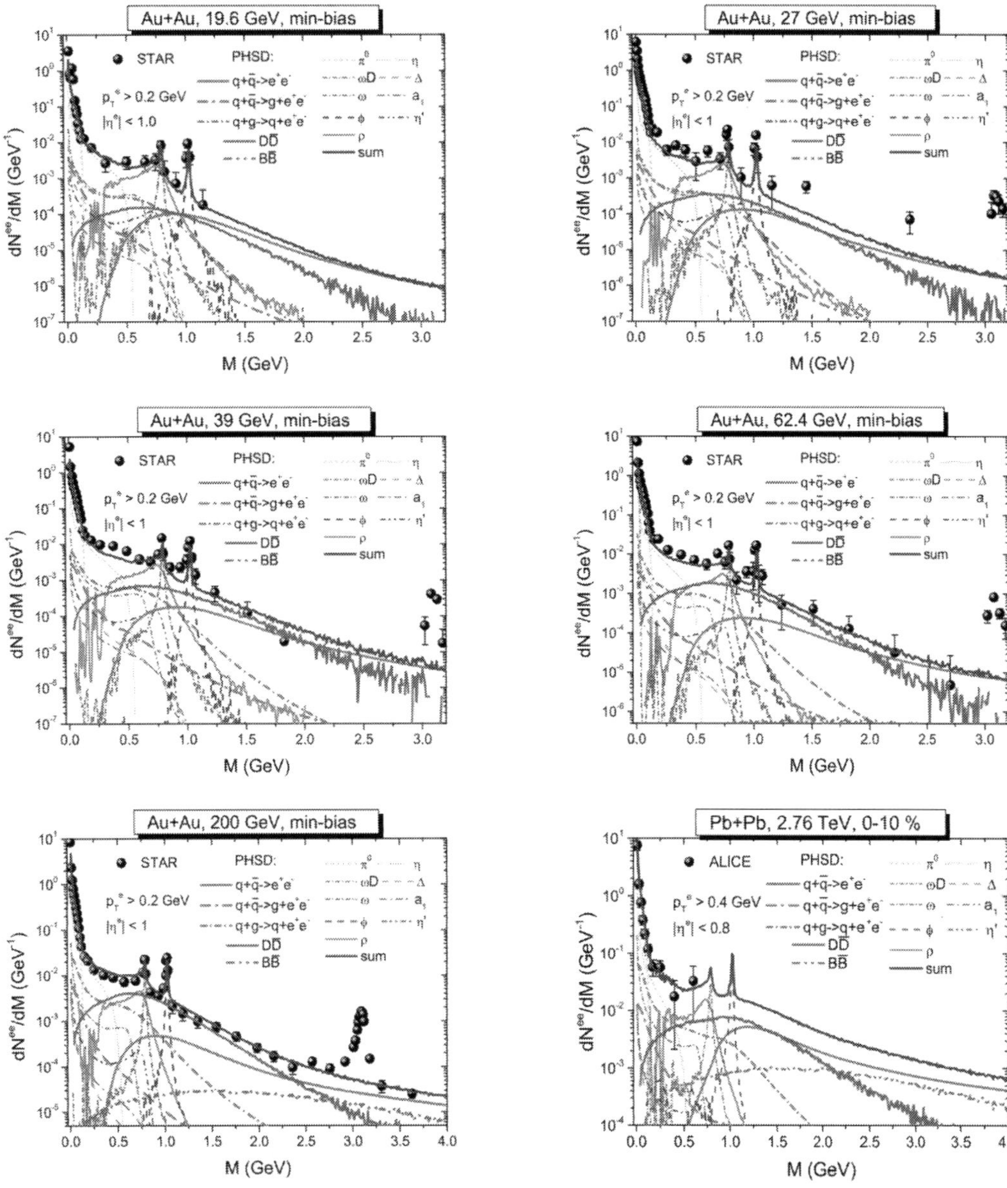

Figure 6. The invariant mass spectra of dielectrons from the PHSD in comparison to the STAR data in Au+Au collisions from $\sqrt{s_{NN}}$ = 19.6 to 200 GeV [55, 56] and to the ALICE data in Pb+Pb collisions at $\sqrt{s_{NN}}$ = 2.76 TeV [57]. The total yield is displayed in terms of the blue lines while the different contributions are specified in the legends. Note that the contribution from J/Ψ and Ψ' decays are not included in the PHSD calculations.

mass region and signals a broadening of the ρ meson spectral function in dense nuclear matter (cf. Ref. [13]. On the other hand, the intermediate-mass range originates predominantly by dielectrons from partonic interactions and those from heavy flavor decays. Similar to the Pb+Pb

IOP Conf. Series: Journal of Physics: Conf. Series **1070** (2018) 012006 doi:10.1088/1742-6596/1070/1/012006

collisions in Fig. 4), the contribution from heavy flavor becomes more and more important with increasing collision energy. The contribution from heavy flavors and from partonic interactions cross around invariant masses $M \approx 1$ GeV in Au+Au collisions at $\sqrt{s_{NN}} = 19.6$ GeV. However, the crossing point shifts to 1.6 GeV at $\sqrt{s_{NN}} = 27$ GeV and to ~2.0 GeV at $\sqrt{s_{NN}} = 39$ and 62.4 GeV. At the top RHIC energy they cross at ~2.4 GeV.

The last panel of Fig. 6 is the invariant mass spectrum of dielectrons in central Pb+Pb collisions at $\sqrt{s_{NN}} = 2.76$ TeV. As in Au+Au collisions at the RHIC energies, the low-mass range is dominated by the dielecrons from hadronic channels and the intermediate-mass region by those from partonic interactions and heavy flavor decays. However, the crossing point of the contribution from partonic interactions and that from heavy flavor is lower than at the top RHIC energy, which is due to a couple of effects: i) the cross section for charm production no longer increases rapidly at the LHC energies as shown in Fig. 1 (a). It is also seen in Fig. 1 (b), which shows the number of produced charm pairs as a function of collision energy. As a result, the growth in the number of produced charm pairs is not faster than the growth of dielectrons from partonic interactions. Additionally the shadowing effect, which is the modification of the parton distribution function in nuclei [58], considerably suppresses charm production at the LHC energies [34]. ii) Another reason is the stronger suppression of the charm four-momentum by partonic scattering at the LHC energies. The strong interaction of heavy flavor with the medium reduces the invariant mass of dielectrons. Since the interaction is stronger at the LHC energies, we can expect a larger suppression of the dielectron spectrum at larger invariant masses. iii) Furthermore, at the LHC energies the contribution from semi-leptonic $B\bar{B}$ decays becomes important. Comparing the lower two panels of Fig. 6, the contribution from $B\bar{B}$ decays is found to be larger than that from $D\bar{D}$ decays above $M \approx 2.2$ GeV in Pb+Pb collisions at $\sqrt{s_{NN}} = 2.76$ TeV, while the contribution from $B\bar{B}$ decays is larger only above $M \approx 3$ GeV in Au+Au collisions $\sqrt{s_{NN}} = 200$ GeV. Since the contribution from $B\bar{B}$ decays amounts to about 50% of the contribution from partonic interactions at the LHC energies, it will distort the information on partonic matter in the intermediate-mass range of the dielectron spectrum.

Besides the interesting points mentioned above, we close this subsection with the comment that the dilepton invariant mass spectra from the PHSD describe reasonably well the available experimental data for collision energies from 19.6 GeV to 2.76 TeV although the experimental data at $\sqrt{s_{NN}} = 2.76$ TeV are available only for invariant masses $M \leq 1$ GeV.

6. Summary

We have studied correlated electron (e^+e^-) production through the semi-leptonic decay of charm hadrons in relativistic heavy-ion collisions from $\sqrt{s_{\rm NN}} = 8$ GeV to 2.76 TeV within the PHSD transport approach in extension of our work on $D-$meson production in relativistic heavy-ion collisions at RHIC and LHC energies [33, 34, 35] and low mass dilepton production from SIS to RHIC energies [13]. In the PHSD the charm partons - produced by the initial hard nucleon-nucleon scattering - interact with the massive quarks and gluons in the QGP by using the scattering cross sections calculated in the Dynamical Quasi-Particle Model (DQPM) which reproduces heavy-quark diffusion coefficients from lattice QCD calculations at temperatures above the deconfinement transition [43]. When approaching the critical energy density for the phase transition from above, the charm (anti)quarks are hadronized into $D-$mesons through the coalescence with light (anti)quarks. Those heavy quarks, which fail in coalescence until the local energy density is below 0.4 GeV/fm^3, hadronize by fragmentation as in p+p collisions [34]. The hadronized $D-$mesons then interact with light hadrons in the hadronic phase with cross sections that have been calculated in an effective lagrangian approach with heavy-quark spin symmetry. Finally, after freeze-out of the $D-$mesons they produce single electrons through semi-leptonic decays with the branching ratios given by the PYTHIA event generator.

The dilepton production from hadronic and partonic channels in central Pb+Pb (or Au+Au) collisions has been calculated including also the contribution from the semi-leptonic decays of heavy flavors in PHSD for the first time on a fully microscopic level. We recall that also the cross sections for dilepton production have been calculated by employing the same propagators and couplings as incorporated in the partonic dynamics in PHSD (cf. [53]). We find that even in central Pb+Pb collisions at $\sqrt{s_{\rm NN}}$ = 8 to 20 GeV the contribution from $D, \bar{D}$ mesons to the intermediate mass dilepton spectra is subleading and one should have a rather clear signal from the QGP radiation whereas at the top RHIC energy this contribution overshines the intermediate mass dileptons from the QGP. It is interesting to note that the dielectrons from $D, \bar{D}$ mesons do not increase any more relative to partonic interactions at the LHC energies for a couple of reasons: i) the cross section for charm production does not grow as fast as at low energies; ii) the shadowing effects, which suppress charm production at low transverse momentum, are stronger at LHC than at RHIC energies; iii) the charm quark pair looses more four-momentum in the partonic medium produced at the LHC, which suppresses the invariant mass of the dielectrons from the semi-leptonic decays. Furthermore, the contribution from $B, \bar{B}$ meson decays becomes more important and superseeds the contribution from $D, \bar{D}$ meson decays above $M = 2.2 \sim 2.3$ GeV at the LHC energies and amounts to about half the contribution from partonic interactions. All these effects strongly distort the information about partonic matter from intermediate-mass dielectrons at the LHC energies. The dilepton spectra at lower masses (0.2 GeV $\leq M \leq$ 0.7 GeV) at SPS, FAIR/NICA and BES RHIC energies show some sensitivity to the medium modification of the ρ meson where the data favor an in-medium broadening as pointed out in the earlier studies on dilepton production reviewed in Refs. [12, 13].

In general the PHSD calculations compare well with the available dilepton data from the BES program at RHIC as well as the LHC energy of $\sqrt{s_{NN}}$ = 2.76 TeV where, unfortunately, only low mass dilepton data are available so far. As noted above, the large background from $D, \bar{D}$ as well as $B, \bar{B}$ correlated semi-leptonic decays - in the intermediate mass range - is by far subleading at lower SPS and FAIR/NICA energies which provides promising perspectives for the future dilepton measurements at these facilities and allows for a fresh look at the electromagnetic radiation from the QGP at finite baryon chemical potential.

Acknowledgements

The authors acknowledge inspiring discussions with J. Butterworth, F. Geurts and O. Linnyk. This work was supported by the LOEWE center "HIC for FAIR", the HGS-HIRe for FAIR and the COST Action THOR, CA15213. Furthermore, PM and EB acknowledge support by DFG through the grant CRC-TR 211 'Strong-interaction matter under extreme conditions'. The computational resources have been provided by the LOEWE-CSC.

References

[1] Bernard C *et al.* [MILC Collaboration] 2005 *Phys. Rev.* D **71** 034504
[2] Aoki Y, Endrodi G, Fodor Z, Katz S and D Szabo K K 2006 *Nature* **443** 675
[3] Aoki Y *et al.* 2009 *JHEP* **0906** 088
[4] Bazavov A *et al.* 2012 *Phys. Rev.* D **85** 054503
[5] Mohanty B [STAR Collaboration] 2011 *J. Phys.* G **38** 124023
[6] Kumar L [STAR Collaboration] 2011 *J. Phys.* G **38** 124145
[7] Cassing W, Palmese A, Moreau P and Bratkovskaya E L 2016 *Phys. Rev.* C **93** 014902
[8] Palmese A Cassing W Seifert E Steinert T, Moreau P and Bratkovskaya E L 2016 *Phys. Rev.* C **94** 044912
[9] Feinberg E L 1970 *Izv. Akad. Nauk Ser. Fiz.* **34** 1987
[10] Shuryak E V 1978 *Phys. Lett.* B **78** 150
[11] Bauer W, Bertsch G F, Cassing W and Mosel U 1986 *Phys. Rev.* C **34** 2127
[12] Rapp R 2013 *Adv. High Energy Phys.* **2013** 148253
[13] Linnyk O, Bratkovskaya E and Cassing W 2016 *Prog. Part. Nucl. Phys.* **87** 50
[14] Moore G D and Teaney D 2005 *Phys. Rev.* C **71** 064904
[15] Zhang B, Chen L W and Ko C M 2005 *Phys. Rev.* C **72** 024906

IOP Conf. Series: Journal of Physics: Conf. Series **1070** (2018) 012006 doi:10.1088/1742-6596/1070/1/012006

[16] Molnar D 2007 *Eur. Phys. J.* C **49** 181
[17] Linnyk O, Bratkovskaya E L and Cassing W 2008 *Int. J. Mod. Phys.* E **17** 1367
[18] Gossiaux P B, Aichelin J, Gousset T and Guiho V 2010 *J. Phys.* G **37** 094019
[19] Cao S and Bass S A 2011 *Phys. Rev.* C **84** 064902
[20] Das S K, Scardina F, Plumari S and Greco V 2015 *Phys. Lett.* B **747** 260
[21] Larionov A B, Cassing W, Greiner C and Mosel U 2000 *Phys. Rev.* C **62** 064611
[22] Sahu P K, Cassing W, Mosel U and Ohnishi A 2000 *Nucl. Phys.* A **672** 376
[23] Cassing W and Bratkovskaya E L 2009 *Nucl. Phys.* A **831** 215
[24] Cassing W 2009 *Eur. Phys. J. ST* **168** 3
[25] Bratkovskaya E L, Cassing, W Konchakovski V P and Linnyk O 2011 *Nucl. Phys.* A **856** 162
[26] Konchakovski V P *et al.* 2015 *J. Phys. G* **42** 055106
[27] Konchakovski V P *et al.* 2012 *Phys. Rev.* C **85** 011902
[28] Konchakovski V P *et al.* 2012 *Phys. Rev.* C **85** 044922
[29] Konchakovski V P *et al.* 2014 *Phys. Rev.* C **90** 014903
[30] Seifert E and Cassing W 2018 *Phys. Rev.* C **97** 024913
[31] Seifert E and Cassing W 2018 *Phys. Rev.* C **97** 044907
[32] Linnyk O, Konchakovski V P, Cassing W and Bratkovskaya E L 2013 *Phys. Rev.* C **88** 034904
[33] Song T *et al.* 2015 *Phys. Rev.* C **92** 014910
[34] Song T *et al.* 2016 *Phys. Rev.* C **93** 034906
[35] Song T *et al.* 2017 *Phys. Rev.* C **96** 014905
[36] Andronic A *et al.* 2016 *Eur. Phys. J.* C **76** 107
[37] Prino F and Rapp R 2016 *J. Phys.* G **43** 093002
[38] Conesa del Valle Z [ALICE Collaboration] 2012 *AIP Conf. Proc.* **1441** 886
[39] Adamczyk L *et al.* [STAR Collaboration] 2012 *Phys. Rev.* D **86** 072013
[40] Adare A *et al.* [PHENIX Collaboration] 2017 *Phys. Rev.* C **96** 024907
[41] Berrehrah H *et al.* 2014 *Phys. Rev.* C **89** 054901
[42] Berrehrah H *et al.* 2014 *Phys. Rev.* C **90** 064906
[43] Berrehrah H *et al.* 2016 *Int. Journal of Mod. Phys. E* **25** 1642003
[44] Peterson C, Schlatter D, Schmitt I and Zerwas P M 1983 *Phys. Rev.* D **27** 105
[45] Tolos L and Torres-Rincon J M 2013 *Phys. Rev.* D **88** 074019
[46] Torres-Rincon, J M Tolos and L Romanets O 2014 *Phys. Rev.* D **89** 074042
[47] Tolos L 2013 *Int. J. Mod. Phys.* E **22** 1330027
[48] Olive K A *et al.* [Particle Data Group Collaboration] 2014 *Chin. Phys.* C **38** 090001
[49] Cassing W and Bratkovskaya E L 1999 *Phys. Rept.* **308** 65
[50] Bratkovskaya E L and Cassing W 2008 *Nucl. Phys.* A **807** 214
[51] Linnyk O, Bratkovskaya E L, Ozvenchuk V, Cassing W and Ko C M 2011 *Phys. Rev.* C **84** 054917
[52] Linnyk O 2011 *J. Phys.* G **38** 025105
[53] Song T, Cassing W, Bratkovskaya E and Moreau P 2018 *arXiv:*1803.02698 *Phys. Rev.* C in print
[54] Sjostrand T, Mrenna S and Skands P Z 2006 *JHEP* **0605** 026
[55] Huck P [STAR Collaboration] 2014 *Nucl. Phys.* A **931** 659
[56] Adamczyk L *et al.* [STAR Collaboration] 2015 *Phys. Rev.* C **92** 024912
[57] Gunji T [ALICE Collaboration] 2017 *Nucl. Part. Phys. Proc.* **289-290** 181
[58] Eskola K J, Paukkunen H and Salgado C A 2009 *JHEP* **0904** 065

IOP Conf. Series: Journal of Physics: Conf. Series **1070** (2018) 012007 doi:10.1088/1742-6596/1070/1/012007

System size dependence of particle production in EPOS

K. Werner[a], **B. Guiot**[b], **Iu. Karpenko**[a], **T. Pierog**[c], **G. Sophys**[a], **M. Stefaniak**[a,d]

[a] SUBATECH, University of Nantes - IN2P3/CNRS - IMT Atlantique, Nantes, France; [b] Universidad Tecnica Federico Santa Maria y Centro Cientifico Tecnologico de Valparaiso, Valparaiso, Chile; [c] Karlsruhe Inst. of Technology, KIT, Campus North, Inst. f. Kernphysik, Germany; [d] Warsaw University of Technology, Warsaw, Poland

Abstract. The aim of this paper is to understand particle production for different collision systems, namely proton-proton (pp), proton-nucleus (pA), and nucleus-nucleus (AA) scattering at the LHC. We will investigate in particular particle yields and ratios versus multiplicity, using the same multiplicity definition for the three different systems, in order to analyse in a compact way the evolution of particle production with the system size and the origin of a very different system size dependence of the different particles.

The main goal of heavy ion physics at very high energies is the proof of existence of the creation of a quark-gluon plasma, and the study of the properties of this exotic state, by analyzing the final state of many thousands of produced hadrons. We will investigate particle production in pp, pA, and AA collisions, our analysis tool being the EPOS model. EPOS3 [1] is a universal model in the sense that for pp, pA, and AA collisions, the same procedure applies, based on several elements:

Initial state. A Gribov-Regge multiple scattering approach is employed ("Parton-Based Gribov-Regge Theory" PBGRT [2]), where the elementary object (by definition called Pomeron) is a DGLAP parton ladder. Starting from a multi-Pomeron structure for the elastic scattering S-matrix, one then uses cutting rules to obtain (partial) cross sections of inelastic processes, by employing Markov chain techniques.

Non-linear effects. High parton density effects like gluon fusion during the parton evolution (see fig. 1(a)) are taken into account by using a dynamical saturation scale for each Pomeron, of the form $Q_s = Q_s(N_{\mathrm{Pom}}, \hat{s})$, depending on the number of Pomerons N_{Pom} connected the Pomeron in question (see

(a) 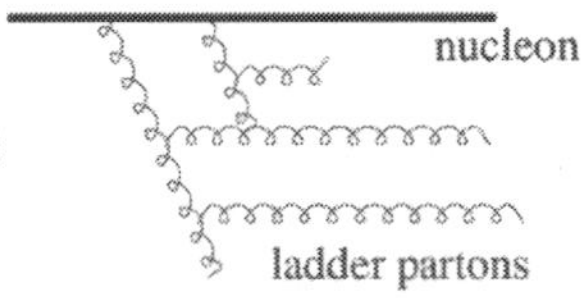

(b) 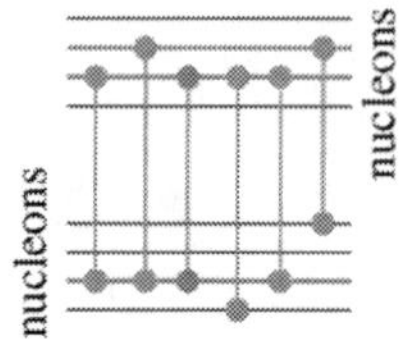

Figure 1. (Color online) (a) Non-linear effect: ladder fusion. (b) Pomerons connected to a given (red) Pomeron

fig. 1(b)), and its energy $\hat{s}$. The functional form of Q_s is obtained from fitting to the energy dependence of basic cross sections and the multiplicity dependence of particle production in pp at very high transverse momenta.

Core-corona approach. The parton ladders corresponding to Pomerons are treated as classical relativistic (kinky) strings. So in general, we have a large number of (partly overlapping) strings. Based on the momenta and the density of string segments, one separates at some early proper time τ_0 the core (going to be treated as fluid) from the corona (escaping hadrons, including jet hadrons). The core-corona procedure has been first described in [3], a more recent discussion is found in [1]. The corresponding energy-momentum tensor of the core part is transformed into an equilibrium one, needed to start the hydrodynamical evolution. This is based on the hypothesis that equilibration happens rapidly and affects essentially the space components of the energy-momentum tensor.

Viscous hydrodynamic expansion. Starting from the initial proper time τ_0, the core part of the system evolves according to the equations of relativistic viscous hydrodynamics [1, 4], where we use presently $\eta/s = 0.08$. A cross-over equation-of-state is used, compatible with lattice QCD [5, 6]. The "core-matter" hadronizes on some hyper-surface defined by a constant temperature T_H, where a so-called Cooper-Frye procedure is employed, using equilibrium hadron distributions, see [6].

Final state hadronic cascade. After hadronization, there occur still hadron-hadron rescatterings, realized via UrQMD [7].

The above procedure is employed for each event (event-by-event procedure).

We will discuss particle production as a function of the average multiplicity per eta interval $\langle dn/d\eta(0)\rangle$ at central pseudorapidity ($\eta = 0$), whereas forward pseudorapidity intervals are used to define the different multiplicity classes. This is the same procedure as used by the ALICE experiment. In the following $\langle dn/d\eta(0)\rangle$ is simply referred to as "multiplicity".

Crucial for the following discussion is the fact that the core-corona separation behaves quite differently for small and big systems (although we employ the same procedure), see fig. 2. Whereas for very large multiplicity (central AA) most of the matter is core, with small corona contributions at

IOP Conf. Series: Journal of Physics: Conf. Series **1070** (2018) 012007 doi:10.1088/1742-6596/1070/1/012007

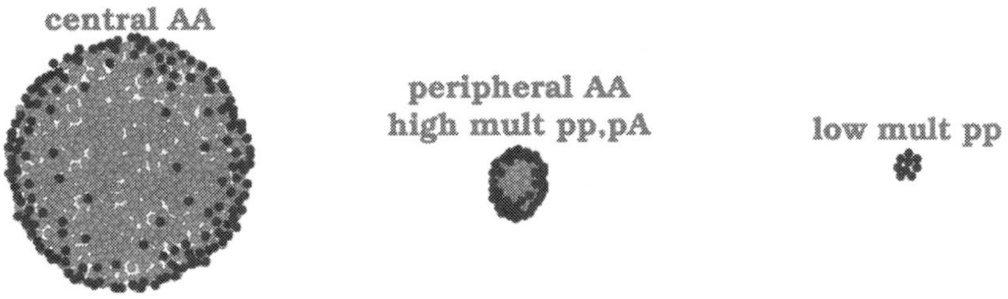

Figure 2. (Color online) Core (red) and corona (blue) in a plane perpendicular to the beam axis, for big (left), small (right) and intermediate (middle) systems.

the surface (or some high pt segments further inside), we get at very low multiplicity (in pp) only corona : the system is not dense enough to form a core. High multiplicity pp and pA, or peripheral AA collisions are in between, both core and corona being important.

core	green dashed-dotted	particle from the core only	no hadronic cascade
corona	blue dotted	particles from corona only	no hadronic cascade
co-co	yellow dashed	particles from core and corona	no hadronic cascade
full	red full	all particles	with hadronic cascade
	blue triangles	particles from pure string decay	no cascade, no hydro
	thin lines	pp simulation	
	intermediate lines	pA (pPb) simulation	
	thick lines	AA (PbPb) simulation	
	open circles	pp data	
	open squares	pA (pPb) data	
	open stars	AA (PbPb) data	

Table 1. Color, line, and symbol codes used for the plots in this paper, to accommodate the different contributions and the different systems.

Another way of seeing the multiplicity dependence of the core-corona separation is shown in fig. 3, where we plot the pion yield versus multiplicity, for only core particles ("core", green dashed-dotted), the contribution from core + corona ("co-co", particles from core and corona, yellow dashed), both contributions from EPOS simulation without hadronic cascade. The ratio "core" over "co-co" would be the relative core fraction, which increases continously from zero (at small multiplicity) to unity (at large multiplicity). We also plot the "full" contribution, referring to all particles, for a simulation with hadronic cascade (A summary of color, line, and symbol codes is found in table 1). In all cases we plot curves for the three systems: pp, pA, and AA. There are substantial overlap regions, where different systems contribute, but we observe unique curves, no system size dependence. We also observe that the

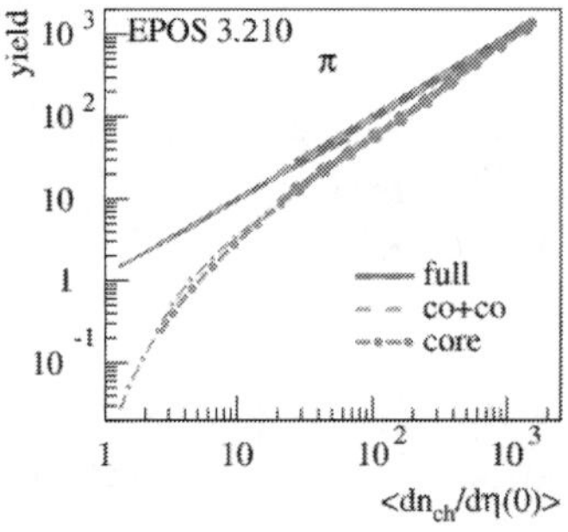

Figure 3. (Color online) Pion yield versus multiplicity. We compare different contributions (core, core plus corona (co-co) and the the full contribution (the latter on including the final state hadronic cascade) for different systems: pp (thin lines), AA (thick lines), and pA (intermediate lines). (A summary of color, line, and symbol codes is found in table 1)

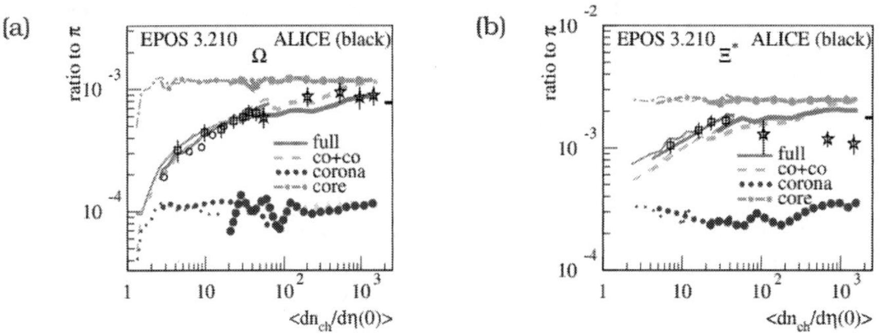

Figure 4. (Color online) Particle to pion ratio for different particle species versus multiplicity, for different contributions from the EPOS simulations, for different systems (pp, pA, AA). We also plot ALICE data. These and data for many more hadrons can be found in [8, 9, 10, 11, 12, 13, 14, 15, 16, 17, 18, 19, 20]. See table 1 for color, line, and symbol codes.

hadronic cascade has no effect on the pion production ("co-co" and "full" gives identical results).

In the following we will investigate the multiplicity dependence (for pp, pA, AA) of some particle ratios to pions, for the different contributions from EPOS simulations, compared to data from the ALICE collaboration, who measured such ratios for many particle species [8, 9, 10, 11, 12, 13, 14, 15, 16, 17]. We always refer to the color, line, and symbol codes of table 1.

In fig. 4(a), we plot the Ω to pion ratio versus multiplicity, for different contributions from the EPOS simulations, for different systems (pp, pA, AA), see table 1 for color, line, and symbol codes. Despite a large overlap for

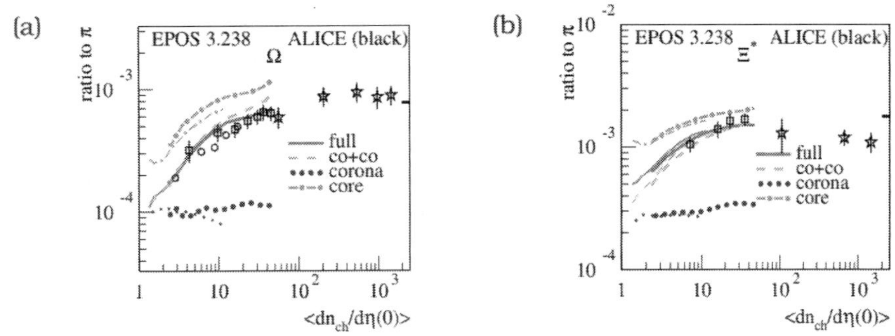

Figure 5. (Color online) Sama as figs. 4, but here we consider phase space decay.

the different systems, we observe universal curves, for all contributions. The core contribution is completely flat, as expected since the particle ratios only depend on the properties of the fluid at freeze out, taken to be the same in all systems. Also the corona contribution (from string segments which escape the fluid) is essentially flat, so we have two flat curves, from core and corona, but here core and corona curves are separated by about a factor of 10, leading to a strong increase of the core+corona (co-co) contribution of a factor of 10 with multiplicity. The full curve is slightly reduced at high multiplicity compared to co-co, due to hadronic final state interactions (baryon-antibaryon annihilation). We observe very similar results for the Ξ^* resonance (lifetime 21.7 fm/c), see fig. 4(b), but some discrepancy compared to the data.

It is not be very realistic to use the (grand canonical) Cooper-Frye prescription for decaying the plasma, in case of very small systems. As a first test, we employ the EPOS LHC model, which uses n-body phase space decay, with complete energy and momentum conservation (but parameterized flow rather than hydrodynamics). The results are shown in fig. 5. Now, the core curves are no longer flat, but they decrease at small multiplicities. But core alone is not enough to decribe the data, still core-corona separation is needed, with a slightly bigger core weight compared to the case of Cooper-Frye prescription.

To summarize, we have shown that the study of the multiplicity dependence of particle ratios for different collision systems (pp, pA, AA) provides useful information concerning the production mecanism of particles. Using phase space decay compared to the Cooper-Frye prescription changes ther core curves, but still core-corona separation is needed to explain the data.

Acknowlegments This research was carried out within the scope of the GDRI EUREA , and was partially supported by COST Action CA15213 (THOR). B. G. gratefully acknowledges generous support from Chilean FONDECYT grants 3160493 and Proyecto Basal FB0821. Iu.K. acknowledges support by Region

Pays de la Loire (France) under contract no. 2015-08473.

[1] K. Werner, B. Guiot, I. Karpenko, and T. Pierog, Phys.Rev. C89 (2014) 064903, arXiv:1312.1233.
[2] H. J. Drescher, M. Hladik, S. Ostapchenko, T. Pierog and K. Werner, Phys. Rept. 350, 93, 2001
[3] K. Werner, Phys. Rev. Lett. 98, 152301 (2007)
[4] Iu. Karpenko, P. Huovinen, M. Bleicher, arXiv:1312.4160
[5] S. Borsanyi et al., JHEP 1011 (2010) 077, arXiv:1007.2580
[6] K. Werner, Iu. Karpenko, T. Pierog, M. Bleicher, K. Mikhailov, arXiv:1010.0400, Phys. Rev. C 83, 044915 (2011)
[7] M. Bleicher et al., J. Phys. G25 (1999) 1859; H. Petersen, J. Steinheimer, G. Burau, M. Bleicher and H. Stocker, Phys. Rev. C78 (2008) 044901
[8] ALICE Collaboration, Phys. Rev. Lett. 106 032301 (2011), Phys. Rev. C 88 044910 (2013)
[9] ALICE Collaboration, Phys. Rev. Lett. 111 222301 (2013)
[10] ALICE Collaboration, Phys. Lett. B 758 389-401 (2016)
[11] ALICE Collaboration, Phys. Lett. B 728 25-38 (2014)
[12] ALICE Collaboration, Eur. Phys. J. C 76 245 (2016)
[13] ALICE Collaboration, Phys. Lett. B 712 309 (2012)
[14] ALICE Collaboration, Phys. Lett. B 728 216–227 (2014)
[15] ALICE Collaboration, Phys. Rev. C 91 024609 (2015)
[16] ALICE Collaboration, Nature Physics 13 (2017) 535-539
[17] digitized from plenary talk of Rafael Derradi de Souza (for the ALICE Collaboration), SQM2016
[18] derived from talk of Jihye Song (for the ALICE Collaboration) at Quark Matter 2017
[19] digitized from A. G. Knospe (for the ALICE Collaboration) at SQM 2016
[20] ALICE, Eur. Phys. J. C 77 (2017) 389

IOP Conf. Series: Journal of Physics: Conf. Series **1070** (2018) 012008 doi:10.1088/1742-6596/1070/1/012008

Quarkonia production in a Langevin approach

N Krenz, H van Hees and C Greiner

Institut für theoretische Physik, Goethe-Universität Frankfurt am Main, Max-von-Laue-Straße 1, 60438 Frankfurt, Germany

E-mail: krenz@th.physik.uni-frankfurt.de

Abstract. We aim to describe the process of dissociation and recombination of quarkonia in the quark-gluon plasma. Therefore we developed a model which allows to observe the time evolution of a system with various numbers of charm-anticharm-quark pairs at different temperatures. The motion of the heavy quarks is realized within a Langevin approach. We use a simplified version of a formalism developed by Blaizot et al. in which an Abelian plasma is considered where the heavy quarks interact over a Coulomb like potential. We have demonstrated, that the system reaches the expected thermal distribution in the equilibrium limit.

1. Introduction

Heavy quarks are an important tool for the investigation of the quark-gluon plasma (QGP). They are primarily produced in the primordial hard collisions of the heavy ion collision, and their number is conserved until the hadronic freezout. Therefore heavy quarks carry information about the whole evolution of the QGP. Especially the surviving probability of heavy-quark bound states such as J/ψ or Υ can give an insight to the medium properties.

The potential between two heavy quarks is screened by the surrounding medium. As suggested long ago, the suppression of J/ψ could be an evidence for the formation of the deconfined state [1]. Higher temperatures should lead to larger screening effects with a full suppression of J/ψ at very high beam energies. The predicted suppression was found at the SPS at CERN [2] but measurements at RHIC at higher beam energies did not show an increase of the suppression [3]. To explain the results the process of recombination of J/ψ inside the medium has been suggested. At higher beam energies the number of initially produced heavy quarks is larger, leading to a higher possibility that a heavy quark which propagates through the medium meets a partner to form a new quarkonium state. The theoretical investigation of recombination processes is therefore necessary to predict the number of J/ψ-mesons measured in the experiments.

The comparatively large masses of heavy quarks makes their motion accessible by Langevin dynamics [4]. The forces that act on the charm quarks by using the Langevin equation are a drag force and random momentum kicks due to collisions with the medium particles. To enable the formation of bound states we add a potential between the heavy quarks, that leads to a attractive force between charm and anti-charm quarks.

In section 2 we will explain the formalism and the parameters that we have used in our simulation. The current results are presented in section 3 while section 4 gives an outlook on further applications for a description of heavy quarkonia in heavy-ion collisions.

IOP Conf. Series: Journal of Physics: Conf. Series **1070** (2018) 012008 doi:10.1088/1742-6596/1070/1/012008

2. Formalism

For the realization of the heavy-quark motion we adopt the formalism by Blaizot et al. [5]. In this description the heavy-quark interaction is reduced to an Abelian model, which means that confinement and color effects are neglected. The Langevin update rules for the coordinate $\boldsymbol{x}$ and the momentum $\boldsymbol{p}$ of a heavy quark with mass M read

$$\frac{\mathrm{d}\boldsymbol{r}}{\mathrm{d}t} = \frac{1}{2M}\boldsymbol{p}, \tag{1}$$

$$\frac{\mathrm{d}\boldsymbol{p}}{\mathrm{d}t} = -\gamma\boldsymbol{p} + F\left(\boldsymbol{r} - \bar{\boldsymbol{r}}\right) + \sqrt{2MT\gamma\Delta t}\boldsymbol{\rho}, \tag{2}$$

where γ is the friction coefficient due to the interaction with the medium, $F\left(\boldsymbol{r} - \bar{\boldsymbol{r}}\right)$ is the force resulting from the heavy-quark potential, T is the temperature of the medium, and $\boldsymbol{\rho}$ are Gaussian normal-distributed random numbers. For the pertinent diffusion coefficient the usual Einstein dissipation-fluctuation relation has been employed. The quark-anti-quark potential is given by a screened Coulomb potential with a cut-off for large momenta at small distances. Following [5] the cut-off is taken to be $\Lambda = 4$ GeV. The potential for different temperatures is displayed in Figure 1. The drag force in [5] contains a dependence on the distance between the heavy quarks. For simplicity we neglect this dependence in our simulation and use a constant drag value. In a numerical calculation a cut-off is also necessary for the friction. With the same cut-off as for the potential the drag-coefficient is given by

$$\gamma = \frac{m_{\mathrm{D}}^2 g^2}{24\pi M}\left[\ln\left(1 + \frac{\Lambda^2}{m_{\mathrm{D}}^2}\right) - \frac{\frac{\Lambda^2}{m_{\mathrm{D}}^2}}{\frac{\Lambda^2}{m_{\mathrm{D}}^2} + 1}\right], \tag{3}$$

where m_{D} is the Debye screening mass, defined as $m_{\mathrm{D}}^2 = \frac{4}{3}g^2T^2$, which is the perturbative expression for a two-flavor quark-gluon plasma. The gauge coupling g is given by the relation [6]

$$g^2 = 4\pi\alpha_s = \frac{4\pi\alpha_s(T_C)}{1 + C\ln\left(\frac{T}{T_C}\right)}, \tag{4}$$

with

$$C = 0.76, \quad T_C = 160\,\mathrm{MeV}, \quad \alpha_s(T_C) = 0.5,$$

where T_C is the critical temperature. The value of γ at different temperatures can be seen in Figure 2. The charm-quark mass is set to $M = 1.8\,\mathrm{GeV}$, which within this model results in binding energies of $\bar{c}c$ pairs leading to bound-state masses close to typical charmonium masses.

3. Results

First the simulation was tested with a single charm-anticharm pair in the medium. The following simulations are done inside a cubic box with side-length 8 fm. At the box boundary the particles are reflected. In case a bound state hits the boundary, we reflect the center-of-mass because we do not want to destroy the binding. To define bound states we use the classical condition that two objects are bound if the energy of the pair is negative. To check this we calculate the relative energy E_{rel} of the pair, which means subtracting the center-of-mass energy from the total energy. After the system reaches equilibrium the distribution of the relative energy should be given by the classical density of states

$$\frac{\mathrm{d}N}{\mathrm{d}E_{\mathrm{rel}}} = C\int_{\mathbb{R}^3}\mathrm{d}^3\boldsymbol{r}\int_{\mathbb{R}^3}\mathrm{d}^3\boldsymbol{p}_{\mathrm{rel}}\delta(E_{\mathrm{rel}} - H_{\mathrm{rel}})\exp\left(-\frac{H_{\mathrm{rel}}}{T}\right), \tag{5}$$

IOP Conf. Series: Journal of Physics: Conf. Series **1070** (2018) 012008 doi:10.1088/1742-6596/1070/1/012008

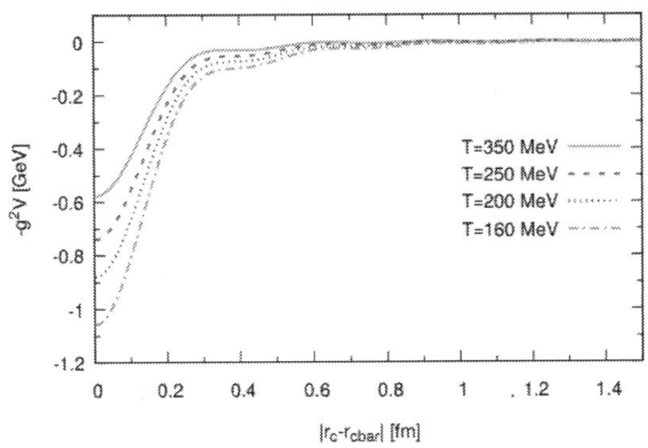

Figure 1. The charm-anti-charm-quark pair potential for different temperatures.

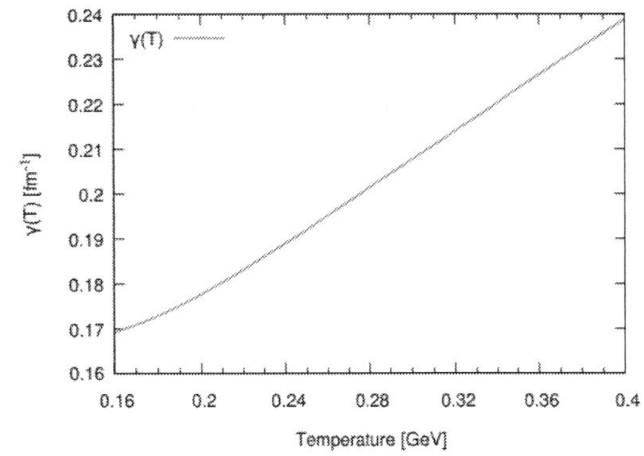

Figure 2. The friction coefficient for different temperatures.

where $H_{\rm rel}$ is the Hamiltonian of the pair and C is a normalization constant. We have solved these integrals numerically for a sphere of radius R with the same volume as our box which leads to

$$\frac{\mathrm{d}N}{\mathrm{d}E_{\rm rel}} = (4\pi)^2(2\mu)^{3/2}C\int_0^R \mathrm{d}r r^2\sqrt{E_{\rm rel}-V(r)}\exp\left(-\frac{E_{\rm rel}}{T}\right), \tag{6}$$

and compare the results with those obtained in the simulation. The results are shown in Figure 3. For this plot both curves are normalized to one. We see that the numerical calculation perfectly fits to the analytic function. We also found this agreement for different temperatures,

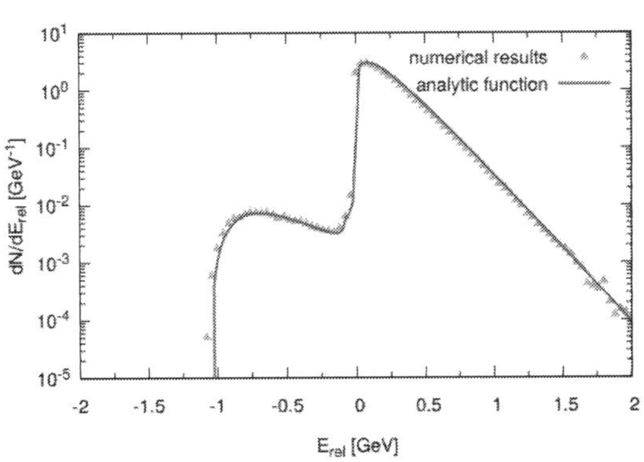

Figure 3. Pair-energy distribution at equilibrium. The simulation is done assuming a cubic box with boxsize 8 fm at a temperature of $T = 160\,\mathrm{MeV}$.

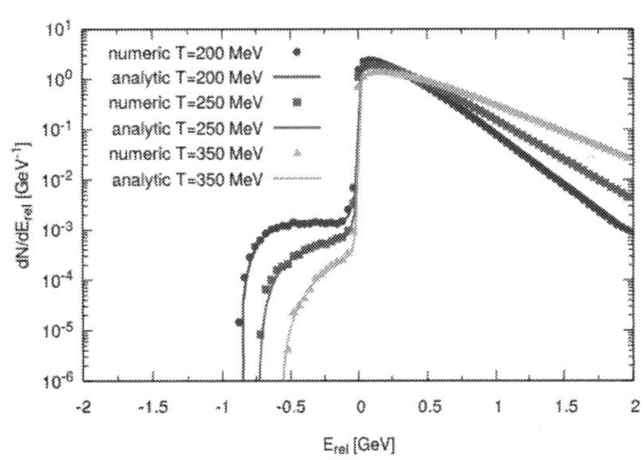

Figure 4. Change of the equilibrium distribution for different temperatures. Due to stronger screening effects higher temperatures lead to a smaller yield of bound states.

as can be seen in Figure 4. For higher temperatures the number of bound states in equilibrium decreases due to stronger screening effects. The different shapes of the box volume, which is a cube in the numerical simulation and a sphere in the analytic calculation seems not to change the result. This indicates that we choose the volume large enough so that the boundary effects can be neglected.

To illustrate the attraction between the charm and the anti-charm quark we plot the distance distribution of a single pair at temperature $T = 160\,\mathrm{MeV}$. The results are shown in Figure

5, where the green (solid) line represents the numerical simulation. The data are taken in the long-time limit when the system is equilibrated. Analytically the distribution can be calculated by evaluating

$$P(r) = \frac{1}{R^6} \int_{R^3} \mathrm{d}^3\boldsymbol{r}_1 \int_{R^3} \mathrm{d}^3\boldsymbol{r}_2 \, \delta\left(r - |\boldsymbol{r}_1 - \boldsymbol{r}_2|\right) \mathrm{e}^{-\frac{V(r)}{T}}, \tag{7}$$

where R^3 is the volume of the box and $V(r)$ is the potential between the heavy quarks. The violet (dashed) line represents a Monte-Carlo evaluation of (7). The numerical and the analytic

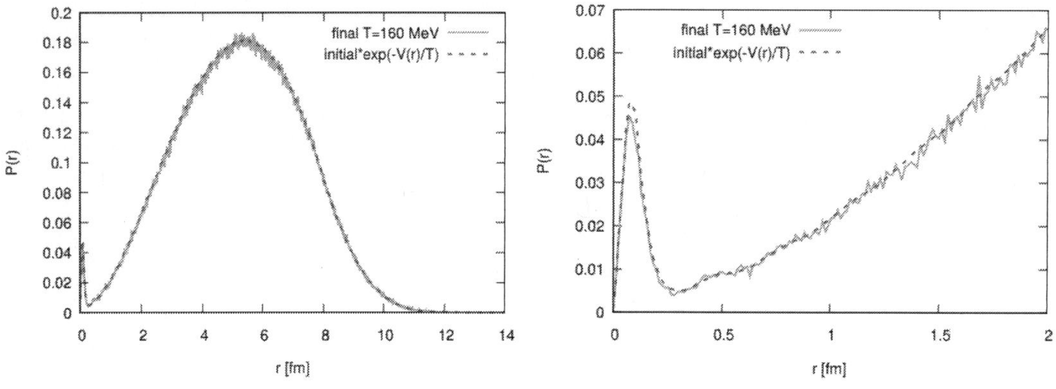

Figure 5. Probability distribution for a single charm anti-charm pair to have the distance r when the system is in equilibrium. The distribution is calculated in the numerical simulation for a single pair in a cubic box with side length 8 fm at temperature $T = 160$ MeV (green solid line) compared to the analytic expectation cf. Eq. (7) (violet dashed line). The right panel zooms into the small-distance region to illustrate the effect of the formation of bound states, leading to a pronounced peak.

function are in good agreement.

We have also investigated the time evolution of the bound states at a temperature of $T = 160\,\mathrm{MeV}$ for two initial conditions. In the first simulation the charm and anticharm quarks are initially randomly placed inside the box to obtain the time evolution for the formation of bound states. For the second simulation the heavy quarks are initially created as bound states with a pair energy of $-700\,\mathrm{MeV}$, which approximately corresponds to the maximum of the peak on the left panel of Figure 3. The momenta are set back-to-back with a relative velocity that is taken from a Maxwell distribution with its center at the average value of typical relative velocities of charmonium $v_0^2 \sim 0.3$ [7]. As shown in Fig. 6, in both cases the fraction of bound states in the system equilibrates to the same value as expected. At this point the dissociation and recombination rates are equal. The time till the equilibrium value is reached is much larger if the pairs are initially created as bound states. This is due to the strong binding between the quarks. Separating an existing pair requires a large energy transfer from the medium. We notice, that in both cases we found very long timescales for the equilibration.

To see the influence of the medium's temperature on the equilibration time we calculated the time evolution at different temperatures. In this simulation we produced five charm-anticharm pairs, initially created as bound states. The results are displayed in Figure 7.

Larger screening effects at higher temperatures lead to a smaller fraction of bound states at equilibrium. Also the equilibration time decreases. This is expected, because the friction coefficient increases with temperature. The momentum and therefore also the energy transfer in the collisions with the light medium constituents is higher, leading to a faster dissociation of a bound state.

IOP Conf. Series: Journal of Physics: Conf. Series **1070** (2018) 012008 doi:10.1088/1742-6596/1070/1/012008

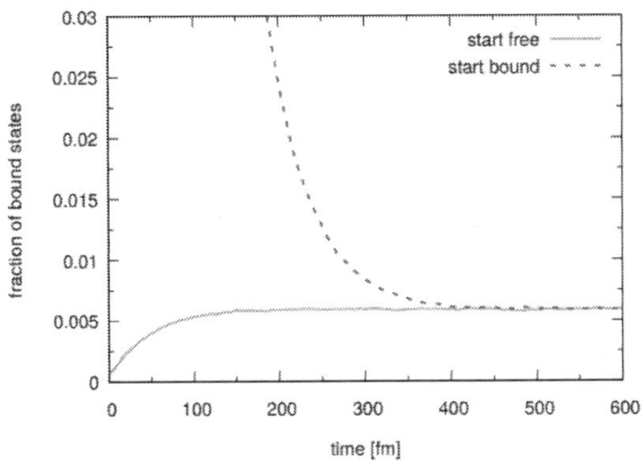

Figure 6. Time evolution of the fraction of bound states with the charm-anticharm pairs produced initially bound (violet dashed line) or placed randomly inside the box (green solid line). Both curves lead to the same equilibrium limit but the equilibration time is longer in case of initial bound states.

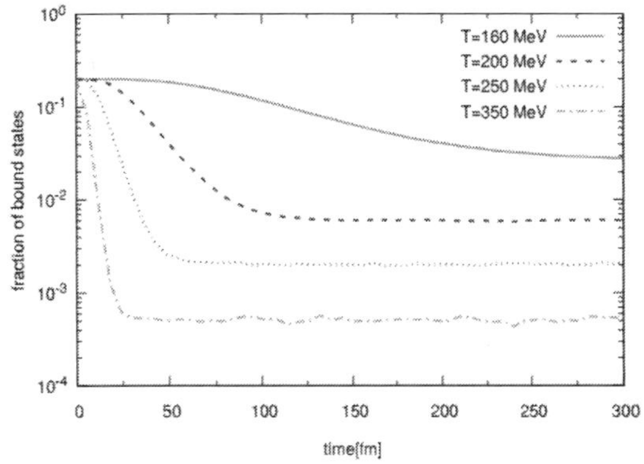

Figure 7. Time evolution of the fraction of bound states for different temperatures. Higher temperatures lead to a less amount of bound states and to a faster equilibration.

To investigate the relation between the relaxation time and the friction coefficient we have run the simulation for different multiples $k\gamma$. For this simulation we created a single pair as an initial bound state and calculated the time until there is only a fraction of $1/e$ of the initial pairs left. The results are shown in Figure 8.

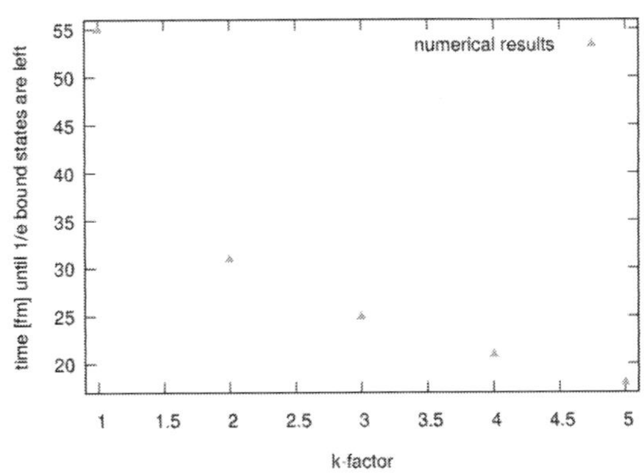

Figure 8. Equilibration time as a function of different multiples k of the friction coefficient γ, defined as the time after which the fraction of bound states has decreased to $1/e$ of its initial value.

According to the kinetic equilibration time $1/\gamma$ we expect that the decrease of the time also depends on $1/\gamma$. In our simulation the decrease is a bit slower but has the expected shape. To explain the large time scales we need to include chemical effects on the equilibration time in addition to the kinetic scales. Still the behavior of the relaxation time for different friction coefficients is approximately consistent with the underlying theory.

To check if our simulation is also in accordance with an equilibrated thermodynamic system, we calculated the particle number using the grand canonical partition function. In a classical approximation the particle number is given by

$$N = gV \left(\frac{mT}{2\pi}\right)^{\frac{3}{2}} \mathrm{e}^{-m/T} \mathrm{e}^{\mu/T}, \tag{8}$$

IOP Conf. Series: Journal of Physics: Conf. Series **1070** (2018) 012008 doi:10.1088/1742-6596/1070/1/012008

where g is the degeneracy factor of the particles, V is the volume of the system, T is the temperature, m is the mass of the particle, and μ the chemical potential. The factor $e^{\mu/T}$ is also called the fugacity and will in the following be denoted with λ. Since the dissociation and recombination of a J/ψ-meson is a process of the kind

$$J/\psi \rightleftarrows c + \bar{c},$$

the chemical potentials are connected by $\mu_{J/\psi} = 2\mu_c$ which means $\lambda_{J/\psi} = \lambda_c^2$. The number of J/ψ in the system can then be calculated by

$$\begin{aligned} N_{J/\psi} &= \lambda_c^2 g_{J/\psi} V \left(\frac{m_{J/\psi} T}{2\pi}\right)^{\frac{3}{2}} e^{-m_{J/\psi}/T} \\ &= N_c^2 \left(\frac{g_{J/\psi}}{g_c^2}\right) \frac{1}{V} \left(\frac{m_{J/\psi}}{m_c^2}\right)^{\frac{3}{2}} \left(\frac{2\pi}{T}\right)^{\frac{3}{2}} e^{(2m_c - m_{J/\psi})/T} \end{aligned} \tag{9}$$

Divided by the initial number of $c\bar{c}$ pairs equation (9) gives the fraction of bound states in the equilibrated system. We compared the results from our simulation with those obtained by using equation (9) for three different box sizes in Table 1. We find that both values are in the same order of magnitude.

Table 1. Comparison between the fraction of bound states obtained with our simulation and that calculated using equation (9).

Volume	grand canonical fraction of J/ψ	numerical fraction of J/ψ
8^3 fm^3	0.0066	0.0059
10^3 fm^3	0.0035	0.0029
12^3 fm^3	0.002	0.0017

4. Conclusion and Outlook

We have developed a model that allows to investigate the time evolution of a system with multiple pairs of charm and anti-charm quarks using the Langevin equation. We could show that our model passes all equilibrium tests and that the bound-state properties in a box are consistent with the expectations for a grand canonical ensemble.

Various extensions are possible to improve the model. First we want to include a distance dependent friction coefficient as used in [5] to replace our constant γ in equation (3) that we chose for simplicity. To simulate a heavy-ion collision with a hot and dense state at the beginning that expands and cools down until the hadronic freezout we want to describe the medium evolution as an expanding fireball. This enables us to observe the number of J/ψ's at the different stages of a collision. We also want to explore different quarkonia potentials. Especially potentials that include effects from strong interaction such as confinement are important to obtain a more realistic description. The influence of the color charges carried by the charm and anticharm quarks have to be considered too. Also the classical picture should be replaced by quantum mechanical calculations. A first step could be to decide whether a bound state is created by using the Wigner function as in [8]. In addition we have to replace the continuous energy distribution of the charm-anti-charm pairs with a quantized one that allows only certain energy levels. The energy levels could be comparable with the different excitation states of charmonium to calculate the respective abundance. The long-time goal of this project is to obtain a full in-medium quantum Langevin treatment of quarkonia.

IOP Conf. Series: Journal of Physics: Conf. Series **1070** (2018) 012008 doi:10.1088/1742-6596/1070/1/012008

References

[1] Matsui T and Satz H 1986 *Phys. Lett.* B **178** 416.
[2] Baglin C et al. (NA38) 1989 *Phys. Lett.* B **220** 471.
[3] Abelev B I et al. (STAR Collaboration) 2009 *Phys. Rev.* C **80** 041902
[4] Svetitsky B 1988 *Phys. Rev.* D **37** 2484.
[5] Blaizot J P, De Boni D, Faccioli P, and Garberoglio G 2016 *Nucl. Phys.* A **946** 49.
[6] Letessier J and Rafelski J 2004 *Hadrons and Quark-Gluon Plasma* (Cambridge University Press).
[7] Bodwin G T, Kang D, Kim T, Lee J and Yu C 2007 *AIP Conference Proceedings* **892** 315.
[8] Young C and Shuryak 2009 *Phys. Rev.* C **79** 034907.

IOP Conf. Series: Journal of Physics: Conf. Series **1070** (2018) 012009 doi:10.1088/1742-6596/1070/1/012009

Nuclear modification factors of strange and multi-strange particles in pPb collisions with the CMS experiment

Hong Ni (for the CMS Collaboration)

Department of Physics and Astronomy, Vanderbilt University, Nashville, TN, United States

E-mail: hong.ni@vanderbilt.edu

Abstract. Recent observation of collective effects in high-multiplicity pp and pA collisions raise the question of whether QGP can also be formed in the smaller systems. Systematic studies of the strange particle abundance and nuclear modification factors can shed light on this issue. The CMS experiment has excellent strange-particle reconstruction capabilities over a broad kinematic range in pp and pPb collisions. The spectra of K^0_S, $\Lambda + \overline{\Lambda}$, $\Xi^- + \overline{\Xi}^+$, and $\Omega^- + \overline{\Omega}^+$ hadrons have been measured in various rapidity regions as a function of p_T in pp and pPb collisions at 5.02 TeV. Based on the measurements of these spectra, nuclear modification factors of K^0_S, $\Lambda + \overline{\Lambda}$, $\Xi^- + \overline{\Xi}^+$, and $\Omega^- + \overline{\Omega}^+$ in mid-rapidity are measured. Since pPb is an asymmetric system, the nuclear modification factor of K^0_S, $\Lambda + \overline{\Lambda}$, $\Xi^- + \overline{\Xi}^+$, and $\Omega^- + \overline{\Omega}^+$ in Pb-going direction are compared to those in p-going direction. These final results for nuclear modification factors measured out to high-p_T can be helpful in discussing the implications of collective behavior and energy loss. In addition, the measurement of the forward-backward rapidity yield asymmetries of K^0_S and Λ as a function of p_T provide sensitivity to initial state effects, such as shadowing in the nuclear parton distributions.

1. Introduction

Particle production in pA and dA, as compared to pp collisions has long been considered as a useful tool to understand particle production mechanism. One way to do the study is using the nuclear modification factor. For collisions between two nuclei, A and B, the nuclear modification factor, R_{AB}, is defined as the ratio of particle yield in AB collisions to those in pp collisions scaled by the average number of binary nucleon-nucleon collisions, $\langle N_{coll}\rangle$, in AB collisions. It is given by

$$R_{AB}(p_T) = \frac{d^2N^{AB}/dp_T dy_{CM}}{\langle N_{coll}\rangle d^2N^{pp}/dp_T dy_{CM}} = \frac{d^2N^{AB}/dp_T dy_{CM}}{\langle T_{AB}\rangle d^2\sigma^{pp}/dp_T dy_{CM}}, \quad (1)$$

where $y_{\rm CM}$ is the rapidity computed in the center-of-mass frame of the colliding nucleons, and $\langle T_{\rm AB}\rangle$, the nuclear overlap function, accounts for the nuclear collision geometry and is calculated from a Glauber model [2]. If nuclear collisions behave as incoherent superpositions of nucleon-nucleon collisions, $R_{\rm AB}$ is expected to be unity.

Since pPb collision system is an asymmetrical system, the other quantity that is informative to measure is the particle yield rapidity asymmetry ($Y_{\rm asym}$). The definition of $Y_{\rm asym}$ is:

$$Y_{\rm asym}(p_{\rm T}) = \frac{{\rm d}^2N(p_{\rm T})/{\rm d}y_{\rm CM}{\rm d}p_{\rm T}|_{y_{\rm CM}\in[-b,-a]}}{{\rm d}^2N(p_{\rm T})/{\rm d}y_{\rm CM}{\rm d}p_{\rm T}|_{y_{\rm CM}\in[a,b]}}, \quad (2)$$

where a and b are always positive and refer to the proton beam direction.

With the measurements of these two quantities for different strange particle species, effects such as the Cronin effect [4], nuclear shadowing effect [5], radial flow effect [6] can be thoroughly studied.

This talk discussed results of measurements of strange and multi-strange particles: $\rm K^0_S$, $\Lambda+\overline{\Lambda}$ (hereafter referred to as Λ), $\Xi^-+\overline{\Xi}^+$ (hereafter referred to as Ξ^-), and $\Omega^-+\overline{\Omega}^+$ (hereafter referred to as Ω^-) $p_{\rm T}$ spectra at $-1.8 < y_{\rm CM} < 1.8$, $-1.8 < y_{\rm CM} < 0$, and $0 < y_{\rm CM} < 1.8$ in pp and pPb collisions at $\sqrt{s_{_{NN}}} = 5.02$ TeV [1]. Based on these spectra, $R_{\rm pPb}$ for each particle species is studied as a function of $p_{\rm T}$ in the three rapidity ranges at $\sqrt{s_{_{NN}}} = 5.02$ TeV. The $R_{\rm pPb}$ of the Ω^- baryon is studied only in the large $-1.8 < y_{\rm CM} < 1.8$ range because of the limited statistical precision. The $Y_{\rm asym}$ of $\rm K^0_S$ and Λ as functions of $p_{\rm T}$ for rapidity ranges $0.3 < |y_{\rm CM}| < 0.8$, $0.8 < |y_{\rm CM}| < 1.3$, and $1.3 < |y_{\rm CM}| < 1.8$ are presented in pPbcollisions at $\sqrt{s_{_{NN}}} = 5.02$ TeV. The results are compared to calculations from the EPOS LHC model including collective flow in pp and pPbcollisions.

2. Results

2.1. Invariant mass peaks

The details of event selection and particle reconstruction can be found in [1]. Figure 1 shows some samples of invariant mass peaks of reconstructed $\rm K^0_S$, Λ, Ξ^-, and Ω^- candidates in the range $-1.8 < y_{\rm CM} < 1.8$ for pPb events. Prominent mass peaks are visible, with little background. Raw yields of strange particles are extracted from these invariant mass peaks by integrating the signal function.

2.2. Nuclear modification factors

With the efficiency-corrected strange-particle spectra, the $R_{\rm pPb}$ values of $\rm K^0_S$, Λ, Ξ^-, and Ω^- particles are calculated in different $y_{\rm CM}$ ranges.

Figure 2 shows that the $R_{\rm pPb}$ of each particle species at $-1.8 < y_{\rm CM} < 1.8$. For $p_{\rm T} > 2$ GeV, the $R_{\rm pPb}$ value of $\rm K^0_S$ are consistent with unity. For strange baryons (Λ, Ξ^-, and Ω^-), the $R_{\rm pPb}$ values increase at low $p_{\rm T}$, and reach peaks at around 5 GeV. At higher $p_{\rm T}$ (around 7 GeV), the $R_{\rm pPb}$ values of Λ and Ξ^- drop to unity. At intermediate $p_{\rm T}$ (2 to 7 GeV), the $R_{\rm pPb}$ values of baryons show clear mass dependence, which is consistent with expectations from the radial-flow

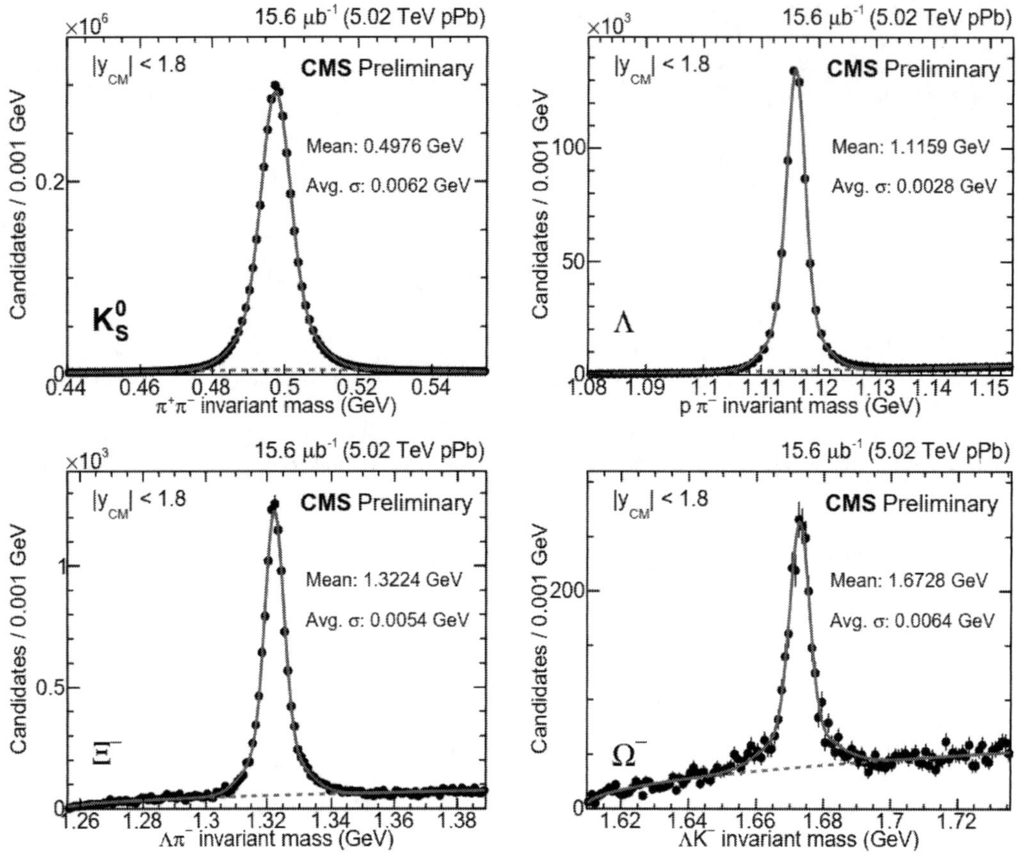

Figure 1. Invariant mass distribution of K_S^0 (upper left), Λ (upper right), Ξ^- (lower left), and Ω^- (lower right) candidates within $|y_{CM}| < 1.8$ in pPb collisions. The solid lines show the results of fits described in the text. The dashed lines indicate the fitted background component. Figure from [1].

effect in hydrodynamic models [6]. Data are compared to Monte Carlo simulations, here EPOS LHC is used. Details about EPOS LHC can be found in [3]. EPOS LHC predicts the mass dependence qualitatively, but with even larger mass dependence.

The R_{pPb} values of K_S^0, Λ, and Ξ^- particles for $-1.8 < y_{CM} < 0$ and $0 < y_{CM} < 1.8$ are presented as functions of p_T in Figure 3. Because of the limited statistical precision, the R_{pPb} of Ω^- baryon is not shown in the p- and Pb-going direction separately. The R_{pPb} of all three species are found to be larger in the Pb-going direction than the p-going direction, with a stronger mass splitting between the heavier and the lighter particles in the Pb-going direction. This trend is consistent with expectations from the radial-flow effect in hydrodynamic models [6].

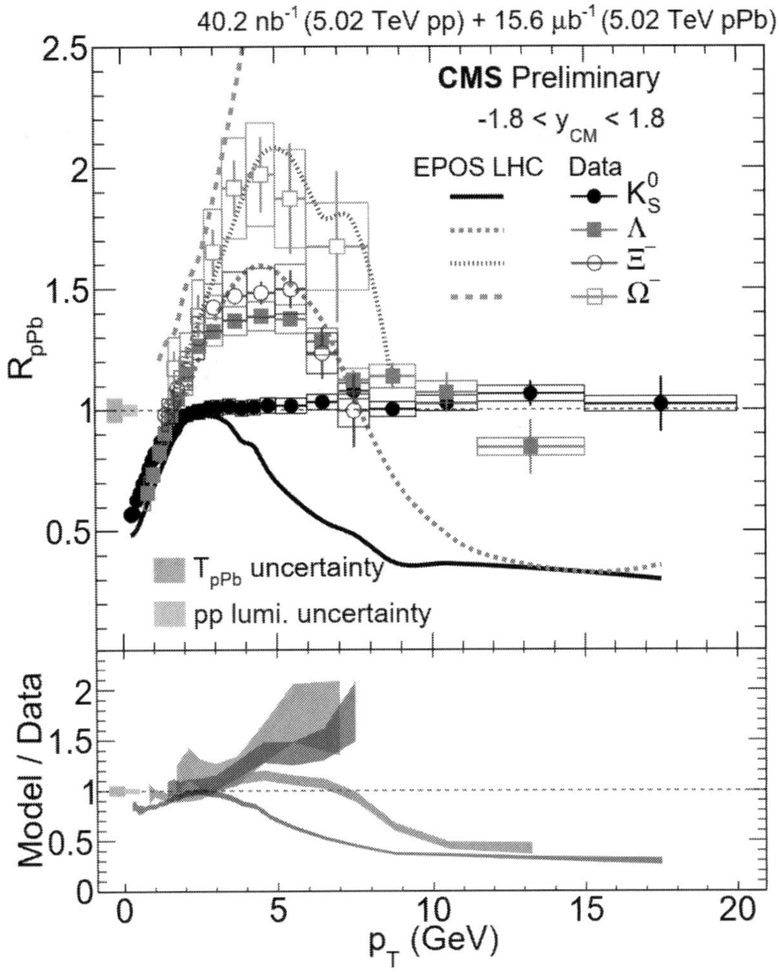

Figure 2. (Upper) Nuclear modification factors for K^0_S (black filled circles), Λ (red filled squares), Ξ^- (blue open circles), and Ω^- (purple open squares) particles for $-1.8 < y_{CM} < 1.8$ in pPbcollisions are presented. The vertical bars correspond to statistical uncertainties, and the horizontal bars represent the bin width, while the boxes around the markers denote the systematic uncertainties. The T_{pPb} and pp integrated luminosity uncertainties are represented by the shaded areas around one. The results are compared to EPOS LHC predictions, that include collective flow in pp and pPb collisions. The data and predictions share the same color for each particle species. (Lower) The ratios of nuclear modification factors for K^0_S, Λ, Ξ^-, and Ω^- particles of EPOS LHC predictions to the measurements are shown. The bands represent the combination of statistical uncertainties and systematic uncertainties. Figure from [1].

2.3. Particle yield rapidity asymmetry

Figure 4 shows the Y_{asym} as functions of p_T for K^0_S and Λ for different rapidity ranges. In forward y_{CM} ranges, the Y_{asym} values of K^0_S and Λ increase to a certain p_T, and drop to unity at higher

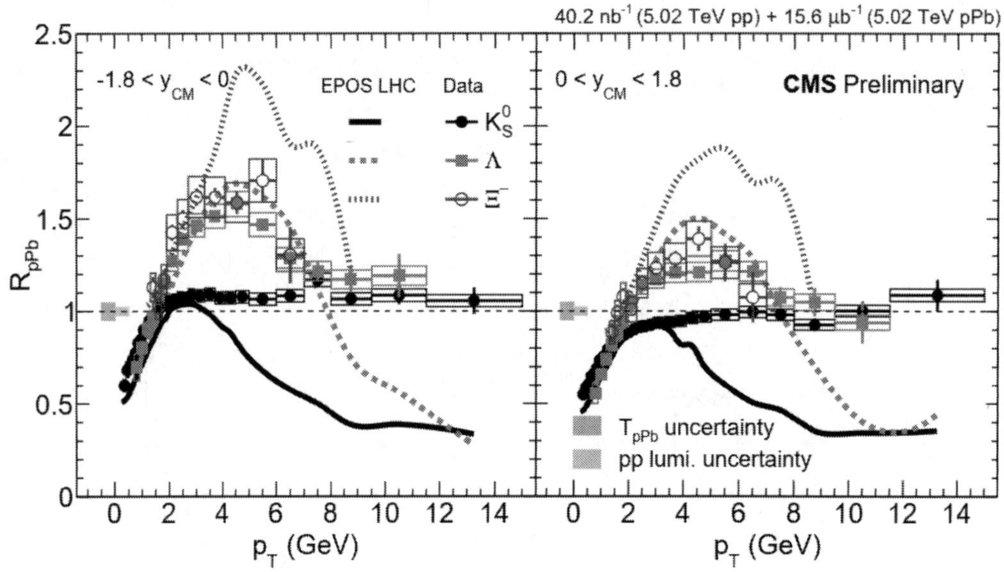

Figure 3. Nuclear modification factors of K^0_S (black filled circles), Λ (red filled squares), and Ξ^- (blue open circles) particles for $-1.8 < y_{CM} < 0$ (left) and $0 < y_{CM} < 1.8$ (right) in pPb collisions are presented. The vertical bars correspond to statistical uncertainties, and the horizontal bars represent the bin width, while the boxes around the markers denote the systematic uncertainties. The T_{pPb} and pp integrated luminosity uncertainties are represented by the shaded areas around one. The results are compared to the EPOS LHC predictions, that include collective flow in pp and pPb collisions [3]. The data and predictions share the same color for each particle species. Figure from [1].

p_T. For all three $|y_{CM}|$ ranges measured, the values of Y_{asym} are larger than unity. The Y_{asym} values increase with $|y_{CM}|$ range, which is consistent with expectations from nuclear shadowing. The Y_{asym} of K^0_S and Λ in the above three $|y_{CM}|$ ranges are compared to the Y_{asym} of charged particles in similar η_{CM} ranges. From the comparison, it is found that the Y_{asym} value of charged particles are smaller than that of K^0_S and Λ, and the peak position of the charged-particle Y_{asym} is between that of K^0_S and Λ in forward $|y_{CM}|$ ranges. The results of Y_{asym} are compared to EPOS LHC calculations in the three rapidity ranges. The calculated Y_{asym} increases from mid-rapidity to forward rapidity, consistent with the trend in data, but fails to describe the particle species dependence at forward rapidities.

3. Summary

The transverse momentum spectra of K^0_S, Λ, Ξ^-, and Ω^- particles in proton-proton and proton-lead collisions are measured in ranges of center-of-mass rapidity. With the efficiency-corrected spectra, the nuclear modification factor and particle yield rapidity asymmetry are calculated. The data are compared to the calculations from EPOS LHC, which predicts the data in a qualitative way. These measurements can help understand particle production mechanism and constrain theoretical models.

IOP Conf. Series: Journal of Physics: Conf. Series **1070** (2018) 012009 doi:10.1088/1742-6596/1070/1/012009

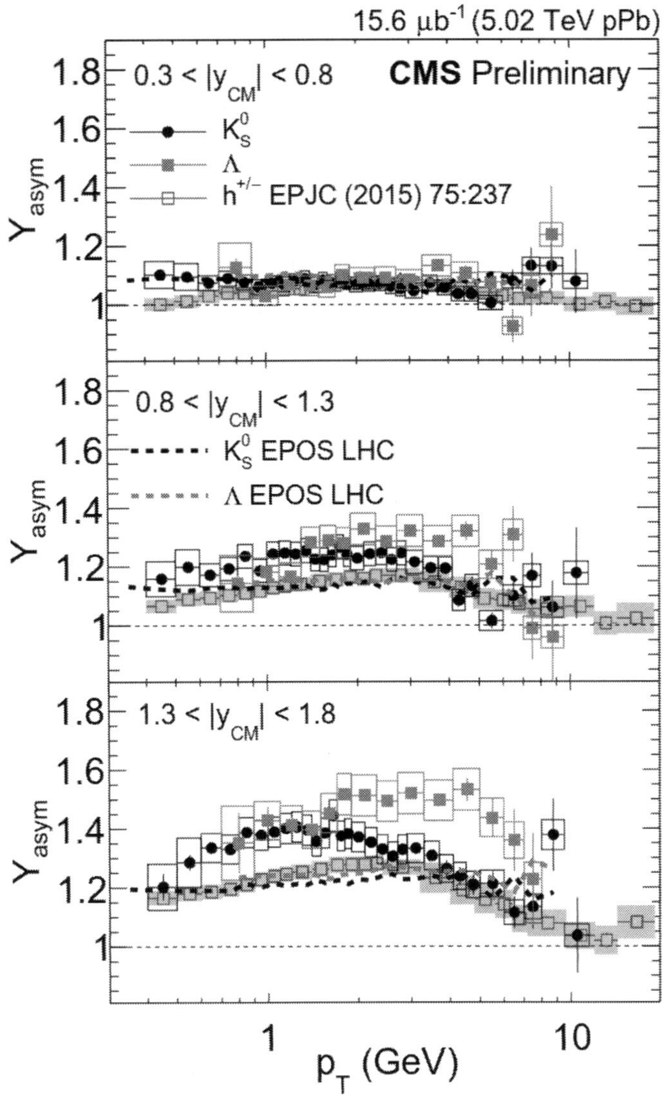

Figure 4. The $Y_{\rm asym}$ of K^0_S (black filled circles), Λ (red filled squares), and charged particles (blue open squares) at $0.3 < |y_{\rm CM}| < 0.8$, $0.8 < |y_{\rm CM}| < 1.3$, and $1.3 < |y_{\rm CM}| < 1.8$ ($|\eta_{\rm CM}|$ rangesfor charged particles) in pPb collisions at $\sqrt{s_{NN}} = 5.02$ TeV. The vertical bars correspond to statistical uncertainties, and the horizontal bars represent the bin width, while the boxes around the markers denote the systematic uncertainties. The results are compared to the EPOS LHC predictions, that include collective flow in pp and pPb collisions [3]. The data and predictions share the same color for each particle species. Figure from [1].

IOP Conf. Series: Journal of Physics: Conf. Series **1070** (2018) 012009 doi:10.1088/1742-6596/1070/1/012009

References

[1] The CMS Collaboration, "Measurement of the nuclear modification factor of identified strange and multi-strange particles in pPb collisions at sqrt(sNN) = 5.02 TeV with CMS experiment", CMS-PAS-HIN-16-013, http://cds.cern.ch/record/2310484

[2] M. L. Miller, K. reygers, S.J. Sanders, and P. Steinberg, "Glauber modeling in high energy nuclear collisions", Ann. Rev. Nucl. Part. Sci., **57** (2007) 205

[3] T. Pierog et al., "EPOS LHC: Test of collective hadronization with data measured at the CERN Large Hadron Collider", Phys. Rev. C **92** (2015) 034906

[4] M.J.S.J.W.Cronin, H.J.Frisch et al., "Production of hadrons at large transverse momentum at 200, 300, and 400 GeV", Phys. Rev. D **11** (1975) 3105

[5] M. Arneodo, "Nuclear effects in structure functions", Phys. Rept. **240** (1994) 301

[6] P. Bozek, A. Bzdak, and V. Skokov, The rapidity dependence of the average transverse momentum in p+Pb collisions at the LHC: the Color Glass Condensate versus hydrodynamics, Phys. Lett. B **728** (2014) 662

IOP Conf. Series: Journal of Physics: Conf. Series **1070** (2018) 012010 doi:10.1088/1742-6596/1070/1/012010

PHENIX results on collectivity in small systems

Weizhuang Peng

Vanderbilt University, 2201 West End Ave, Nashville, TN 37235, USA

E-mail: `weizhuang.peng@vanderbilt.edu`

Abstract. Recent results from small collision systems at RHIC and LHC indicate that many of the signatures of collective behavior observed in AA collisions are also present in small systems in high-multiplicity events. Using the extraordinary versatility of RHIC in selecting different colliding species, the PHENIX experiment has collected data in p+Al, p+Au, d+Au, and ^{3}He+Au at a nucleon-nucleon center-of-mass energy of 200 GeV and conducted a comprehensive set of anisotropic flow measurements. These geometry-controlled experiments provide a unique testing ground for theoretical models that produce azimuthal particle correlations based on initial and/or final state effects. In this paper, we present measurments of elliptic and triangular flow for inclusive charged particles, and $v_2(p_T)$ measurements for identified pions, and protons. Detailed model comparisons and the implications for the origin of collectivity in p/d/ ^{3}He+Au collisions are discussed.

1. Introduction

Collective behavior, such as the azimuthal momentum anisotropy of the particles produced in collisions between large nuclei, such as Au+Au, is considered to be evidence for the formation of a strongly interacting, nearly-perfect fluid [1]. However, in recent years, similar phenomena have been observed in various small systems (e.g., p/d/^{3}He+Au collisions at RHIC [2, 3, 4, 5], and in pp and p+Pb collisions at the LHC [6, 7, 8, 9]. Many explanations have been proposed to explain the anisotropy signals in small systems. One explanation is that the initial-state geometry is translated into final-state flow under the gradient pressure. We can test this by changing the projectile/target of the collision system and see if the final-state flow strength corresponds to the initial-state geometric eccentricity. We should also see a mass ordering of $v_2(p_T)$ for identified final-state hadrons.

2. Methods

The PHENIX collaboration has measured the $v_2(p_T)$ of charged particles and identified pions, kaons, and protons in p+Au, d+Au, and ^{3}He+Au collisions and $v_3(p_T)$ of charged in d+Au and ^{3}He+Au collisions [2, 4, 5, 10]. The v_2 values are measured with the event-plane method. The event plane is determined using the south section (in the direction of the Au beam) of the forward vertex detector (FVTX), which is a silicon detector comprised of two identical end-cap assemblies symmetrically arranged in the longitudinal direction around the interaction region. The pseudorapitidy coverage of the FVTX is $1 < |\eta| < 3$. The pions, kaons, and protons are identified by measuring their mass using time-of-flight measurements from the two Time-of-flight detectors located in the East and West arms of PHENIX in conjuction with the momentum and path-lenth measurements from the tracking detectors.

3. Results

3.1. Measurements of $v_2(p_T)$ and $v_3(p_T)$ of charged hadrons in geometry-controlled experiments

Figure 1 shows the transverse momentum dependence of v_2 for charged hadrons in central p+Au, d+Au, and ^{3}He+Au collisions [2] and the theory predicitons from the SONIC hydrodynamics model [2]. From this figure, we see that the values of $v_2(p_T)$ in d+Au and ^{3}He+Au collisions are larger than those measured in p+Au collisions, which corresponds to the ordering of the initial geometric eccentricity $\epsilon_2(p+Au) < \epsilon_2(^3He+Au) \approx \epsilon_2(d+Au)$. The SONIC model [11] describes the data well. It uses the Glauber model to set-up the initial energy density of the system, which then evolves with QGP equation of state. Table 1 shows the average spatial eccentricities for p+Au, d+Au and ^{3}He+Au collisions calculated from a Monte Carlo Glauber model for the top 5% central collisions in each system.

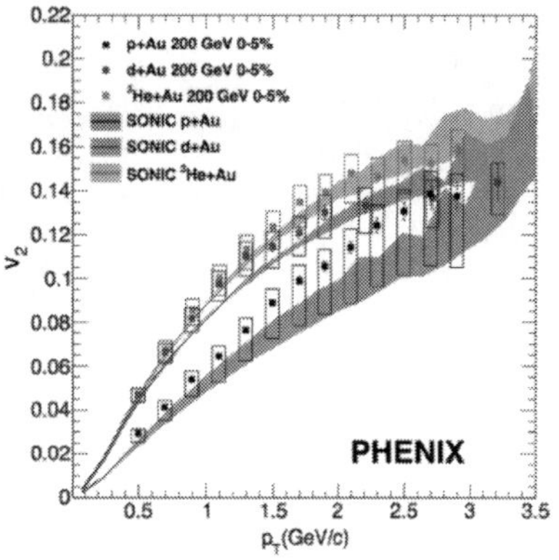

Figure 1. Measurement of v_2 for charged hadrons produced at mid-rapidity in central p+Au, d+Au, and ^{3}He+Au collisions [2] in comparision to SONIC predictions.

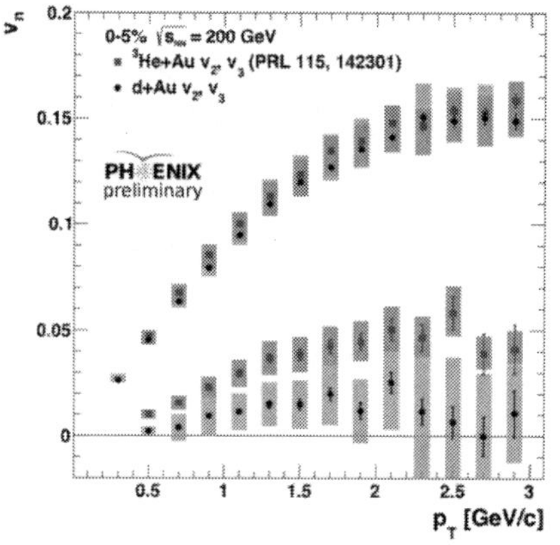

Figure 2. Transverse momentum dependence of v_2 and v_3 for charged hadrons produced at mid-rapidity in central d+Au, and ^{3}He+Au collisions.

Table 1. Average spatial eccentricities for p+Au, d+Au and ^{3}He+Au calculated from a Monte Carlo Glauber model for 0-5% central collisions.

Spatial eccentricity	p+Au	d+Au	^{3}He+Au
$\langle\epsilon_2\rangle$	0.23±0.01	0.54±0.04	0.50±0.02
$\langle\epsilon_3\rangle$	0.16±0.01	0.18±0.01	0.28±0.02

Figure 2 shows both $v_2(p_T)$ and $v_3(p_T)$ in d+Au and ^{3}He+Au collisions. The v_2 values are similar in these two systems, while the v_3 values are definitely different, as expected from the geometry.

Figure 3 shows the comparision of v_2 and v_3 to the iEBE-VISHNU hydrodynamics model [12]. The v_2 and v_3 are well described simultaneously with the same model parameters. Since the presentation at the conference, the PHENIX collaboration has submitted for publication the

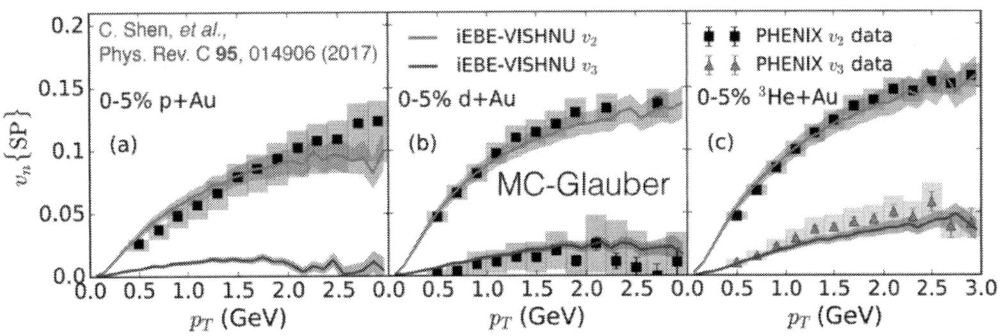

Figure 3. Transverse momentum dependence of v_2 and v_3 for hadrons in central p+Au, d+Au, and ^{3}He+Au collisions and theory predictions from iEBE-VISHNU model.

measurements of $v_3(p_T)$ in p+Au and d+Au collisions [13]. The measurement in d+Au has significantly reduced systematic uncertainty in comparison to the preliminary results presented here. With the complete set of v_2 and v_3 measurements in three small systems with distinct initial geometry, the geometric origin of the observed azimuthal anisotropies is further established.

3.2. Measurements of $v_2(p_T)$ for identified particles in central p/d/^{3}He+Au collisions
Mass-ordering is expected in the $v_2(p_T)$ distributions of identified particles, if the particles emerge from a common velocity field. Figure 4 shows $v_2(p_T)$ for identified pions and protons in 0%-5% central p+Au, d+Au, and ^{3}He+Au collisions [10]. We can see a clear seperation between the pion and proton v_2 values in the d+Au and ^{3}He+Au systems, with the pion v_2 values being larger than those of the protons for $p_T < 1.5$ GeV/c, and the order being reversed at higher p_T. In the p+Au system, the splitting between pion and proton v_2 is smaller.

Figure 4 also compares the measured $v_2(p_T)$ with hydrodynamic calculations using the iEBE-VISHNU model. This model includes event-by-event fluctuating initial conditions via Monte Carlo Galuber simulation and then viscous hydrodynamics starting at $\tau_0 = 0.6$ fm/c. The hydrodynamic evolution utilizes an $\eta/s = 0.08$ and ends at $T = 155$ MeV. After that point, hadronization occurs and hadronic rescattering is implemented using URQMD. This model shows good agreement with the measurments at low p_T, but does not describe the mass-dependence of the measured v_2 values at high p_T. This figure also shows the theory curves with no hadronic rescattering. For $p_T < 1.5$ GeV, there is little change in the model calculations with or without hadronic rescattering. This means that the mass splitting at low p_T in the hydrodynamic models mostly comes from the early-stage interactions.

Figure 5 compares the expermental data to the theory calculations from a-multiphase-transport model (AMPT) [14]. AMPT uses Monte Carlo Glauber initial conditions, and it models the evolution of the system by a succession of partonic scattering, spatial coalescence, and late-stage hadronic scattering as implemented in ART. The results shown in Fig. 5 are calculated from the AMPT time evolution with a partonic cross section $\sigma_{part} = 1.5$ mb. Also shown are the results with the hadronic rescattering turned off. We observed that the full AMPT describes the mass-splitting in all three collision systems for $p_T < 1.5$ GeV/c. At higher p_T, the v_2 from the theory becomes lower than the data points. The splitting of pion and proton v_2 at higher p_T comes from the implementation of recombination model for hadronization. The comparison of the calculations with and without hadronic rescattering reveals that in AMPT the mass-splitting between v_2 of pions and protons at low p_T comes mainly from the hadronic rescattering stage, which is different than the results from iEBE-VISHNU. These differences in the role of the

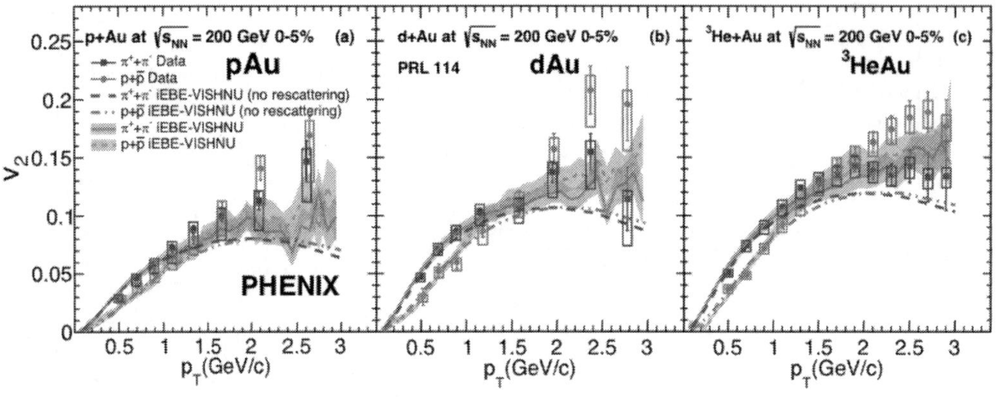

Figure 4. Transverse momentum dependence of v_2 for identified pions and protons within $|\eta|$ < 0.35 in 0%-5% central p+Au, d+Au, and ^{3}He+Au collisions. The measurments are compared to hydrodynamic calculations using the iEBE-VISHNU hydrodynamic model, illustrating the effect of hadronic rescattering on the mass-dependent v_2 values.

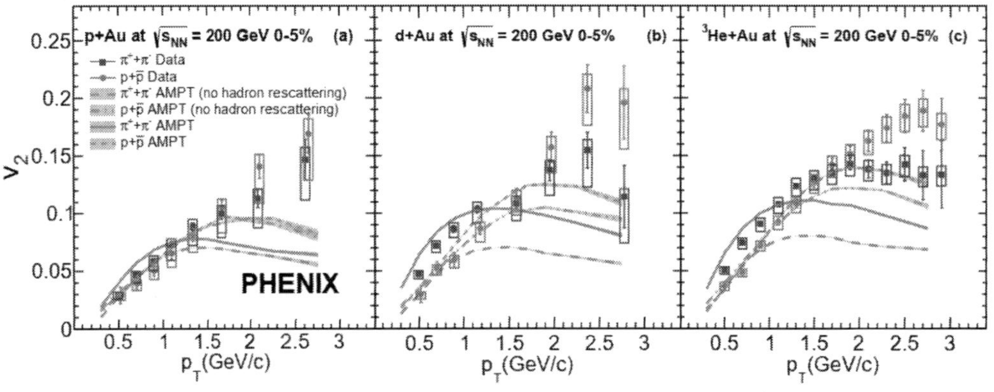

Figure 5. Same as Figure 4, but also shown are v_2 transport model calculations using AMPT.

hadronic rescattering may be due to differences in the hadronic cascade implimentation and are worth investigating further.

Figure 6 shows the ratio of pion to proton $v_2(p_T)$ in all collision systems, together with the theory curves. In the ratio, many systematic uncertainties are canceled and thus we can see more clearly that the data in the three systems exhibit a similar trend, where pion v_2 is larger than the proton v_2 for $p_T < 1.5$ GeV/c, with the order reversed at higher p_T. Linear fit of the ratio within the p_T range from 0.5 GeV/c to 3.0 GeV/c gives nozero values for all three systems: -0.22 $\pm$ 0.07 in p+Au collisions, -0.40 $\pm$ 0.07 in d+Au collisions, and -0.34 $\pm$ 0.03 in ^{3}He+Au collisions. While both the hydrodynamics models and AMPT describe the ratio at low p_T, only AMPT successfully describes the ratio for $p_T > 1.5$ GeV/c, which emphasizes the importance of hadronization by recombination.

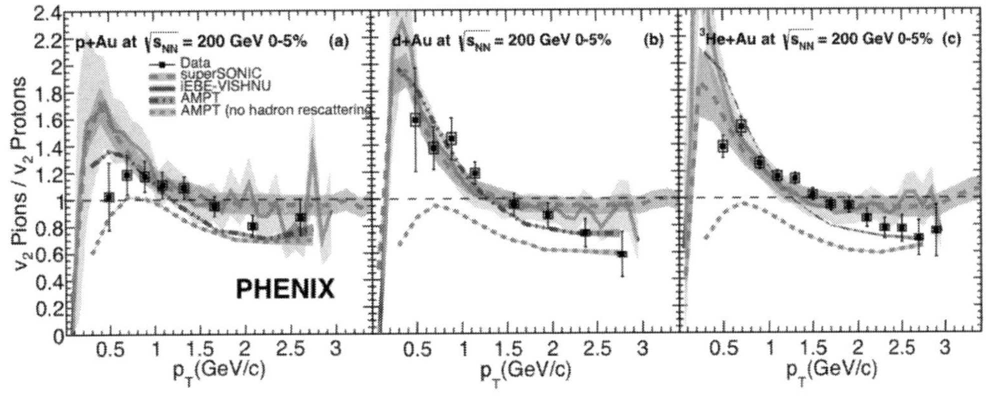

Figure 6. Ratio of pion v_2 over proton v_2 in central 0%-5% (a) p+Au, (b) d+Au, (c) ^{3}He+Au collisions at $\sqrt{s_{NN}}$ = 200 GeV. Theoretical calculations from SUPERSONIC and AMPT are also shown.

4. Summary

We have shown $v_2(p_T)$ of inclusive charged hadrons for p+Au, d+Au,^{3}He+Au collisions and $v_3(p_T)$ of inclusive hadrons produced in d+Au and ^{3}He+Au collision. These measurements support the idea that the final-state flow observed in small systems arises from the intial-state geometry. We have also measured $v_2(p_T)$ for identified pions and proton in p+Au, d+Au, ^{3}He+Au collisions. We observe mass-splitting in all three systems, with the heavier particles having smaller v_2 at a given p_T in the range $p_T < 1.5$ GeV/c, and the reverse order at higher p_T. These observations are qualitatively similar to previously measured effects in A+A collisions.

References

[1] U.Heinz and R. Snellings, "Collective flow and viscosity in relativistic heavy-ion collisions," Ann. Rev. Nucl. Part. Sci.**63**, 123 (2013).
[2] C. Aidala *et al.* (PHENIX Collaboration) "Measurement of long-range angular correlations and azimuthal anisotropies in high-multiplicity p+Au collisions at $\sqrt{s_{NN}}$ = 200 GeV," Phys. Rev. C **95**, 034910 (2017).
[3] A. Adare *et al.* (PHENIX Collaboration), "Quadrupole anisotropy in dihadron azimuthal correlations in central d+Au collisions at $\sqrt{s_{NN}}$ = 200 GeV," Phys. Rev. Lett. **111**, 212301 (2013).
[4] A. Adare *et al.* (PHENIX Collaboration) "Measurements of long-range angular correlation and quadrupole anisotropy of pions and (anti)protons in central d+Au collisions at $\sqrt{s_{NN}}$ =200 GeV" Phys. Rev. Lett. **114**, 192301 (2015).
[5] A. Adare *et al.* (PHENIX Collaboration), "Meassurements of elliptic and triangular flow in high-multiplicity ^{3}He+Au collisions at $\sqrt{s_{NN}}$ = 200 GeV," Phys. Rev. Lett. **115**, 142301 (2015).
[6] Khachatryan *et al.* (CMS Collaboration), Observation of long-range near-side angular correlations in proton-proton collisions at the LHC, J. High Energy Phys. 09 (2010) 091.
[7] S. Chatrchyan *et al.*(CMS Collaboration),"Multiplicity and transverse momentum dependence of two- and four-particle correlations in pPb and PbPb collisions," Phys. Lett. B 724, 213 (2013).
[8] B. Abelev *et al.* (ALICE Collaboration), Long-range angular correlations on the near and away side in p-Pb collisions at $\sqrt{s_{NN}}$ = 5.02 TeV, Phys. Lett. B 719, 29 (2013).
[9] G. Aad *et al.* (ATLAS Collaboration), Observation of associated near-side and away-side long-range correlations in $\sqrt{s_{NN}}$ = 5.02 TeV proton-lead collisions with the ATLAS detector, Phys. Rev. Lett. **110**, 182302 (2013).
[10] A. Adare *et al.* (PHENIX Collaboration) "Measurments of mass-dependent azimuthal anisotropy in central p+Au, d+Au, and ^{3}He+Au collisions at $\sqrt{s_{NN}}$ = 200 GeV," arXiv:1710.09736, accepted in Phys. Rev. C
[11] M. Habich, J. L. Nagle, and P. Romatschke, Particle spectra and HBT radii for simulated central nuclear

collisions of C+C, Al+Al, Cu+Cu, Au+Au, and Pb+Pb from $\sqrt{s_{NN}} = 62.4 \quad 2760$ GeV, Eur. Phys. J. C 75, 15 (2015)

[12] C. Shen *et al.*, "Collectivity and electromagnetic radiation in small systems," Phys. Rev. C 95, 014906 (2017).

[13] C.Aidala *et al.* (PHENIX Collaboration) "Creating small circular, elliptical, and triangular droplets of quark-gluon plasma," arXiv:1805.02973.

[14] Z. Lin, C. Ko, B. Li, B. Zhang, and S. Pal, "Multiphase transport model for relativistic heavy ion collisions," Phys. Rev. C **72**, 064901 (2005).

IOP Conf. Series: Journal of Physics: Conf. Series **1070** (2018) 012011 doi:10.1088/1742-6596/1070/1/012011

Recent PHENIX Results on High-p_T π^0 and η Production in Cu+Au and U+U Collisions

Murad Sarsour (for the PHENIX Collaboration)
Georgia State University, Atlanta, Georgia 30303, USA
E-mail: msar@gsu.edu

Abstract. One of the key signatures of the Quark-Gluon Plasma (QGP), is the modification of hadron transverse momentum differential cross-sections in heavy-ion collisions as compared to p+p collisions. Suppression of hadron production at high transverse momenta (p_T) in heavy ion collisions has been explained by the energy loss of the partons produced in the hard scattering processes that traverse the deconfined quantum chromodynamics (QCD) matter. Recent RHIC runs with asymmetric Cu+Au collisions and non-spherical nuclear U+U collisions provide the means to systematically study suppression pattern of hadrons in different nuclear overlap geometry and different energy density and surface biases compared to Au+Au collisions. These studies provide an important input to improve theoretical description of parton energy loss in QGP. In this proceeding, we present results from π^0 and η production in Cu+Au and U+U systems and discuss how these results further our understanding of parton energy loss.

1. Introduction

Ever since RHIC announced the discovery of the hot and dense state of strongly interacting matter called Quark-Gluon Plasma (QGP) [1, 2] the main objective has been to characterize its properties. Jet quenching, manifested in suppression of high-p_T hadron production in A+A collisions relative to that in p+p collisions, was one of the early evidences to the existence of QGP [3] and provides an excellent way to characterize its properties. The suppression, which is explained by the energy loss of the hard scattered partons in the QGP medium, is studied by measuring the nuclear modification factor, R_{AB}. R_{AB} is defined as the ratio of the hadron yields in A+B collisions to p+p collisions scaled by the number of nucleon-nucleon collisions in the A+B system, $N_{\rm coll}$ [4], and is calculated as:

$$R_{AB} = \frac{d^2N_{AB}/dydp_T}{\langle N_{\rm coll}\rangle \times d^2N_{pp}/dydp_T}, \tag{1}$$

where $d^2N_{AB}/dydp_T$ is the per-event yield of particle production in heavy ion collisions and $d^2N_{pp}/dydp_T$ is the per-event yield of the same process in p+p collisions. In PHENIX, we use leading hadrons, π^0 and η, as proxy for jets. π^0 is abundantly produced and can be measured to a high p_T while η has hidden strangeness and allows studying hadron suppression as a function of flavor and mass. At RHIC, jet quenching was studied extensively in symmetric collisions such as Au+Au and Cu+Cu, recent asymmetric Cu+Au collisions at $\sqrt{s_{NN}}$ = 200 GeV and non-spherical nuclear U+U collisions at $\sqrt{s_{NN}}$ = 193 GeV provided the opportunity to study jet quenching in different nuclear overlap geometry and different energy density and surface biases. Cu+Au collision is the first asymmetric system and differs from symmetric systems by special form of nuclear overlap region while U+U collision is the largest heavy collision system with the largest density.

2. Experimental setup

The PHENIX detector [5] has a high rate capability utilizing a fast DAQ and specialized triggers, high granularity detectors, and good mass resolution and particle ID. Detection of γ, π^0 and η utilizes the finely grained electromagnetic calorimeter (EMCal). The very forward beam-beam counters (BBC) are used to determine the collision vertex position and time, the beam luminosity and form a minimum bias trigger.

In 2012, the PHENIX collaboration collected data from Cu+Au and U+U collisions provided by the Relativistic Heavy Ion Collider (RHIC) at Brookhaven National Laboratory (BNL). Results from these data sets are presented in this proceeding.

3. Results

π^0 and η are identified through their photon decay channel. The photons are reconstructed in the finely segmented EMCal with high efficiency. The invariant mass spectrum is formed from

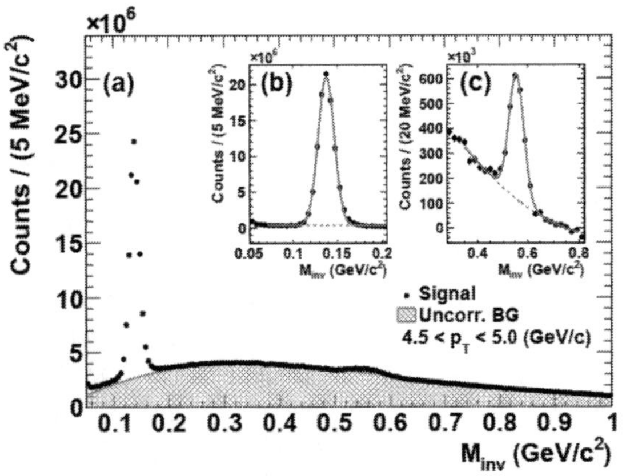

Figure 1. The invariant mass spectrum at $4.5 < p_T < 5.0$ GeV/c for all photon pairs is shown in solid black points while the shaded distribution shows the normalized mixed events background. Insert (b) shows the background subtracted invariant mass spectrum in the π^0 region fitted with a Gaussian and a first order polynomial while insert (c) shows the background subtracted invariant mass spectrum in the η region fitted with a Gaussian and a second order polynomial.

all pairs of the reconstructed photons, as shown in Figure 1, and the combinatorial background is estimated using mixed-event technique and subtracted from it. The π^0 yield is then extracted by fitting the π^0 region (0.10 - 0.17 GeV/c^2) by a Gaussian and a first order polynomial while the η yield is extracted by fitting the η mass region (0.48 - 0.62 GeV/c^2) by a Gaussian and a second order polynomial, as shown in inserts (b) and (c) of Figure 1.

3.1. Cu+Au collisions

To extract R_{AB} of π^0 and η in Cu+Au collisions, the PHENIX collaboration measured the π^0 and η p_T-spectra over a wide p_T range in different centrality classes, as shown in Figure 2.

The ratio of η to π^0, shown in Figure 3, is consistent between all centrality classes and consistent with that measured in p+p, p+A and A+A collisions at different energies [6, 7].

Figure 4 shows R_{AB} of π^0 and η versus p_T using the spectra shown in Figure 2. This figure shows that in central and semi central Cu+Au collisions π^0 and η are suppressed while in peripheral Cu+Au collisions we observe a hint of π^0 and η enhancement. In addition, π^0

IOP Conf. Series: Journal of Physics: Conf. Series **1070** (2018) 012011 doi:10.1088/1742-6596/1070/1/012011

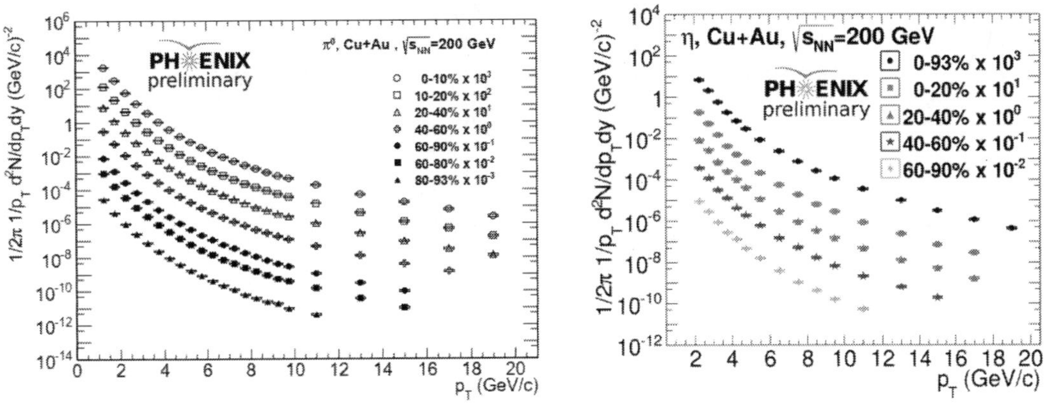

Figure 2. The invariant p_T-spectra of π^0 (left) and η (right) measured in different centrality classes of Cu+Au collisions at $\sqrt{s_{NN}} = 200$ GeV. The spectra are scaled by arbitrary factors for clarity.

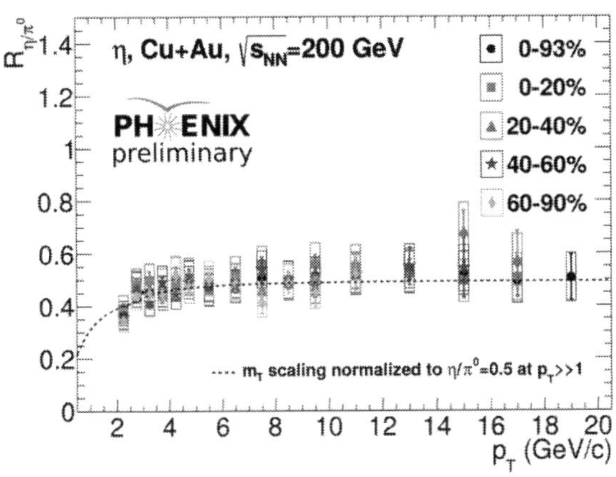

Figure 3. Ratios of η to π^0 yields as a function of p_T in different centrality classes of Cu+Au collisions at $\sqrt{s_{NN}} = 200$ GeV. The dashed curve shows the m_T scaling curve normalized to 0.5 at high momentum.

and η results show similar suppression/enhancement pattern which could be an indication that they have similar fragmentation function modification by the medium in the accessed p_T range. Furthermore, this suppression/enhancement pattern shows very similar centrality dependence to that observed in jets [8], which supports using leading hadrons to study jet quenching.

In central and semi-central Cu+Au collisions π^0 and η yields show similar suppression as in Cu+Cu and Au+Au collisions at similar number of participants which lead us to conclude that π^0 and η production depends on the size of the nuclear overlap but not on its shape. However, in peripheral Cu+Au collisions π^0 and η yields show a hint of enhancement while a suppression is in Au+Au collisions and Cu+Cu collisions is in between.

3.2. U+U collisions

Figure 5 shows the π^0 and η p_T-spectra over a wide range in all centrality classes. However, to extract R_{AA} one needs to carefully select the collision geometry since the Uranium nucleus is deformed. Therefore in the Monte Carlo Glauber based approach [9], the nucleon density

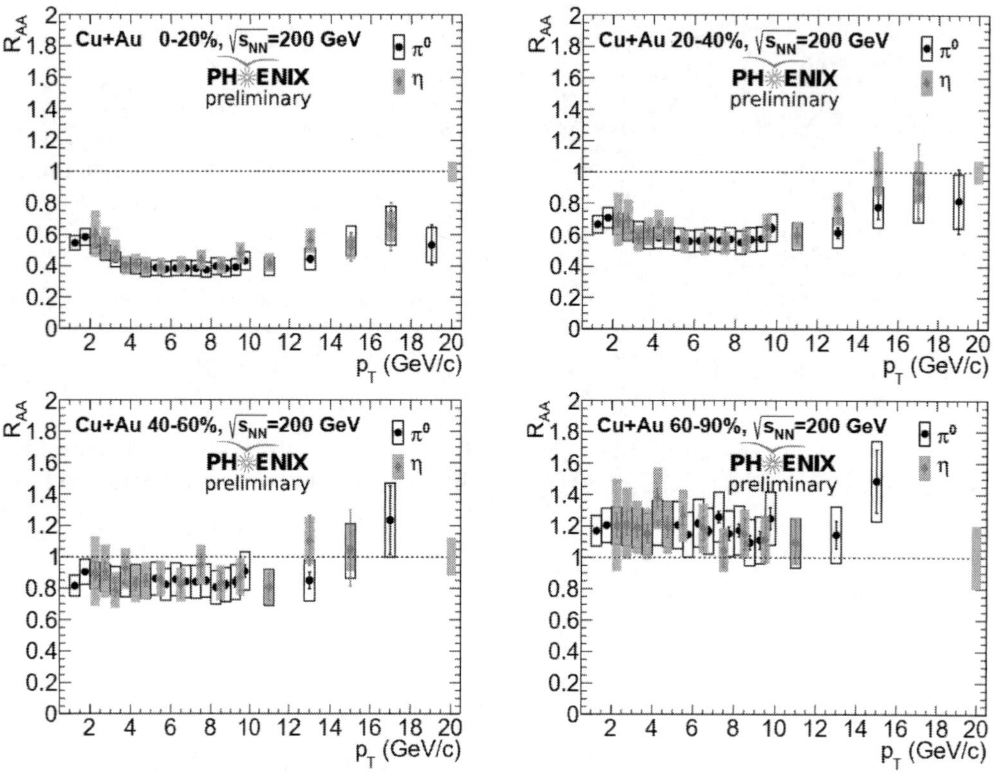

Figure 4. R_{AB} of π^0 (black) and η (red) as a function of p_T in different centrality classes of Cu+Au collisions at $\sqrt{s_{NN}} = 200$ GeV.

distribution is parameterized by a deformed Woods-Saxon profile [10],

$$\rho = \frac{\rho_0}{1 + \exp([r - R']/a)}, \tag{2}$$

$$R' = R[1 + \beta_2 Y_2^0(\theta) + \beta_4 Y_4^0(\theta)], \tag{3}$$

where ρ_0 is the normal nuclear density, R and a denote the radius of nucleus and the surface diffuseness parameter, respectively. The spherical harmonic functions and the β values introduce the deformation from spherical shape in the Uranium nucleus. Table 1 summarizes the two sets of Wood-Saxon parameters that were used in Glauber MC.

Table 1. Parameter sets for Uranium nucleus [11, 12].

Parameter	Set1/Glauber1 [11]	Set2/Glauber2 [12]
$R(fm)$	6.81	6.86
$a(fm)$	0.61	0.42
β_2	0.286	0.265
β_4	0.093	0

IOP Conf. Series: Journal of Physics: Conf. Series **1070** (2018) 012011 doi:10.1088/1742-6596/1070/1/012011

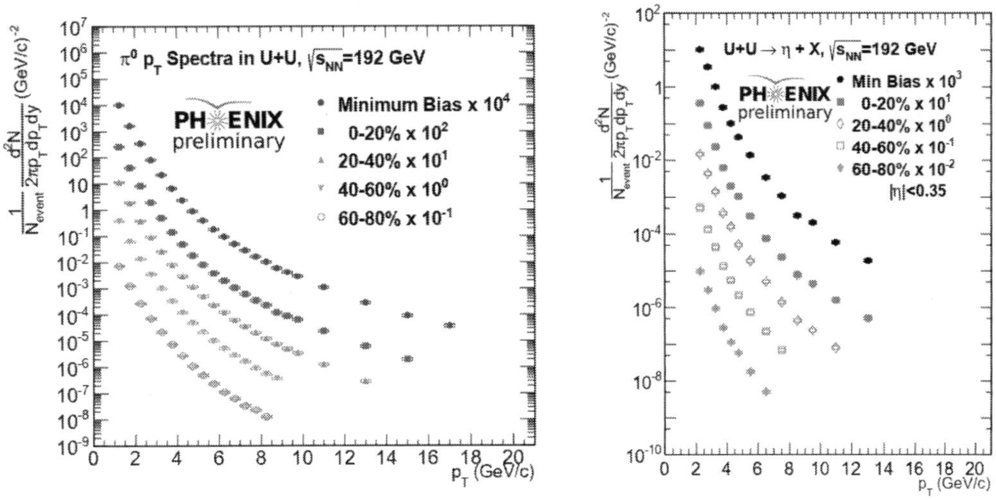

Figure 5. The invariant p_T-spectra of π^0 (left) and η (right) measured in different centrality classes of U+U collisions at $\sqrt{s_{NN}} = 193$ GeV. The spectra are scaled by arbitrary factors for clarity.

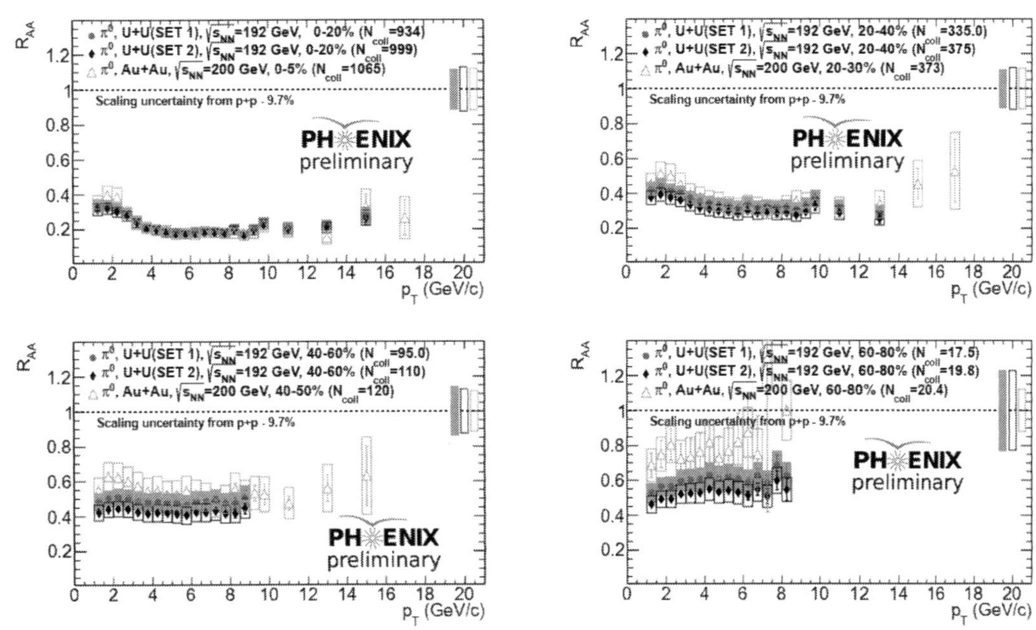

Figure 6. R_{AA} of π^0 as a function of p_T in different centrality classes of U+U collisions at $\sqrt{s_{NN}} = 193$ GeV. The U+U results are compared to those of Au+Au at $\sqrt{s_{NN}} = 200$ GeV.

Figure 6 shows π^0 R_{AA} in U+U collisions as a function of p_T for both sets of parameters at different centralities. The data are also compared to π^0 R_{AA} in Au+Au collisions. At the same N_{coll} values, R_{AA} in Au+Au and U+U collisions are consistent in most- to mid-central

IOP Conf. Series: Journal of Physics: Conf. Series **1070** (2018) 012011 doi:10.1088/1742-6596/1070/1/012011

collisions which implies that π^0 production depends on the size of the nuclear overlap but not on its density. However, the most peripheral collisions show larger suppression in U+U collisions.

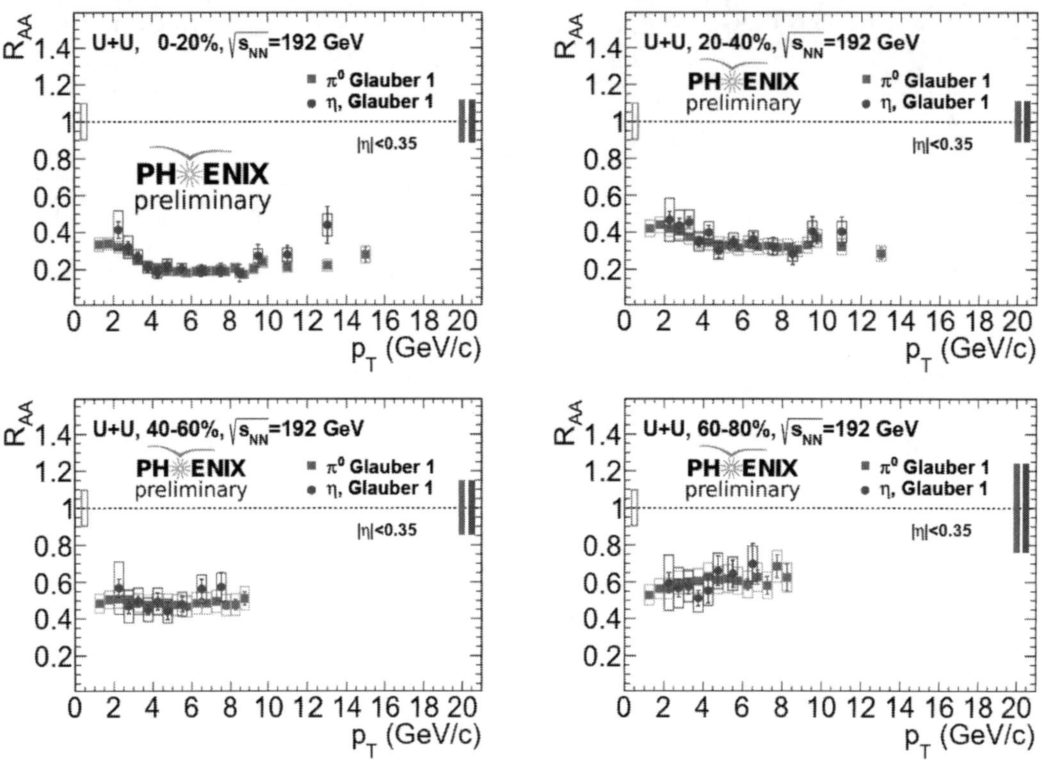

Figure 7. R_{AA} of π^0 (magenta) and η (blue) as a function of p_T in different centrality classes of U+U collisions at $\sqrt{s_{NN}} = 193$ GeV.

Figure 7 shows η R_{AA} compared to that of π^0 for different centrality classes. Similar to the case of Cu+Au collisions, R_{AA} of η and π^0 are consistent over all centralities.

Comparisons of the integrated R_{AB} versus N_{part} between the different collision systems, Cu+Cu, Cu+Au, Au+Au and U+U, are shown in Figures 8. Additional comparison of R_{AB} versus p_T for the most peripheral collisions in Cu+Cu, Cu+Au, and Au+Au is shown in Figure 9. These comparisons show that at $N_{part} > 50$ all studied systems are consistent with each other. However, at $N_{part} < 50$, although consistent within uncertainties, there seems to be ordering in the medium modification, Cu+Au > Cu+Cu > Au+Au > U+U.

4. Summary

PHENIX has measured p_T spectra and nuclear modification factors for π^0 and η in Cu+Au and U+U collisions at 200 and 193 GeV, respectively. R_{AB} factors for π^0 and η are consistent within uncertainties at all momenta and centralities which indicates that they have similar fragmentation function modification by the medium in the accessed p_T range. We also observed very consistent centrality dependent medium modification to previous jet measurements.

In central and semi-central A+B collisions: π^0 and η production depend on the size of the nuclear overlap but not on its shape or density. However, there is a hint of dependence on both in most peripheral, Cu+Au > Cu+Cu > Au+Au > U+U.

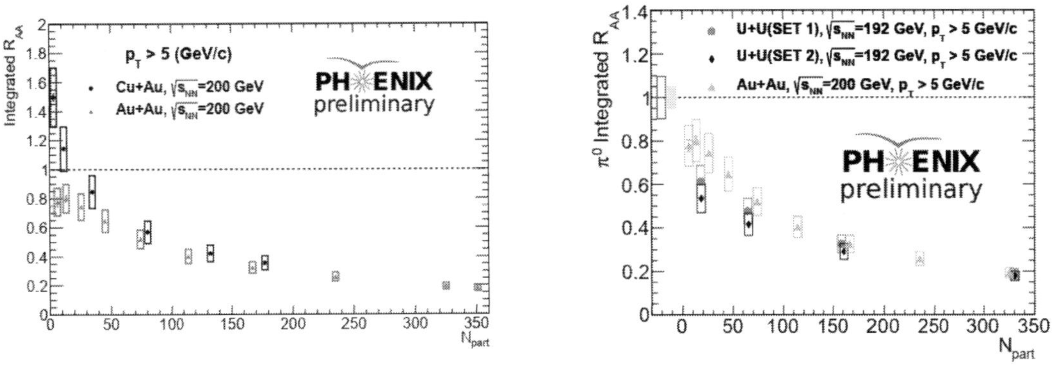

Figure 8. R_{AB} as a function of N_{part} for Cu+Au collisions compared with Au+Au collisions (left) and for U+U collisions (two sets) compared with Au+Au collisions (right).

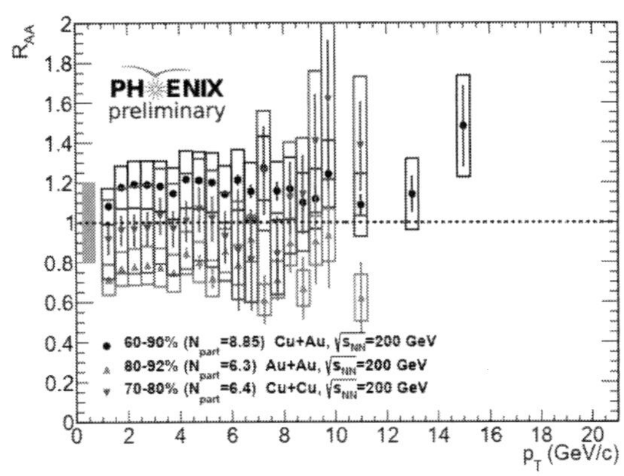

Figure 9. R_{AB} as a function of p_T for peripheral Cu+Au collisions compared with peripheral Cu+Cu and Au+Au collisions.

References

[1] Adcox K *et al.* 2005 *Nucl. Phys. A* **757** 184 – 283
[2] Adams J and other 2005 *Nucl. Phys. A* **757** 102 – 183
[3] Afanasiev S *et al.* 2012 *Phys. Rev. Lett.* **109** 152302
[4] Adare A *et al.* 2013 *Phys. Rev. C* **87**(3) 034904
[5] Adcox K *et al.* (PHENIX Collaboration) 2003 *Nucl. Instrum. Methods Phys. Res., Sec. A* **499** 469
[6] Adler S S, *et al.* 2007 *Phys. Rev. C* **75** 024909
[7] Bsser F *et al.* 1975 *Phys. Lett. B* **55** 232 – 236
[8] Timilsina A 2016 *Nucl. Phys. A* **956** 637 – 640
[9] Miller M L, Reygers K, Sanders S J and Steinberg P 2007 *Annu. Rev. Nucl. Part. Sci.* **57** 205–243
[10] Hagino K, Lwin N W and Yamagami M 2006 *Phys. Rev. C* **74**(1) 017310
[11] Masui H, Mohanty B and Xu N 2009 *Phys. Lett. B* **679** 440 – 444
[12] Shou Q, Ma Y, Sorensen P, Tang A, Videbk F and Wang H 2015 *Phys. Lett. B* **749** 215 – 220

IOP Conf. Series: Journal of Physics: Conf. Series **1070** (2018) 012012 doi:10.1088/1742-6596/1070/1/012012

Overview of low p_T photon results from PHENIX at RHIC

Takao Sakaguchi
Physics Department, Brookhaven National Laboratory, Upton, NY 11973-5000, USA
E-mail: takao@bnl.gov

Abstract. Overview of the latest low p_T direct photon measurements by the PHENIX experiment at RHIC is presented. We found a smooth change of the effective temperature as a function of $\sqrt{s_{NN}}$. We studied the scaling properties of the direct photon yield with various global observables. When plotting the direct photon yield as a function of $dN_{ch}/d\eta$, a universal scaling independent of the $\sqrt{s_{NN}}$ was observed.

1. Introduction

Direct photons are of special interest in the relativistic heavy ion collisions. They are emitted from all the stages of the collisions with wide range of Q^2 and don't interact strongly with medium once produced, thus they carry thermodynamical information of the stages they are emitted [1, 2]. The production rate of the photons can be written as:

$$E\frac{dR_\gamma}{d^3p} = -\frac{\alpha_{em}}{\pi^2}\mathrm{Im}\Pi_{em}(\omega,k)\frac{1}{e^{E/T}-1}, \quad \mathrm{Im}\Pi_{em}(\omega,k) \approx \ln\left\{\frac{\omega T}{(m_{th}(\approx gT))^2}\right\}$$

where Π_{em} is the photon self energy, and is process dependent. When we measure the inverse slope of the distribution at a high energy region ($E \gg T$) the inverse slope will tell the temperature of the system. However, the source of photons is not unique as mentioned above, therefore the measured photon spectra are convolution of various sources and so for the inverse slope. Figure 1 shows the photon production sources proposed by now, outlined as a function of their production times and energy ranges that they manifest. We define the direct photons as photons not originating from the hadron decays (π^0, η etc). The general trend is that the earlier the photons are produced, the higher their typical energy ranges will be. The Jet-Bremsstralung and Jet-photon conversion are those produced by the second interaction of hard scattered partons with a hot dense medium [3].

In order to disentangle the various photon sources, we have to perform as many differential measurements as the number of possible photon sources. One of the recently proposed and actually realized measurements is collective flow (v_n). Practically, this measurement is possible for only first couple orders (2, 3 ...) because of limited statistics. We list the expected behavior of yield, v_2 and v_3 for various proposed photon sources in the Table 1.

Comparing to the high p_T photons ($p_T > 5\,\mathrm{GeV}/c$) that have been successfully measured using calorimetric method [4, 5], the measurement of low to mid p_T photons ($p_T \lesssim 4\,\mathrm{GeV}/c$) called for new methods. By now, two methods have been developed; one is called internal

IOP Conf. Series: Journal of Physics: Conf. Series **1070** (2018) 012012 doi:10.1088/1742-6596/1070/1/012012

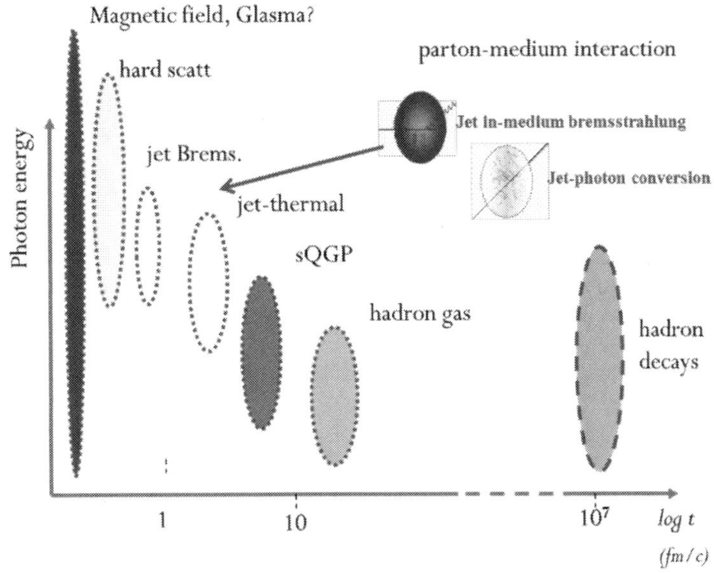

Figure 1. Schematic representation of the photon production sources as a function of their production times and energy ranges.

conversion method [6, 7], and the other is external conversion method [8]. The internal conversion method utilizes the fact that the process of producing real photons (zero mass) always has some probability of producing virtual photons with a finite mass that decay into electron-positron pairs. Then, the analysis method similar to dilepton measurement can be applied. The advantage of this method is that one can select the virtual mass region where π^0 contribution which is the primary background for photon measurement becomes very small. A restriction exists in this method, however, which is that the p_T to measure has to meet $p_\mathrm{T} \gg m_\mathrm{ee}$. The typical lower limit in p_T is 1 GeV/c. The external conversion method measures photons converted to the electron-positron pairs at materials. In this method, there is no lower p_T limit, but one has to deal with a large background from π^0 decay. Both methods have good energy resolution and small inefficiency of hadron rejection, thanks to the particle tracking system and the Ring Imaging Cherenkov detector (RICH) [9].

Table 1. List of photon production sources, and their manifesting p_T regions and the typical behavior of their v_2 and v_3.

Production source	p_T	v_2	v_3	t-dependence of v_n
Hadron gas	low p_T	positive	positive	⟹
QGP	Mid p_T	positive	positive	⟹
Primordial (hard)	High p_T	~zero	~zero	⟹
Jet-Bremsstrahlung	Mid p_T	positive	?	⟹
Jet-photon conversion	Mid p_T	negative	?	⟹
Initial B-field	All p_T	positive	~zero	⇒

2. Direct photon production in Au+Au collisions at $\sqrt{s_{NN}}$ =200 GeV

PHENIX has successfully measured low p_T direct photons with both internal and external conversion methods. A huge excess of photon yield over the primordial hard photon production scaled from the photon spectra in $p+p$ collisions was observed. The inverse slopes (effective temperature, $T_{\rm eff}$) were measured for the photon spectra for $p_T > 1.0(0.5)$ GeV/c by internal (external) conversions, and were found consistently to be ~220 MeV which are much higher than the transition temperature (170 MeV) [6]. On the other hand, a sizable v_2 for low p_T photons was found [10], which is controversy to the spectra results. A large photon yield implies that the photons are emitted from initial stages of the collisions, while the large v_2 implies that the photons are emitted from late stages where the flow is sufficiently built up. The v_3 was recently measured and was also found to be sizable [11]. Figure 2 shows the latest measurement of the spectra, v_2 and v_3 for low p_T direct photons [8, 11] together with representative calculations from several different models [12, 13, 14, 15, 16]. When the v_2 result was presented for the

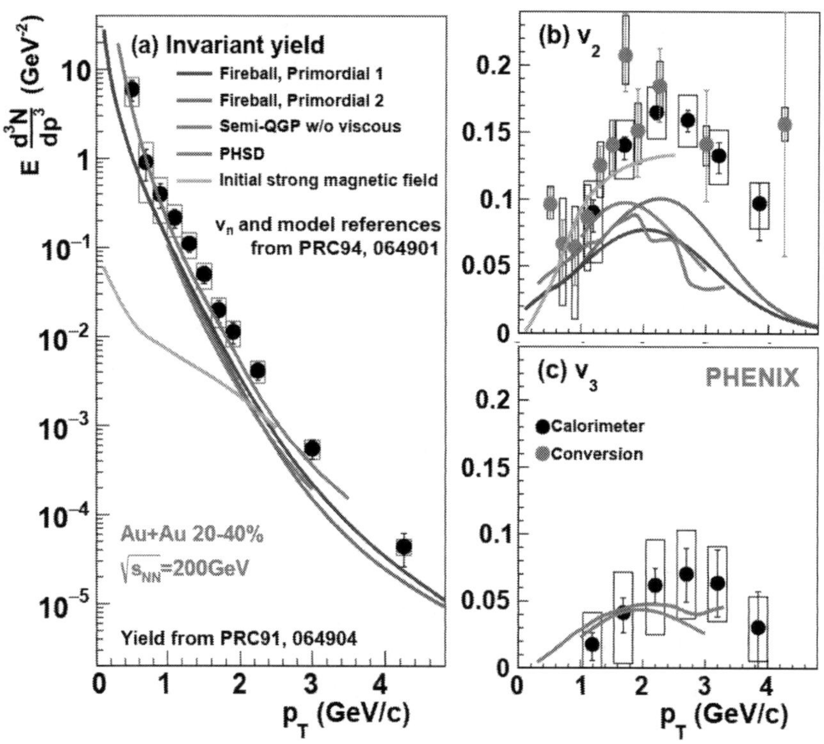

Figure 2. Low p_T direct photon spectra, v_2 and v_3 in 20-40 % Au+Au collisions at $\sqrt{s_{NN}}$ =200 GeV [8]. The lines show representative calculations from several different models.

first time, none of the model calculations was able to produce spectra and v_2. However the considerable theory efforts since then improved the model calculations significantly and the data are reasonably described by now. One of the key ingredients in the model calculations is the blue shift of the spectra [17, 18, 19]; as the collision system expands, the photons are boosted (blue shifted) and give a large inverse slope. This implies that the majority of the photons may have been produced around the hadronization stage.

3. Direct photon production in new collision systems

In order to investigate the origin of photons observed, PHENIX has extended the photon measurement to different systems and different collision energies. In the RHIC Year-2005 run, PHENIX has collected Cu+Cu collision data at $\sqrt{s_{NN}}$ =200 GeV. Figure 3 shows the photon spectra in Cu+Cu collisions at $\sqrt{s_{NN}}$ =200 GeV together with the ones in Au+Au collisions previously published [8]. The inverse slope parameters were obtained as $288 \pm 49(stat) \pm$

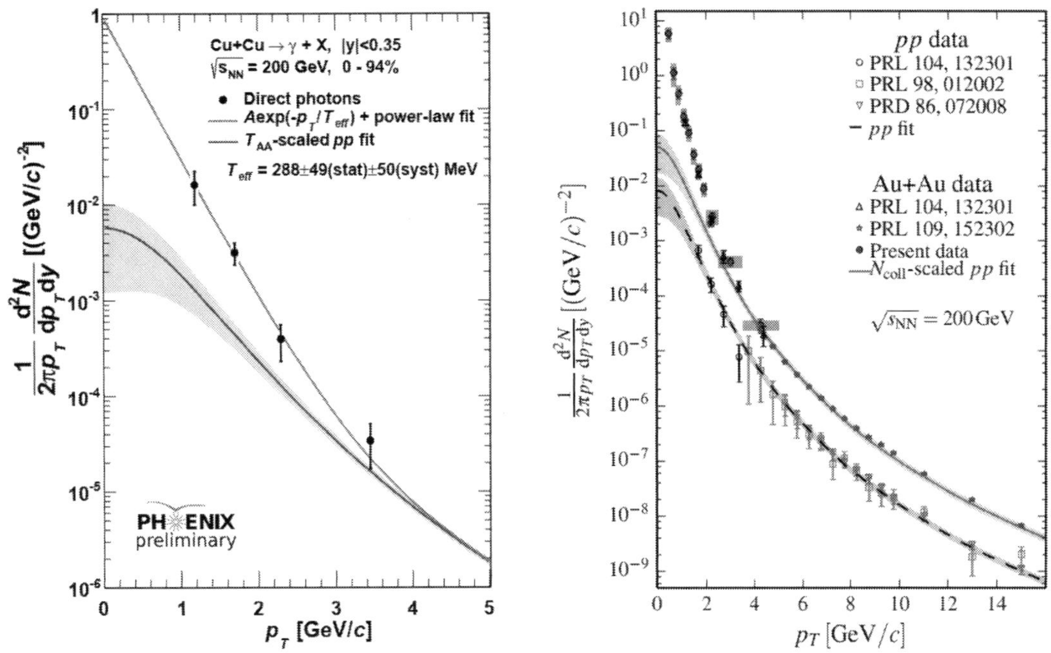

Figure 3. Direct photon spectra in minimum bias Cu+Cu collisions (left) and in minimum bias Au+Au collisions (right) at $\sqrt{s_{NN}}$ =200 GeV.

$50(syst)$ MeV for minimum bias Cu+Cu collisions, and $242 \pm 28(stat) \pm 7(syst)$ MeV for minimum bias Au+Au collisions, respectively. The parameters for both systems are agreeing each other within quoted uncertainties.

PHENIX has recently measured the direct photons in Au+Au collisions at lower energies as well. Figure 4 shows the low p_T direct photon spectra in Au+Au collisions at $\sqrt{s_{NN}}$ =62.4 GeV and 39 GeV. Unfortunately, the photons from initial hard scattering cannot be quantified due to the lack of statistics at high p_T. However, from a pQCD calculation that is validated by results from past experiments shows that the hard photon contribution at $\sqrt{s_{NN}}$ =62.4 GeV is small. It is reasonable to also assume that the hard photon contribution at $\sqrt{s_{NN}}$ =39 GeV is even smaller and negligible, since the hard scattering probability decreases much rapidly compared to the bulk particle production.

We summarized the inverse slope parameters, T_{eff} in the Figure 5, for Au+Au and Cu+Cu collisions at $\sqrt{s_{NN}}$ =200 GeV and Au+Au collisions at $\sqrt{s_{NN}}$ =62.4 and 39 GeV, together with the measurement in Pb+Pb collisions at $\sqrt{s_{NN}}$ =2.76 TeV performed by the ALICE experiment at the LHC [20]. It is seen that the T_{eff} increases systematically as $\sqrt{s_{NN}}$ increases, but not linearly. This observation is reasonable since the temperature of the system increases by the energy density to the power of $\sim 1/4$.

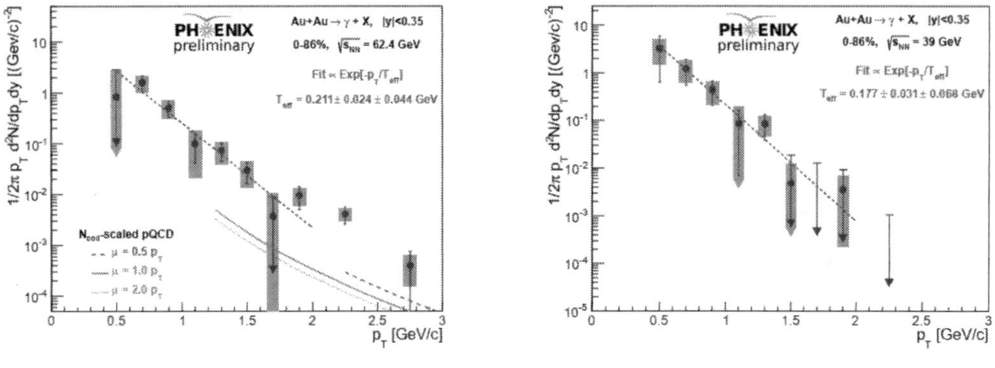

Figure 4. Direct photon spectra in Au+Au collisions at $\sqrt{s_{NN}}$ =62.4 GeV (left) and 39 GeV (right). The dotted lines going through the data points are exponential fits. The three lines far below the data points in the left show pQCD calculations with three different mass scales.

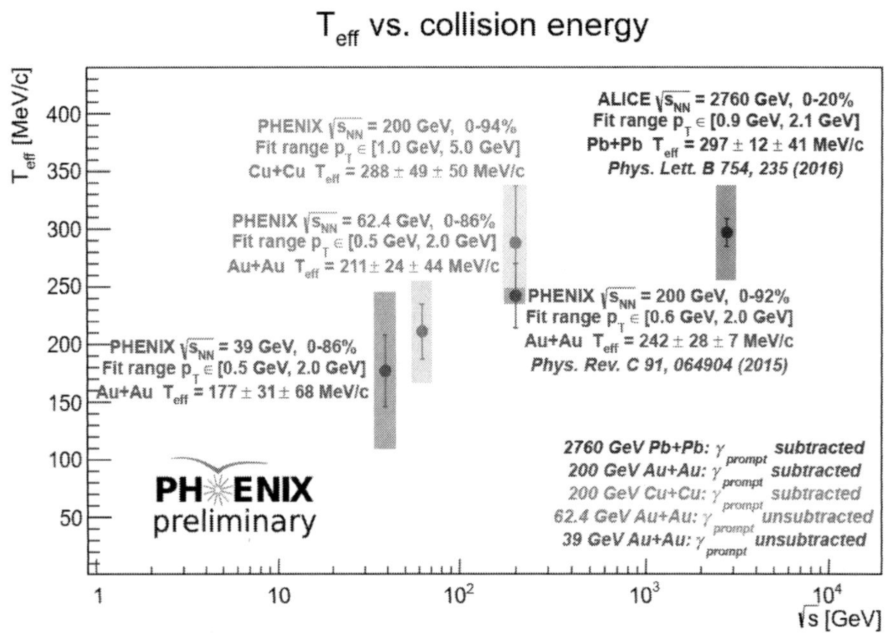

Figure 5. Summary of the inverse slope parameters, $T_{\rm eff}$ for various collision systems and energies. The 2.76 TeV point is from ALICE measurement at the LHC.

4. Scaling property of direct photon yield

With variety of datasets we collected, we studied the scaling properties of the direct photon yields. In the previous publication, we plotted the integrated direct photon yield (dN/dy) as a function of $N_{\rm part}$ for various integration range [8]. This is reproduced in the left plot of the Figure 6. Surprisingly, $N_{\rm part}$ dependence of the photon dN/dy is independent of the lower p_T limit of the integration, suggesting that the source of these photons are similar. The data points are fitted with a power-law function, $dN/dy = \beta {N_{\rm part}}^{\alpha}$, and obtained the α as $\alpha = 1.48 \pm 0.08(stat) \pm 0.04(syst)$. This number can be compared with a model calculation

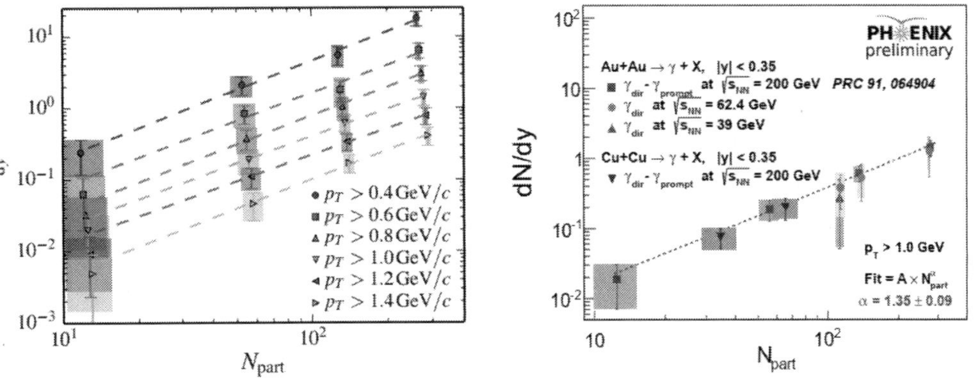

Figure 6. (Left) Integrated direct photon yield (dN/dy) as a function of N_{part} in Au+Au collisions at $\sqrt{s_{NN}}$ =200 GeV. (Right) Compilation of photon dN/dy (p_{T} >1 GeV/c) for all the available data at PHENIX, again plotted against N_{part}.

where α is predicted to change as a function of the lower p_{T} limit [18]. We have extended this scaling study to all the available data as shown in the right plot of the Figure 6 [21]. It is interesting to see that all the available data are on the same power-law function. We performed simultaneous fit over the data and obtained as $\alpha = 1.35 \pm 0.09$, which is consistent with the previous publication.

Lastly, we studied the scaling properties with another global observables, that is charged particle multiplicity ($dN_{\mathrm{ch}}/d\eta$). Figure 7 shows the photon dN/dy plotted against $dN_{\mathrm{ch}}/d\eta$ for various collisions systems and energies [22, 21, 23]. Interesting enough, all the data points except for those from p+p and d+Au are found to be nicely fitted with a power-law function of $dN/dy = \beta(dN_{\mathrm{ch}}/dy)^{\alpha}$ with $\alpha \sim 1.2$. This shows that the photons are emitted similarly over the wide range of the energies in terms of temperature and space time evolution. This may be natural, if most of the photons are emitted near the transition from QGP to hadronic gas. In the same plot, we showed the dN/dy for the pQCD calculations scaled by the number of collisions corresponding to that of A+A collisions. It is interesting to see that the pQCD lines are parallel to the A+A points except for an order of magnitude difference. There looks to be a similar production mechanism playing in p+p collisions as well.

We performed a separate power-law fit only for the p+p, d+Au and peripheral Au+Au collision data and showed in Figure 8. It is found that the α is $\sim$ 2.2 for this fit and is quite different from the one for the A+A points. It is possible that there is a turn-on of the hot medium production somewhere in this range of $dN_{\mathrm{ch}}/d\eta$.

We here note that some of the plots shown here became final and submitted to journals after the conference [24, 25].

5. Summary

We have shown and reviewed the low p_{T} direct photon measurements performed by the PHENIX experiment at RHIC. PHENIX has measured low p_{T} direct photons in Au+Au and Cu+Cu collisions at $\sqrt{s_{NN}}$ =200 GeV and in Au+Au collisions at $\sqrt{s_{NN}}$ =62.4 and 39 GeV, either using internal conversion or external conversion methods that reduced the background significantly. The inverse slope was found to change smoothly as a function of $\sqrt{s_{NN}}$ from 39 GeV to 2.76 TeV. It was found that the direct photon yields scale very well with $dN_{\mathrm{ch}}/d\eta$ with a power of $\alpha \sim 1.2$. This implies that the photons are emitted similarly over the wide range of the energies in terms

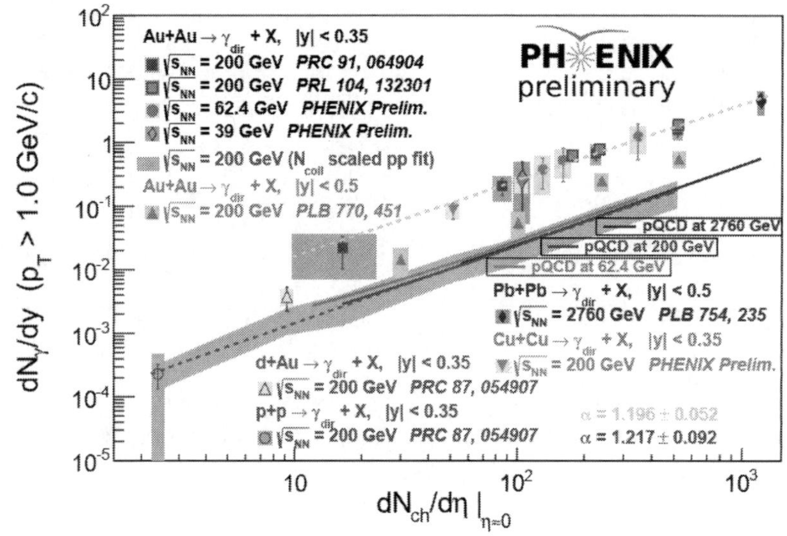

Figure 7. Compilation of photon dN/dy (p_T >1 GeV/c) for all the available data at PHENIX as well as the one from ALICE at the LHC, plotted against $dN_{ch}/d\eta$. pQCD calculations scaled by the number of collisions corresponding to that of A+A collisions are also plotted.

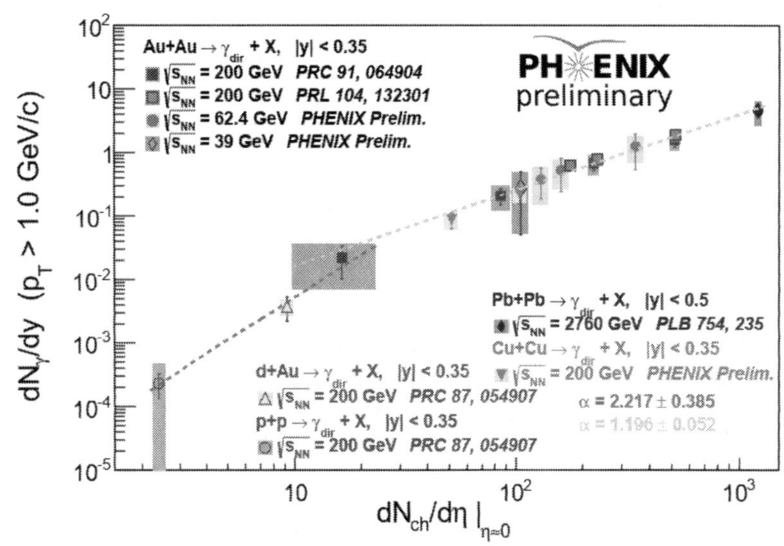

Figure 8. Same as Figure 7 but without the pQCD calculations shown. In this plot, another power-law fit performed only for p+p, d+Au and peripheral Au+Au collision data is shown.

of temperature and space time evolution. This may be natural, if most of the photons are emitted near the transition from QGP to hadronic gas.

IOP Conf. Series: Journal of Physics: Conf. Series **1070** (2018) 012012 doi:10.1088/1742-6596/1070/1/012012

References

[1] Shuryak E V 1978 *Phys. Lett. B* **78** 150
[2] Alam J, Sinha B and Raha S 1996 *Phys. Rept.* **273** 243–362
[3] Fries R J, Muller B and Srivastava D K 2005 *Phys. Rev.* **C72** 041902 (*Preprint* nucl-th/0507018)
[4] Adler S S *et al.* (PHENIX) 2005 *Phys. Rev. Lett.* **94** 232301 (*Preprint* nucl-ex/0503003)
[5] Afanasiev S *et al.* (PHENIX Collaboration) 2012 *Phys. Rev. Lett.* **109** 152302
[6] Adare A *et al.* (PHENIX Collaboration) 2010 *Phys. Rev. Lett.* **104** 132301
[7] Adare A *et al.* (PHENIX) 2013 *Phys. Rev.* **C87** 054907 (*Preprint* 1208.1234)
[8] Adare A *et al.* (PHENIX Collaboration) 2015 *Phys. Rev. C* **91** 064904
[9] Adcox K *et al.* 2003 *Nucl. Inst. Methods Phys. Res., Sect. A* **499** 469
[10] Adare A *et al.* (PHENIX Collaboration) 2012 *Phys. Rev. Lett.* **109** 122302
[11] Adare A *et al.* (PHENIX) 2016 *Phys. Rev.* **C94** 064901 (*Preprint* 1509.07758)
[12] van Hees H, He M and Rapp R 2015 *Nucl. Phys. A* **933** 256 and private communication
[13] Gale C, Hidaka Y, Jeon S, Lin S, Paquet J F, Pisarski R D, Satow D, Skokov V V and Vujanovic G 2015 *Phys. Rev. Lett.* **114** 072301
[14] Linnyk O, Cassing W and Bratkovskaya E L 2014 *Phys. Rev. C* **89** 034908
[15] Linnyk O, Konchakovski V, Steinert T, Cassing W and Bratkovskaya E L 2015 *Phys. Rev. C* **92** 054914 and private communication
[16] Basar G, Kharzeev D E and Skokov V 2012 *Phys. Rev. Lett.* **109** 202302
[17] van Hees H, Gale C and Rapp R 2011 *Phys. Rev. C* **84** 054906
[18] Shen C, Heinz U, Paquet J F and Gale C 2014 *Phys. Rev. C* **89** 044910
[19] Shen C, Paquet J F, Heinz U and Gale C 2015 *Phys. Rev. C* **91** 014908
[20] Adam J *et al.* (ALICE Collaboration) 2016 *Phys. Lett. B* **754** 235
[21] Adare A *et al.* (PHENIX Collaboration) 2016 *Phys. Rev. C* **93** 024901
[22] Adler S S *et al.* (PHENIX Collaboration) 2014 *Phys. Rev. C* **89** 044905
[23] Aamodt K *et al.* (ALICE Collaboration) 2011 *Phys. Rev. Lett.* **106** 032301
[24] Adare A *et al.* (PHENIX) 2018 *Submitted to: Phys. Rev. Lett.* (*Preprint* 1805.04066)
[25] Adare A *et al.* (PHENIX) 2018 *Submitted to: Phys. Rev. Lett.* (*Preprint* 1805.04084)

IOP Conf. Series: Journal of Physics: Conf. Series **1070** (2018) 012013 doi:10.1088/1742-6596/1070/1/012013

Quarkonium in A–A collisions with the ALICE experiment

Audrey Francisco on behalf of the ALICE Collaboration

SUBATECH, UMR 6457, IMT Atlantique, Université de Nantes, IN2P3/CNRS,
4 rue Alfred Kastler, 44307 Nantes cedex 3, France

E-mail: `audrey.francisco@cern.ch`

Abstract. Quarkonia, i.e. bound states of heavy quarks (charm/bottom), provide remarkable probes of the hot and strongly-interacting medium, the quark gluon plasma (QGP), which is created during heavy-ion collisions. Heavy quarks are produced during initial hard-scattering processes prior to the QGP formation and their number is conserved throughout the partonic and hadronic phases of the collision. At LHC, experimental observations of quarkonium states in A-A collisions are reproduced through two antagonist mechanisms: the sequential suppression and quarkonium production by (re)combination of deconfined quarks. Recent measurements of a significant J/ψ elliptic flow (v_2) in Pb-Pb collisions at $\sqrt{s_{\rm NN}}$ = 5.02 TeV suggest a participation of the charm quarks to the collectivity of the medium. However, quarkonium interactions in the medium are not yet fully understood and several points should still be clarified. The ALICE experiment at the LHC measures quarkonia at mid-rapidity in the dielectron decay channel and at forward rapidity in the dimuon one, both down to zero transverse momentum. Latest measurements of quarkonium nuclear modification factor and elliptic flow in Pb-Pb collisions will be presented and compared to lower energy results and theoretical predictions.

1. Introduction

Measurements of quarkonia in heavy-ion collisions have been performed for several decades. Heavy quarks are produced prior to the formation of the QGP and experience the full medium evolution, providing crucial probes of the QGP properties. Quarkonium mesons formed of either a charm and anti-charm pair(J/ψ , ψ(2S)) or a beauty and anti-beauty pair (Υ(1S),Υ(2S),Υ(3S)) can be experimentally measured through their dilepton decay. In the QGP they experience a suppression mechanism by color-screening in the medium and (re)generation by quark (re)combination at a later stage of the collision. The relative extent of both mechanisms depends on the system energy density, the heavy quark density and the specific bound state. Whereas light quarks typically reflect medium thermalization and expansion properties, heavy quarks carry information about their early interactions in the medium. Since the thermalization time of heavy quarks is longer, a relevant question is their degree of thermalization in the QGP. Substantial anisotropies are observed in the heavy-flavour sector and if charm quarks interact in the medium, their flow should be transmitted to quarkonium states formed through (re)combination.

2. Apparatus

The ALICE detector [1] measures quarkonium down to zero transverse momentum in two rapidity ranges: at mid-rapidity ($|y| < 0.9$) with the central barrel through the dielectron decay channel and at forward rapidity ($2.5 < y < 4$) with the muon arm through the dimuon decay channel. Other detectors [2] are involved in the presented results for triggering, centrality determination and luminosity measurement. The 2015 Pb–Pb data campaign provided unprecedented statistics for quarkonium measurements, with an integrated luminosity reaching $\sim 13\ \mu b^{-1}$ in the central barrel and $\sim 225\ \mu b^{-1}$ in the muon spectrometer. All quarkonium measurements are inclusive, i.e. they include direct production and contributions from heavier charmonium states and B decays.

3. Quarkonium production in Pb–Pb collisions

The measurement of quarkonium suppression with respect to the corresponding pp yield is quantified through the nuclear modification factor (R_{AA}). Results for J/ψ are presented on Fig.1 as a function of the number of participants in the collision. Compatible R_{AA} values are observed between both J/ψ decay channels. After a strong decrease in peripheral collisions, the J/ψ production reaches a plateau around $R_{AA}\sim 0.6$ [3]. Mid-rapidity results exhibit a hint of production enhancement for the most central collisions, in compliance with theoretical expectations including a (re)generation contribution dominant at mid-rapidity for most central events. The results are compatible with previous observations at $\sqrt{s_{NN}} = 2.76$ TeV but benefit from an increased precision. These observations are interpreted as an indication of the interplay between the suppression and (re)combination mechanisms. A precise discrimination based on theory models [4, 5, 6, 7] is not possible because of substantial uncertainties on the determination of charm cross-section and the estimation of cold nuclear matter effects [8, 9]. The J/ψ R_{AA} is enhanced at low transverse momentum (p_T) for the most central collisions, this effect being expected from the (re)combination contribution, while a stronger suppression is observed as the J/ψ p_T increases, for both forward and mid-rapidity results. Available predictions [4] from a transport model predict a similar trend.

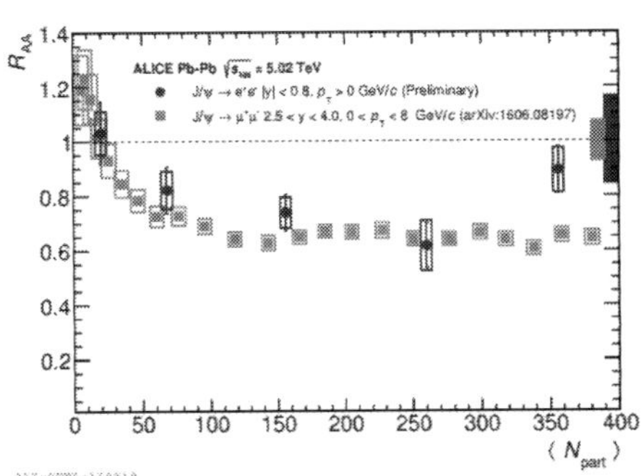

Figure 1. Comparison of J/ψ R_{AA} measurements as a function of the number of participants in Pb–Pb collisions at $\sqrt{s_{NN}} = 5.02$ TeV in the dielectron and dimuon decay channels

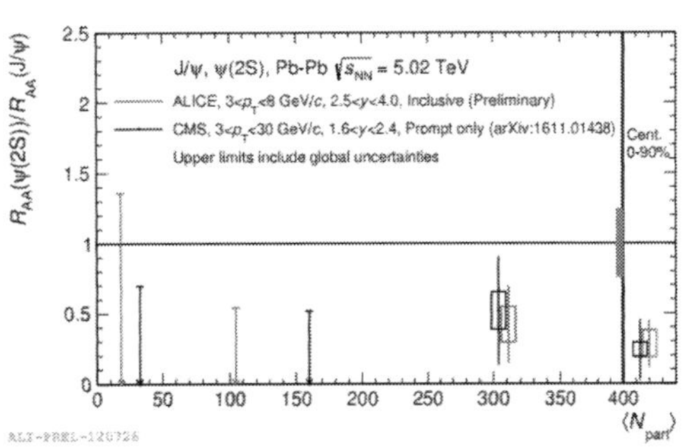

Figure 2. Double ratio of ψ(2S) and J/ψ R_{AA} as a function of the number of participants in Pb–Pb collisions at $\sqrt{s_{NN}} = 5.02$ TeV

More statistics is required to better study ψ(2S) production in A–A collisions. A relative suppression to J/ψ measurement is observed at forward rapidity (as seen on Fig. 2), as expected from the sequential suppression picture since the ψ(2S) is more loosely bound. Data show a

IOP Conf. Series: Journal of Physics: Conf. Series **1070** (2018) 012013 doi:10.1088/1742-6596/1070/1/012013

stronger suppression in semi-central and central collisions. An upper limit with a confidence level of 95% was calculated for low signal significances. More precise (semi-)peripheral measurements would greatly contribute to describe the suppression and (re)combination mechanisms for charmonium states. An overall improved measurement of the ψ(2S) over J/ψ ratio would particularly help model discrimination. During the LHC run 3 a substantial improvement of the collected statistics and of the signal over background ratio is expected for this challenging measurement.

Inclusive measurements of Υ(1S) are also available at forward rapidity in Pb–Pb collisions at $\sqrt{s_{\mathrm{NN}}} = 5.02$ TeV. Because of the high mass of the bottom quark, a lower density of bottom quarks is expected in medium compared to charm. R_{AA} values around ~ 0.5 [10] are observed, similar to $\sqrt{s_{\mathrm{NN}}} = 2.76$ TeV results [11]. The centrality dependence is more pronounced than in the charmonium case and results exhibit almost no variation with p_{T}.

4. Quarkonium azimuthal anisotropy in Pb–Pb collisions

The elliptic shape of the medium created in non-central collisions causes an anisotropic matter distribution, passed to the particle momentum distribution if the system is interacting. The decomposition of particle momentum distribution into a Fourier series [12] allows for a precise study of each harmonic. The second coefficient, the elliptic flow (v_2), reflects the anisotropy in the transverse plane and is very sensitive to the early times of the collision. Because of their large mass, the thermalization degree of heavy quarks might be partial and needs to be ascertained. Heavy flavour hadrons and in particular recent D mesons measurements of ALICE [13] exhibit significant azimuthal anisotropies. Event shape engineering results give strong hints that charm follows the collective expansion of the medium.

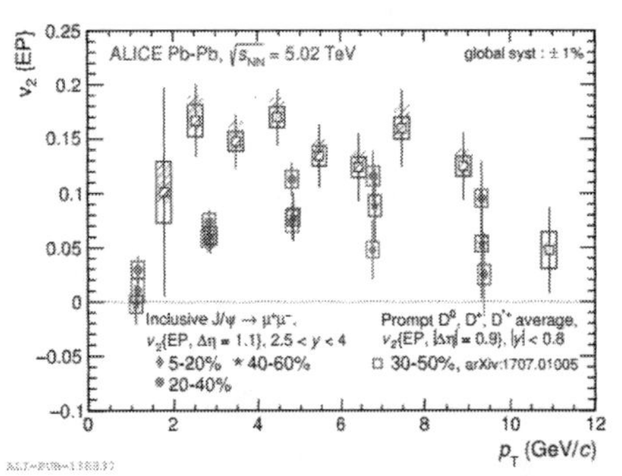

Figure 3. J/ψ elliptic flow vs p_{T} at forward rapidity in Pb–Pb collisions at $\sqrt{s_{\mathrm{NN}}} = 5.02$ TeV in several centrality ranges compared to D mesons results

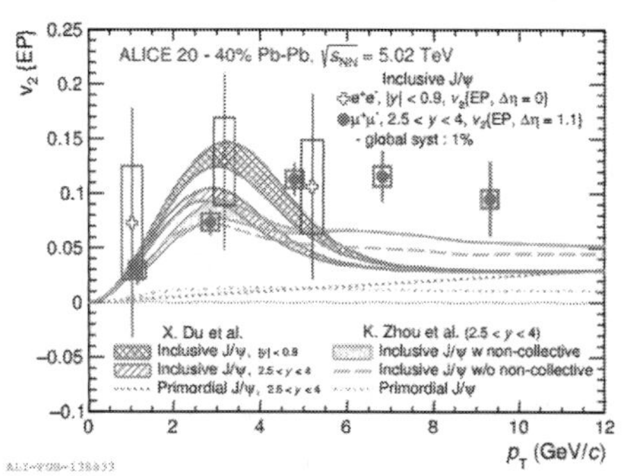

Figure 4. J/ψ elliptic flow vs p_{T} in semi-central Pb–Pb collisions at $\sqrt{s_{\mathrm{NN}}} = 5.02$ TeV compared to theoretical predictions [14, 7]

Previous studies at RHIC energies [15, 16] of quarkonium elliptic flow led to small anisotropies compatible with zero but dominated by large uncertainties. LHC first collisions at $\sqrt{s_{\mathrm{NN}}} = 2.76$ TeV revealed first positive J/ψ v_2 values measured both by CMS [17] and ALICE [10] in different kinematic ranges. The analysis at $\sqrt{s_{\mathrm{NN}}} = 5.02$ TeV [18] features a sizeable azimuthal anisotropy for several transverse momentum and centrality ranges. For the first time, mid-rapidity measurements are also performed. The comparison to open heavy flavour on Fig.3 reveals similar amplitudes, pointing towards charm thermalization in the medium.

Theoretical calculations reproducing R_{AA} measurements can reproduce the measured amplitude of the elliptic flow if a strong (re)combination component is included (Fig.4). Therefore J/ψ elliptic flow measurements weight in favour of quarkonium formation through charm quarks (re)combination before or during the hadronization stage. At high p_{T} a small anisotropy is expected from primordial J/ψ, arising from path length differences through the elliptic shape of the fireball. A second contribution is considered by [7] from the early magnetic field [19]. However the measured amplitude lies far above theory predictions. A missing mechanism should possibly be invoked explain this discrepancy.

5. Conclusion

ALICE measurements in Pb–Pb collisions at $\sqrt{s_{\mathrm{NN}}} = 5.02$ TeV provide deeper understanding of quarkonium physics. The study of the nuclear modification factor allows an investigation of the interplay between suppression and (re)combination mechanisms. J/ψ R_{AA} and elliptic flow require that a significant proportion of J/ψ are produced through charm quark (re)combination. The J/ψ azimuthal anisotropy is a sign of charm interaction with the medium constituents but the unexpected amplitude at high p_{T} has not been explained yet. Bottomonium results show a suppression which increases towards most central collisions, with a negligible p_{T} dependence and little room for a (re)generation contribution. The additional statistics of the next LHC run, also thanks to the ALICE detector upgrade, should bring new insights to quarkonium formation and evolution in the QGP.

References

[1] K. Aamodt *et al.* [ALICE Collaboration], JINST **3**, S08002 (2008). doi:10.1088/1748-0221/3/08/S08002
[2] B. B. Abelev *et al.* [ALICE Collaboration], Int. J. Mod. Phys. A **29**, 1430044 (2014) doi:10.1142/S0217751X14300440 [arXiv:1402.4476 [nucl-ex]].
[3] J. Adam *et al.* [ALICE Collaboration], Phys. Lett. B **766**, 212 (2017) doi:10.1016/j.physletb.2016.12.064 [arXiv:1606.08197 [nucl-ex]].
[4] X. Zhao and R. Rapp, Nucl. Phys. A **859**, 114 (2011) doi:10.1016/j.nuclphysa.2011.05.001 [arXiv:1102.2194 [hep-ph]].
[5] A. Andronic, P. Braun-Munzinger, K. Redlich and J. Stachel, Nucl. Phys. A **904-905**, 535c (2013) doi:10.1016/j.nuclphysa.2013.02.070 [arXiv:1210.7724 [nucl-th]].
[6] E. G. Ferreiro, Phys. Lett. B **731**, 57 (2014) doi:10.1016/j.physletb.2014.02.011 [arXiv:1210.3209 [hep-ph]].
[7] K. Zhou, N. Xu, Z. Xu and P. Zhuang, Phys. Rev. C **89**, no. 5, 054911 (2014) doi:10.1103/PhysRevC.89.054911 [arXiv:1401.5845 [nucl-th]].
[8] J. Adam *et al.* [ALICE Collaboration], JHEP **1506**, 055 (2015) doi:10.1007/JHEP06(2015)055 [arXiv:1503.07179 [nucl-ex]].
[9] S. Acharya *et al.* [ALICE Collaboration], arXiv:1802.00765 [nucl-ex].
[10] S. Acharya *et al.* [ALICE Collaboration], arXiv:1805.04387 [nucl-ex].
[11] B. B. Abelev *et al.* [ALICE Collaboration], Phys. Lett. B **738**, 361 (2014) doi:10.1016/j.physletb.2014.10.001 [arXiv:1405.4493 [nucl-ex]].
[12] S. Voloshin and Y. Zhang, Z. Phys. C **70**, 665 (1996) doi:10.1007/s002880050141 [hep-ph/9407282].
[13] S. Acharya *et al.* [ALICE Collaboration], Phys. Rev. Lett. **120**, no. 10, 102301 (2018) doi:10.1103/PhysRevLett.120.102301 [arXiv:1707.01005 [nucl-ex]].
[14] X. Du and R. Rapp, Nucl. Phys. A **943**, 147 (2015) doi:10.1016/j.nuclphysa.2015.09.006 [arXiv:1504.00670 [hep-ph]].
[15] C. Silvestre [PHENIX Collaboration], J. Phys. G **35**, 104136 (2008) doi:10.1088/0954-3899/35/10/104136 [arXiv:0806.0475 [nucl-ex]].
[16] C. Zhou [STAR Collaboration], J. Phys. Conf. Ser. **832**, no. 1, 012021 (2017). doi:10.1088/1742-6596/832/1/012021
[17] V. Khachatryan *et al.* [CMS Collaboration], Eur. Phys. J. C **77**, no. 4, 252 (2017) doi:10.1140/epjc/s10052-017-4781-1 [arXiv:1610.00613 [nucl-ex]].
[18] S. Acharya *et al.* [ALICE Collaboration], Phys. Rev. Lett. **119**, no. 24, 242301 (2017) doi:10.1103/PhysRevLett.119.242301 [arXiv:1709.05260 [nucl-ex]].
[19] X. Guo, S. Shi, N. Xu, Z. Xu and P. Zhuang, Phys. Lett. B **751**, 215 (2015) doi:10.1016/j.physletb.2015.10.038 [arXiv:1502.04407 [hep-ph]].

IOP Conf. Series: Journal of Physics: Conf. Series **1070** (2018) 012014 doi:10.1088/1742-6596/1070/1/012014

Open heavy flavour production in heavy–ion collisions with ALICE

Martin Völkl for the ALICE Collaboration
Eberhard Karls Universität Tübingen, Geschwister-Scholl-Platz, 72074 Tübingen, Germany
E-mail: martin.andreas.volkl@cern.ch

Abstract. Heavy–flavour measurements give an important contribution to the understanding of the Quark–Gluon Plasma (QGP) produced in heavy–ion collisions. Heavy quarks are effective probes for the QGP and understanding their interaction with the medium can give insights into its transport properties. The ALICE Collaboration investigates heavy–flavour production by measuring leptons from heavy–flavour decays and by reconstructing hadronic decays of D mesons and charmed baryons. In this proceeding, recent results from ALICE as well as the direction of future heavy–flavour measurements are discussed.

1. Introduction

Heavy quarks – charm and beauty – are important tools to study the properties of the hot and dense matter produced in relativistic heavy–ion collisions. The study of their interaction with the medium can yield insight into its properties. Due to their large mass ($m_{c/b} \approx 1.5/4.5$ GeV/c), heavy quarks are produced almost exclusively in the initial hard scattering processes of the collision. Therefore, they witness the full evolution of the collision. Additionally, the large quark masses make perturbative calculations applicable even to a production at rest. In the vast majority of models, the interaction is modeled as discrete scatterings of the heavy quark off the medium constituents. In this framework, collisional ($2 \to 2$) and radiative ($2 \to 3$) processes can be distinguished. The relative contribution of these processes differs between models. It depends on the momentum and mass of the heavy quark. As a consequence, measurements over a large momentum range for both flavours are needed to disentangle the contributions.

After the initial hard scatterings, the number of heavy quarks in the system stays constant throughout the interaction. Thus, the effect of the medium is a redistribution of the produced particles in momentum space. Fast heavy quarks are slowed down, while slow ones may be accelerated due to the collective flow of the medium around them. The nuclear modification factor is frequently used to quantify the effects due to the formation of a medium. It compares the effect of a sample of heavy–ion collisions with an estimated mean value of binary (nucleon–nucleon) collisions of $\langle N_{\mathrm{coll}} \rangle$ to a superposition of the same number of proton–proton (pp) collisions:

$$R_{\mathrm{AA}}(p_{\mathrm{T}}) = \frac{1}{\langle N_{\mathrm{coll}} \rangle} \frac{\mathrm{d}N_{\mathrm{AA}}/\mathrm{d}p_{\mathrm{T}}}{\mathrm{d}N_{\mathrm{pp}}/\mathrm{d}p_{\mathrm{T}}} = \frac{1}{\langle T_{\mathrm{AA}} \rangle} \frac{\mathrm{d}N_{\mathrm{AA}}/\mathrm{d}p_{\mathrm{T}}}{\mathrm{d}\sigma_{\mathrm{pp}}/\mathrm{d}p_{\mathrm{T}}} , \quad (1)$$

where T_{AA} is the nuclear overlap function. This approach assumes that the individual nucleon–nucleon collisions are well represented by proton–proton collisions in the vacuum. In reality, the fact that the nucleons are part of a large nucleus may give rise to cold–nuclear–matter (CNM) effects which have to be estimated to interpret the measured nuclear modification factor. Additional measurements are performed in p–Pb collisions to disentangle CNM effects from those due to the medium. Another important observable to quantify the interaction of heavy quarks with the medium is the elliptic flow coefficient v_2. It is defined as $v_2 = \langle \cos[2(\phi - \Psi_2)] \rangle$, the second Fourier coefficient with respect to the second order symmetry plane of the azimuthal distribution. The v_2 quantifies the degree of participation in the collective motion of the surrounding medium.

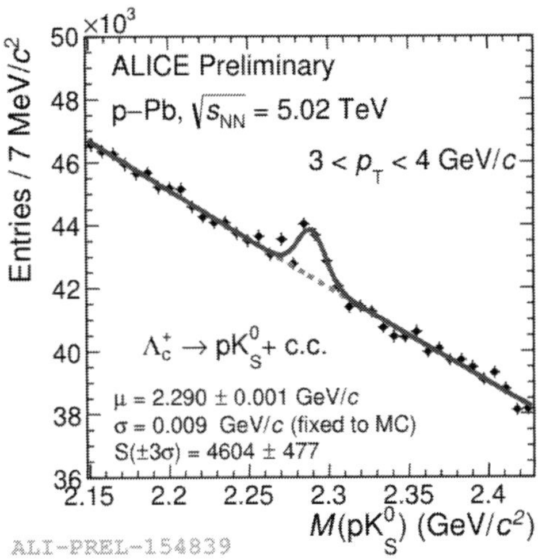

Figure 1. Example of an invariant mass peak in the fully hadronic reconstruction of Λ_c^+.

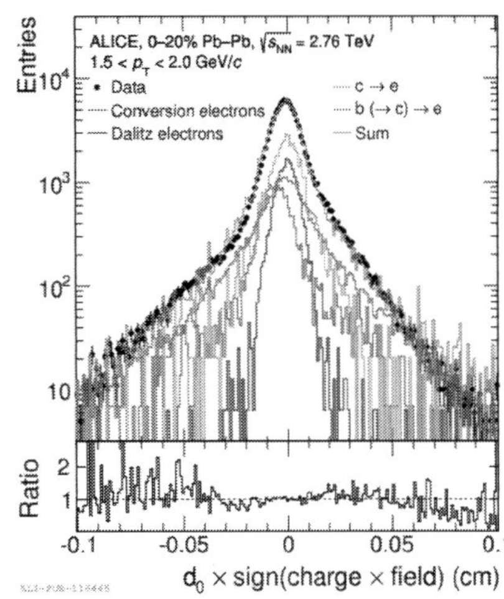

Figure 2. Example of the transverse impact parameter template fit for extracting the beauty contribution to the electrons [1].

2. Experimental methods

The measurement at low transverse momenta and particle identification are particular strengths of the ALICE experiment. At mid–rapidity, open charm hadrons are measured via their hadronic decays (an example is shown in Figure 1). This includes $D^0 \to K^-\pi^+$ (branching ratio $3.93 \pm 0.04\%$), $D^+ \to K^-\pi^+\pi^+$ ($9.46 \pm 0.24\%$), $D^{*+} \to D^0\pi^+$ ($67.7 \pm 0.5\%$), $D_s^+ \to \phi^-\pi^+$ ($2.27 \pm 0.08\%$). The Λ_c^+ is reconstructed via the $pK^-\pi^+$ and pK^0 channels and via the semi–leptonic $e^-\nu_e\Lambda$ channel. The Ξ_c^0 is reconstructed via the semi–leptonic $e^+\Xi^-\nu_e$ channel. The decay products can be identified using the signals of the Time Projection Chamber (TPC) and the Time Of Flight detector (TOF). For measurements at forward rapidity and to gain an access to beauty, semileptonic decays of heavy flavour hadrons are used. These are decays of the type $b \to l + X$, where only the lepton l is measured. Such a measurement has the advantage that just by using particle identification, a large amount of the background can be removed. Due to the fairly large branching ratios and the softer p_T–distributions of electrons from background sources, electrons with transverse momenta above a few GeV/c mostly come from heavy flavour sources. The branching ratios of charmed hadrons to a final state containing electrons is about 10%. For beauty hadrons, the ratio is about 20%, with about half of the contribution coming from decays with intermediate charmed states. At forward rapidity ($2.5 < \eta < 4$ for symmetric collisions), heavy flavour muons are measured by the Muon Spectrometer. The central barrel measurements ($|\eta| < 0.8$) are based on electrons identified using the TPC and TOF with additional PID information from the Inner Tracking System (ITS) at low p_T and the Electromagnetic Calorimeter (EMCal) at high p_T. To extract the contribution of heavy flavours, an estimate for the light–flavour contribution is subtracted from the measured leptons. To separate the beauty contribution, the impact parameter of the electron in the transverse plane is used as additional information. The different electron sources are then separated by a template fit of the transverse impact parameter distributions (example in Figure 2).

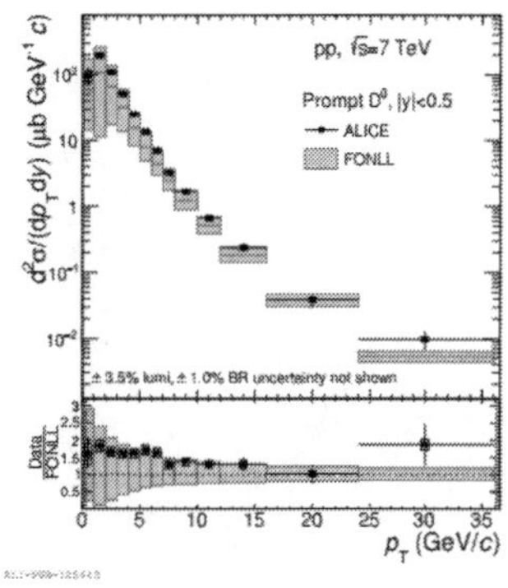

Figure 3. D^0 cross section measurement as function of transverse momentum [2].

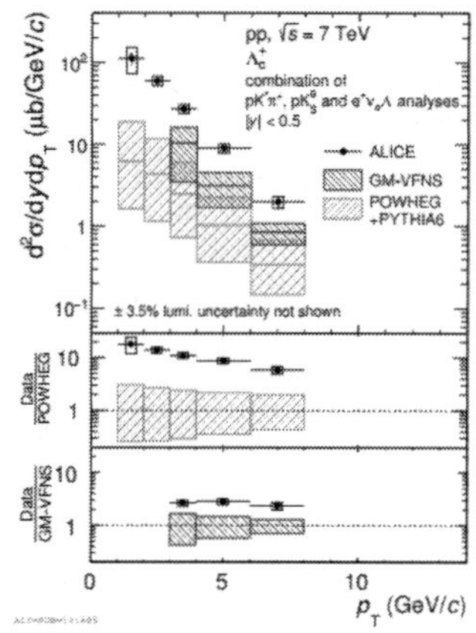

Figure 4. Λ_c^+ measurement in pp collisions at $\sqrt{s}$ = 7 TeV[3].

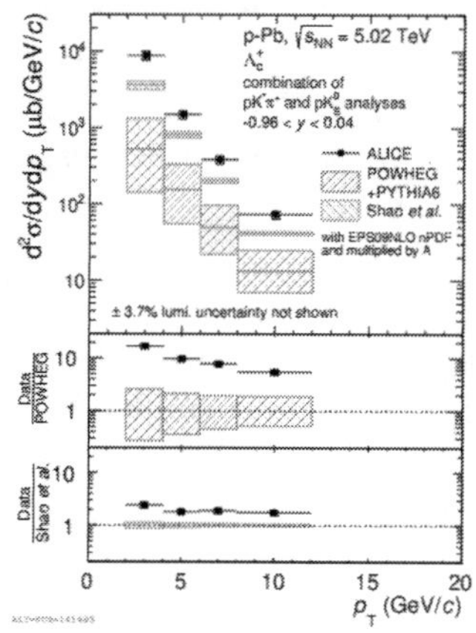

Figure 5. Λ_c^+ measurement in p–Pb collisions at $\sqrt{s_{\mathrm{NN}}}$ = 5.02 TeV[3].

3. Results in small systems

Measurements in pp give the needed baseline to calculate the nuclear modification factor R_{AA}. In addition they can test the quality of our theoretical understanding of the production processes. Figure 3 shows a measurement of the production cross sections of D^0 mesons in pp collisions at $\sqrt{s} = 7$ TeV. The measurement uncertainties are much smaller than those from the theoretical calculations. The central points are near the upper edge of the calculations. Another important physical process is the hadronization. It is inherently a nonperturbative process, thus the experimental input is particularly important for its understanding. Additionally, it is of interest to study further not just how heavy flavours are distributed in phase space but also how they are distributed among the possible hadron species. In particular, the charmed meson to baryon ratio can give insight into the charm hadronization mechanism. Figures 4 and 5 show the measured yield of Λ_c^+ in pp and p–Pb collisions together with model calculations. The models underestimate the production of Λ_c^+ by about a factor of 10 (for the central values) in both collision systems [3]. This also provides an important baseline to compare the ratios in Pb–Pb with. In addition, the yield of Ξ_c^0 was measured. Even considering the large uncertainties in the branching ratio, the measured yields are much larger than suggested by the models.

Comparison of the yields in pp and p–Pb collisions can aid in the understanding of cold nuclear matter effects. While the results for the R_{pA} of reconstructed D mesons shown in Figure 6 broadly follow the trend of the calculations based on nuclear parton distribution functions (nPDFs), no strong deviation from unity is found either. To gain a more thorough understanding of CNM effects it has proven useful to compare measurements at forward and backward rapidity. This is possible by using the Muon Spectrometer in p–Pb collisions and Pb–p collisions, which can be achieved by switching the beams. Both in the model calculations and in the measurements, the ratio of the forward and backward nuclear modification factors

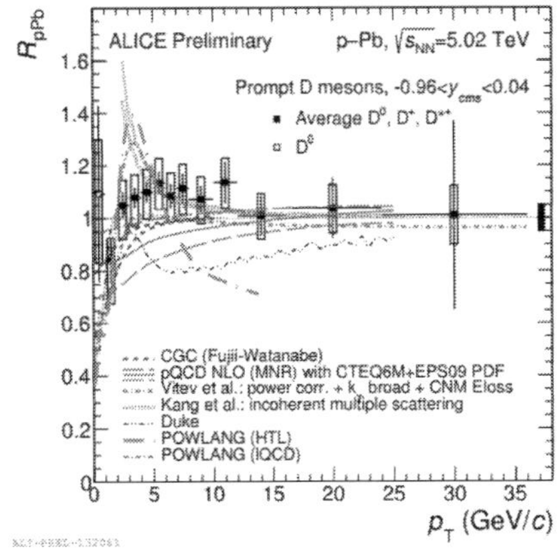

Figure 6. $R_{\rm pA}$ of D mesons in p–Pb collisions at $\sqrt{s_{\rm NN}} = 5.02$ TeV.

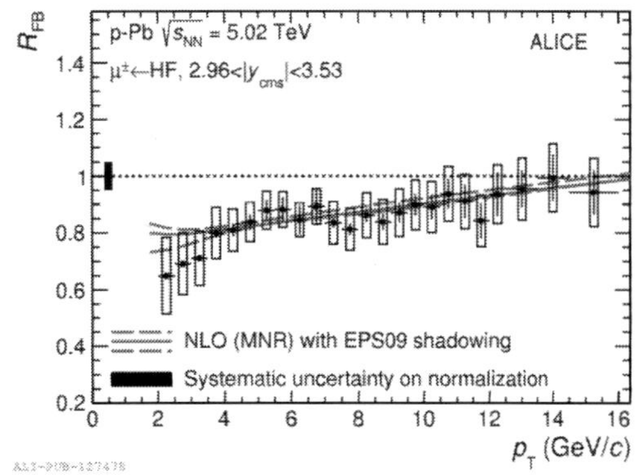

Figure 7. Ratio of $R_{\rm pA}$ of muons from heavy flavour hadron decays measured in p–Pb and Pb–p collisions

$R_{\rm FB}$ has small uncertainties, because many contributions cancel out in the ratio. The $R_{\rm FB}$ of muon from heavy-flavour hadron decays is shown in Figure 7. It deviates from unity and is consistent with the expectations from NLO calculations which include nPDFs [4].

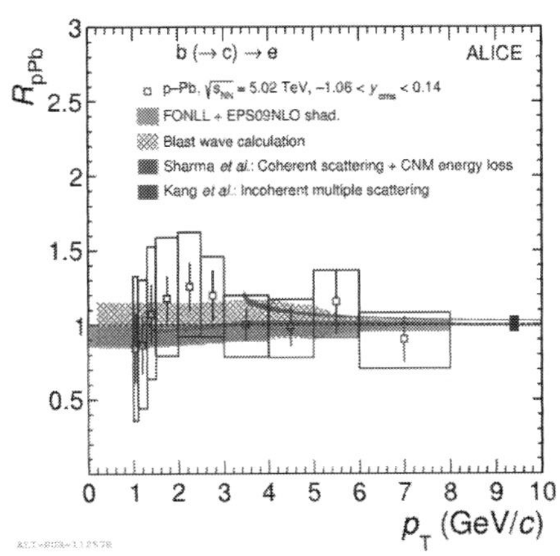

Figure 8. $R_{\rm pA}$ of electrons from beauty–hadron decays [1].

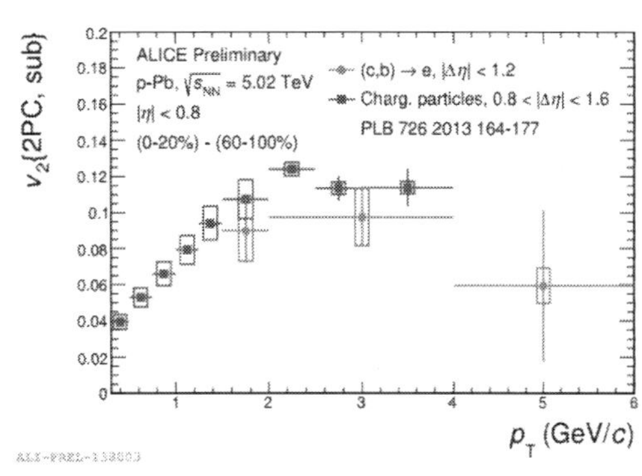

Figure 9. Elliptic flow of electrons from heavy–flavour hadron decays in p–Pb collisions.

Measuring electrons from beauty–hadron decays specifically, also yields an $R_{\rm pA}$ consistent with unity as shown in Figure 8. All these results point towards pp and p–Pb collisions not showing large modifications of the heavy flavour production. As a consequence, strong deviations from the model of vacuum superpositions of nucleon–nucleon collisions in Pb–Pb could be interpreted as being medium induced. However, measurements of collective flow in small systems challenge this simple picture. Figure 9 shows a measured elliptic flow coefficient for electrons form heavy–flavour decays, which is significantly larger than zero. It is similar in maginitude to the measurement of light particles [5]. The measurement of collective behaviour in the angular distributions, simultaneous with essentially unmodified p_T–distributions requires careful additional studies.

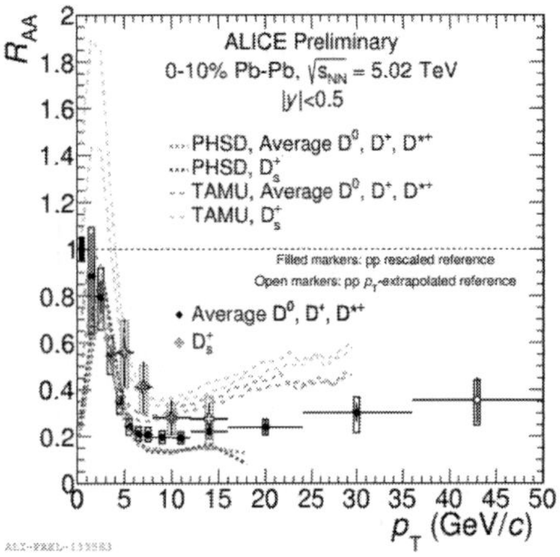

Figure 10. $R_{AA}$of D mesons in central Pb–Pb collisions.

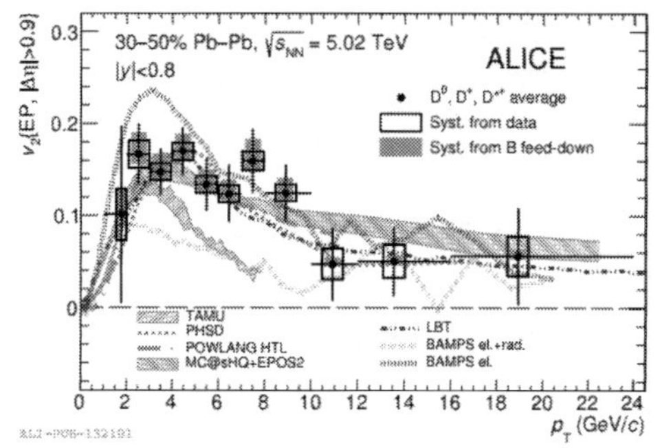

Figure 11. Elliptic flow of D mesons in Pb–Pb collisions.

4. Results in Pb–Pb collisions

The p_T–distributions of D mesons show a strong modification in central Pb–Pb collisions as shown in Figure 10. In addition, the elliptic flow of D mesons (Figure 11) shows a magnitude similar to that of light hadrons. This suggests a thermalization time for charm quarks of the order of the system lifetime. Careful comparison of the most successful models suggests $\tau_c = 3 - 14$ fm/c. The nuclear modification factor of D_s^+ was also measured. The result at low to intermediate p_T hints at a larger $R_{AA}$with respect to non–strange D mesons at the same p_T, consistent with an increased production of strange mesons expected by model predictions.

More differential quantities can be useful to further constrain models. In the case of the elliptic flow, events of the same centrality can produce a strong variation in the final state anisotropy. This can be quantified in by the reduced second–order flow vector q_2^{TPC}. It grows with event multiplicity and flow strength [6]. As shown in figure 12, it fluctuates strongly within centrality classes. Selecting q_2^{TPC} as well as centrality classes of events gives a clear separation in the flow strength measured (Figure 13). Due to the fact that q_2^{TPC} is measured in at mid–rapidity, where the D mesons are measured as well, some correlations can be present. As a result, the measured v_2 decreases when an η gap is introduced between the measurement of the D mesons and the q_2^{TPC} vector. This effect requires some additional careful analysis.

Even with the constraints from the nuclear modification factor and the elliptic flow, several diverse model calculations are able to describe the data reasonably. To complete the picture of heavy flavours in heavy ion collision, information about the mass dependence of the interaction is needed from the measurement of the beauty sector. Figure 14 shows the nuclear modification factor for electrons and muons from the decay of heavy–flavour hadrons. Towards higher p_T, the relative contribution of beauty increases, suggesting a suppression of beauty. This is further refined in the flavour–separated measurement shown in Figure 15. There is an indication for a suppression of electrons from beauty–hadron decays at the higher p_T values measured, with the central points rising towards lower p_T. Several models can describe the measurement within uncertainties.

IOP Conf. Series: Journal of Physics: Conf. Series **1070** (2018) 012014 doi:10.1088/1742-6596/1070/1/012014

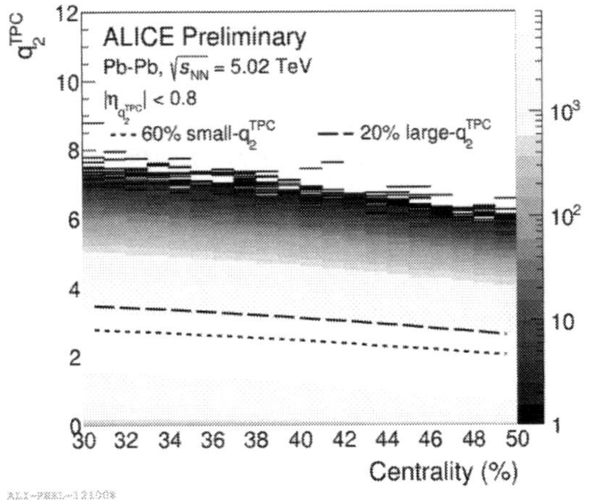

Figure 12. Distribution of reduced second–order flow vector q_2^{TPC} for different centralities.

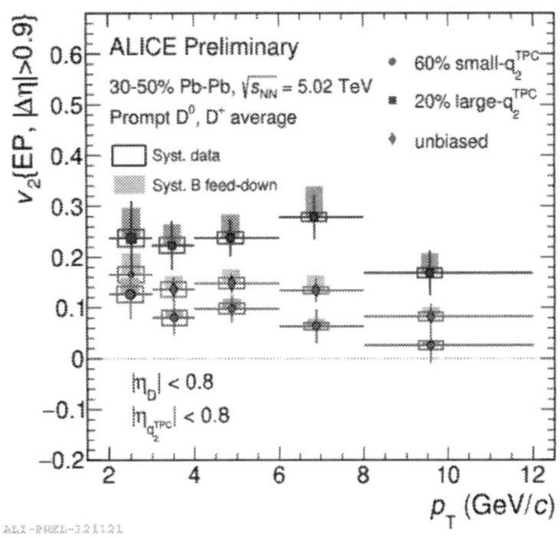

Figure 13. D meson v_2 in different classes of q_2^{TPC} .

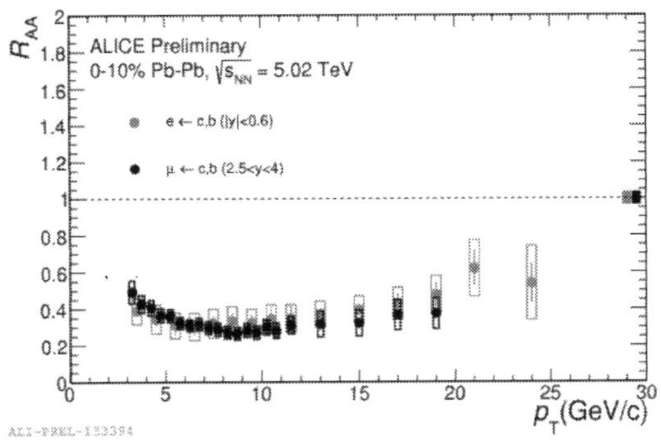

Figure 14. $R_{\mathrm{AA}}$of muons and electrons from heavy–flavour hadron decays at forward and mid–rapidity respectively.

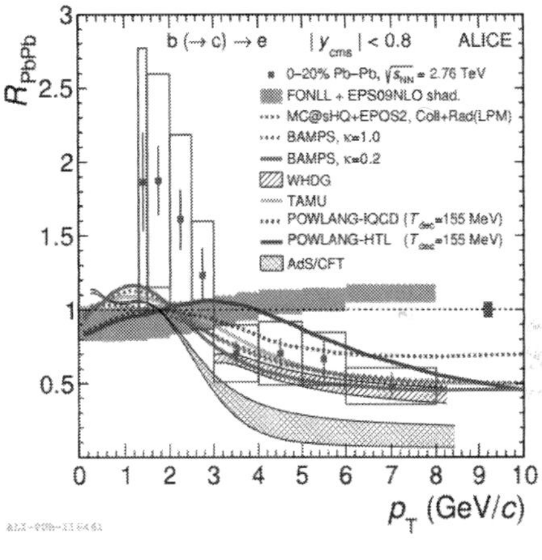

Figure 15. $R_{\mathrm{AA}}$of electrons from heavy–flavour hadron decays [1].

5. Summary

A comparison of the measurements of p_{T}–differential cross sections in pp and p–Pb collisions shows effects consistent with the expectations from cold nuclear matter effects, supporting the interpretation that the measured suppression and elliptic flow of heavy flavours in Pb–Pb collisions is a final state effect. This picture is complicated by the measurement of a significant elliptic flow for electrons from heavy–flavour decays even in p–Pb collisions. In addition, the measured production of charmed baryons in pp and p–Pb collisions is strongly underestimated by the model predictions. Future measurements may use more differential variables to constrain models further.

IOP Conf. Series: Journal of Physics: Conf. Series **1070** (2018) 012014 doi:10.1088/1742-6596/1070/1/012014

References

[1] Adam J *et al:* (ALICE) 2017 *JHEP* **07** 052 (*Preprint* 1609.03898)
[2] Acharya S *et al.* (ALICE) 2017 *Eur. Phys. J.* **C77** 550 (*Preprint* 1702.00766)
[3] Acharya S *et al.* (ALICE) 2017 *Submitted to: JHEP* (*Preprint* 1712.09581)
[4] Acharya S *et al.* (ALICE) 2017 *Phys. Lett.* **B770** 459–472 (*Preprint* 1702.01479)
[5] Abelev B B *et al.* (ALICE) 2013 *Phys. Lett.* **B726** 164–177 (*Preprint* 1307.3237)
[6] Voloshin S A, Poskanzer A M and Snellings R 2010 *Landolt-Bornstein* **23** 293–333 (*Preprint* 0809.2949)

IOP Conf. Series: Journal of Physics: Conf. Series **1070** (2018) 012015 doi:10.1088/1742-6596/1070/1/012015

Event Topology Reconstruction in the CBM Experiment

Ivan Kisel
(for CBM Collaboration)

Goethe University, 60323 Frankfurt am Main, Germany
FIAS Frankfurt Institute for Advanced Studies, 60437 Frankfurt am Main, Germany
GSI Helmholtz Center for Heavy Ion Research, 64291 Darmstadt, Germany

E-mail: i.kisel@gsi.de

Abstract. One of the main purposes of the physics program of the future heavy ion experiment CBM (FAIR, Germany) is to understand the properties of strongly interacting matter at very high baryonic densities and to study the possibility of a phase transition to a deconfined and chirally restored phase of quark matter. The experiment will operate at high interaction rates up to 10 MHz, that requires a full event reconstruction in real time.

In order to make an efficient event selection online a clean sample of particles has to be provided by the reconstruction package called First Level Event Selection (FLES). The FLES package operates in two stages. First, particles registered in the CBM detector system are reconstructed. Then short-lived particles decayed before or inside the setup are searched based on their charged and neutral daughter particles. Since the FLES package is developed to run on many-core computer architectures, the reconstruction of particles is done in parallel that provides a possibility for a global competition between particle candidates. Such a global event topology reconstruction significantly improves suppression of a combinatorial background and provides for further physics analysis a very clean sample of particles produced at different stages of heavy ion collision.

The global event topology reconstruction procedure and the results of its application to simulated collisions in the CBM detector setup are presented and discussed.

1. Introduction

The Facility for Antiproton and Ion Research (FAIR) project is aimed to build a large international center for studying the structure and fundamental properties of matter. It is an accelerating complex of new generation which will provide unique opportunities to perform research in the most exciting fields of modern science: nuclear, hadron and particle physics, atomic and anti-matter physics, high density plasma physics, and applications in condensed matter physics, biology and bio-medical sciences [1].

In the Compressed Baryonic Matter (CBM) experiment [2] with heavy ions highest baryon densities will be created and the properties of super-dense nuclear matter will be explored in various extreme states similar, for instance, to the conditions of matter in the center of neutron stars, where the matter is at the final stage of the evolution before transition into black hole. The CBM experiment will complement the experimental heavy-ion program at the LHC accelerator complex at CERN where the properties of the hot matter similar to that after the Big Bang are investigated.

The scientific program of the CBM experiment includes:

- explore properties of super-dense nuclear matter;
- search for in-medium modifications of hadrons;
- search for the transition from dense hadronic matter to quark-gluon matter;
- search for the critical endpoint in the phase diagram of strongly interacting matter;
- investigate the structure of neutron stars and the dynamics of core-collapse supernovae.

The experiment will measure rare and penetrating probes such as dilepton pairs from light vector mesons and charmonium, open charm, multistrange hyperons, together with collective hadron flow and fluctuations in heavy-ion collisions at rates up to 10^7 collisions per second.

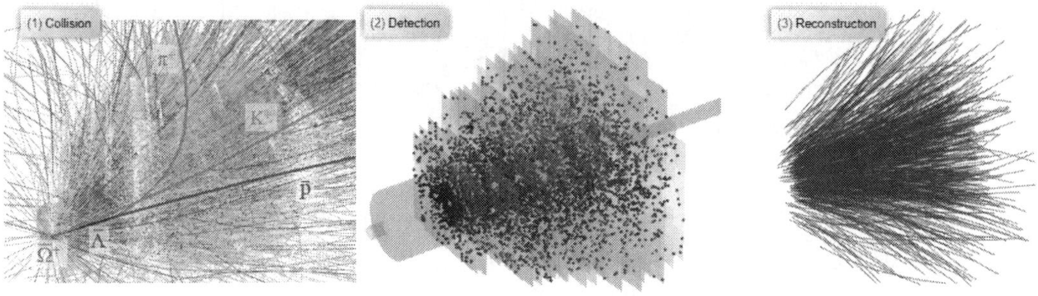

Figure 1. A simulated central Au-Au collision at 25 *A*GeV energy with about 1000 charged particles (left), registered in the CBM detector (middle) and reconstructed by the Cellular Automaton track finder (right).

CBM is characterized by high collision rates, large amount of produced particles, non-homogeneous magnetic fields and a very complex detector system. Event reconstruction is the most complicated and time consuming task of the data analysis in modern high-energy physics experiments. It is a key part of success in the CBM experiment with up to thousand particles per central collision (Fig. 1). An additional complication in CBM is its continuous data stream represented in form of time slices. This makes the reconstruction of such 4-dimensional data with time stamps and the search for interesting physics extremely difficult. All of the above mentioned makes necessary to develop fast and efficient algorithms for data analysis and to optimize them for running on a modern high-performance computer cluster [3].

2. First Level Event Selection

The First Level Event Selection (FLES) package [4, 5] of the CBM experiment is intended to reconstruct online the full event topology including tracks of charged particles and short-lived particles. The FLES package consists of several modules (the block diagram is shown on Fig. 2): CA track finder, KF track fitter, KF Particle Finder and physics selection. In addition, a quality check module is implemented, that allows to monitor and control the reconstruction process at all stages. The FLES package is platform and operating system independent.

The FLES package is portable to different many-core CPU architectures. The package is vectorized using SIMD (Single Instruction, Multiple Data) instructions and parallelized between CPU cores. All algorithms are optimized with respect to the memory usage and the speed. Fig. 3 shows a strong scalability for all many-core systems achieving the reconstruction speed of 1700 events per second on the 80-cores server.

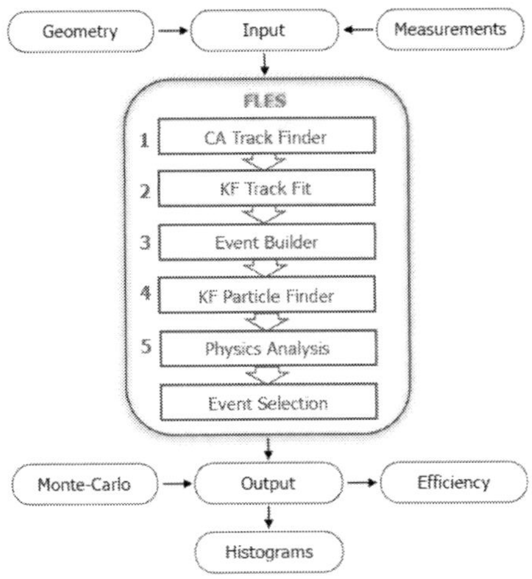

Figure 2. Block diagram of the FLES package, which consists of several modules: track finder, track fitter, particle finder and physics selection.

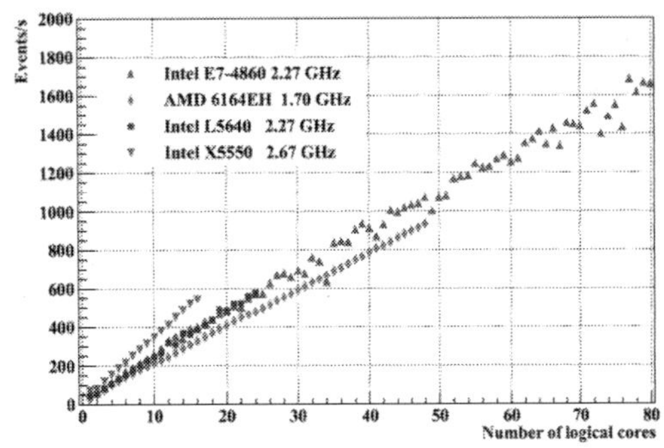

Figure 3. Scalability of the FLES package on many-core servers with 16, 24, 48 and 80 logical cores.

2.1. 4D Cellular Automaton (CA) track finder

The 4-dimensional (4D, space and time) Cellular Automaton (CA) track finder [6] takes as input hit measurements from the tracking detector in the form of a time-slice, which includes time and spacial measurements. The track finding procedure starts with combining the hits into triplets, combination of three hits on adjacent stations. The triplet structure was chosen, since it allows to estimate the momentum of a particle, which could produce it. The triplets with two common hits are combined into track candidates. In order to take into account detector inefficiency, a missing hit in one station is tolerated for a reconstructed track. The track candidates should survive a dedicated selection based on the track length and calculated χ^2-value to be accepted to the reconstructed tracks.

In order to optimize and speed up the process of track reconstruction the track finding procedure is performed in several iterations. At first the track finder algorithm is searching for the tracks, which are easier to reconstruct, for instance, high momentum and primary tracks. Having found these tracks, the algorithm removes from the input information hits, which were used to construct these tracks. Thus, the combinatorics, which is higher for the iterations when searching for low-momenta and secondary tracks gets significantly reduced. Therefore, in the first iteration only high-momentum primary tracks are reconstructed, in the second one — low-momentum primary tracks, and then — all other tracks.

Input time information is used in the algorithm to the same extent and in similar manner as it is done with the spacial coordinates. The same logic is used while constructing triplets: the hits in the triplet should belong to the same particle, therefore they should correlate not only in space, but also in time. Since the estimated time of flight between two consecutive station is negligible in comparison to the detector time precision, the hits, belonging to the same track, should coincide within 3σ of the detector time precision. The resulting track reconstruction efficiencies for the cases of event-by-event analysis (so-called 3D analysis) as well as for the 4D case (with included time measurement, as well as 3-dimensional spacial information) while reconstructing time-slices, produced out of one hundred AuAu minimum bias UrQMD events at 25 AGeV, are presented in Tab. 1.

Table 1. Track reconstruction efficiency performance for 100 minimum bias AuAu UrQMD simulated collisions at 25 AGeV for the cases of event-by-event reconstruction, time-slice-based reconstruction at 0.1 MHz, 1 MHz and 10 MHz interaction rate, presented for different track categories.

Track category, %	E-by-E	10^5 Hz	10^6 Hz	10^7 Hz
All tracks	92.1	92.6	92.6	92.2
Primary high-p	97.9	98.2	98.2	97.9
Primary low-p	93.6	94.1	94.1	93.5
Secondary high-p	92.0	92.7	92.7	92.0
Secondary low-p	65.7	66.7	66.6	65.9
Clone level	2.8	0.3	0.3	3.1
Ghost level	4.9	3.5	3.5	4.2
MC tracks found	145	146	146	145

2.2. *Kalman Filter (KF) track fit library*
High precision of the parameters of particle trajectories (tracks) and their covariance matrices is a prerequisite for finding rare signal events among hundreds of thousands of background events. Such high precision is usually obtained by using the estimation algorithms based on the Kalman filter (KF) method.

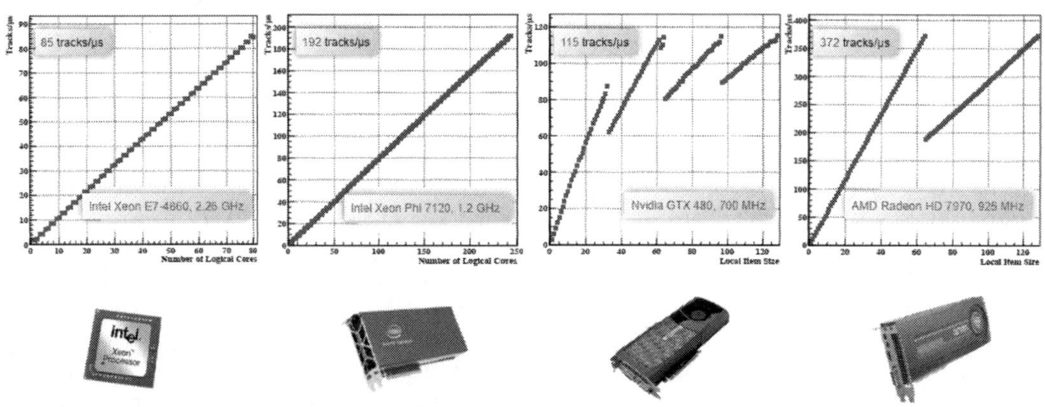

Figure 4. The KF track fitter is optimized and portable with respect to all many-core HPC architectures.

The developed Kalman filter based library for track fitting includes following tracking algorithms:

- track fit based on the conventional Kalman filter;
- track fit based on the square root Kalman filter;
- track fit based on the UD Kalman filter;
- track smoother based on the listed above approaches and
- deterministic annealing filter based on the listed above track smoothers.

High speed of the reconstruction algorithms on modern many-core computer architectures can be accomplished by: (a) optimizing with respect to the computer memory, in particular declaring all variables in single precision, (b) vectorizing in order to use the SIMD instruction set and (c) parallelizing between cores within a compute node.

Several formulations of the Kalman filter method, such as the square root KF and the UD KF, increase its numerical stability in single precision. All algorithms, therefore, can be used either in double or in single precision.

The vectorization and parallelization of the algorithms are currently done by using of: header files, Vc vector classes, Intel TBB, OpenMP, Intel ArBB and OpenCL.

The KF track fitter library is well investigated and developed to run on all many-core CPU/Phi/GPU architectures (Fig. 4).

2.3. KF Particle Finder — a package for reconstruction of short-lived particles

Today the most interesting physics is hidden in the properties of short-lived particles, which are not registered, but can be reconstructed only from their decay products. A fast and efficient KF Particle Finder package [7], based on the Kalman filter (hence KF) method, for reconstruction and selection of short-lived particles is developed to solve this task. A search of more than 100 decay channels has been currently implemented (Fig. 5).

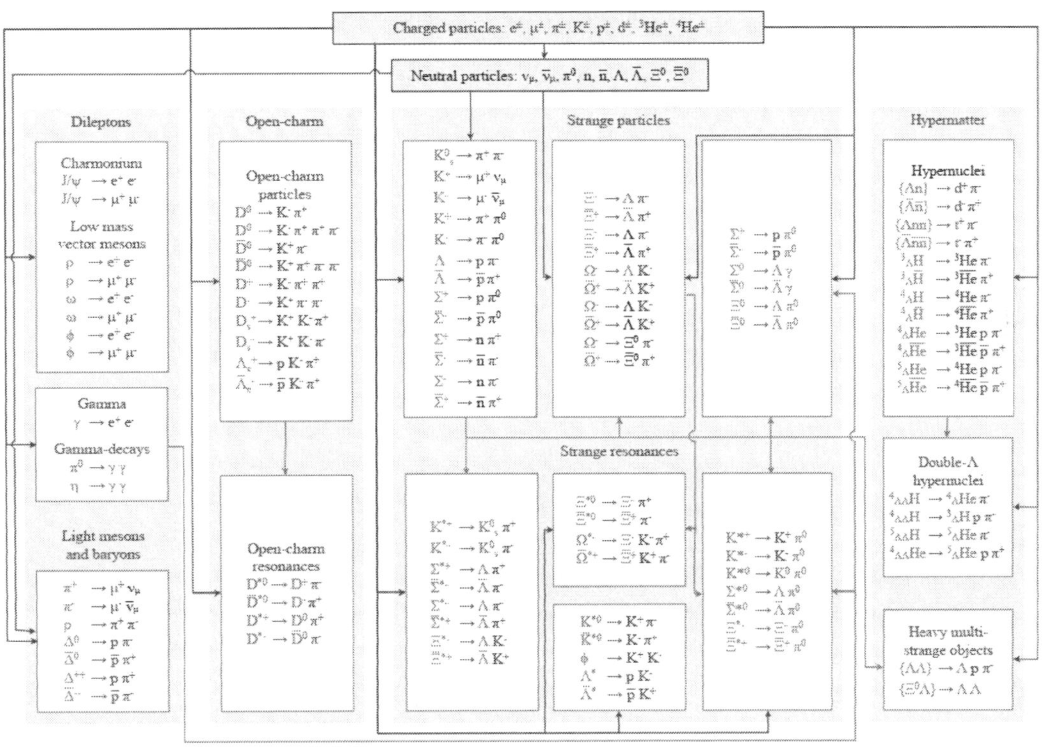

Figure 5. Block diagram of the KF Particle Finder package. The particle parameters, such as decay point, momentum, energy, mass, decay length and lifetime, together with their errors are estimated using the Kalman filter method.

In the package all registered particle trajectories are divided into groups of secondary and primary tracks for further processing. Primary tracks are those, which are produced directly in the collision point. Tracks from decays of resonances (strange, multi-strange and charmed resonances, light vector mesons, charmonium) are also considered as primaries, since they are produced directly at the point of the primary collision. Secondary tracks are produced by the short-lived particles, which decay not in the point of the primary collision and can be clearly separated. These particles include strange particles (K_s^0 and Λ), multi-strange hyperons (Ξ and Ω) and charmed particles (D_0, $D^{\pm}$, $D_s^{\pm}$ and Λ_c). After that tracks are combined according to the block diagram in Fig. 5. The package estimates the particle parameters, such as decay point, momentum, energy, mass, decay length and lifetime, together with their errors. The package has a rich functionality, including particle transport, calculation of a distance to a point or another particle, calculation of a deviation from a point or another particle, constraints on mass, decay length and production point. All particles produced in the collision are reconstructed at once, that makes the algorithm local with respect to the data and therefore extremely fast.

In addition, simultaneous reconstruction in the KF Particle Finder of different decay channels of the same particle, including also decays with a neutral particle in the final state, makes it possible to calculate the efficiency of the reconstruction of rare particles and reliably estimate their systematic errors.

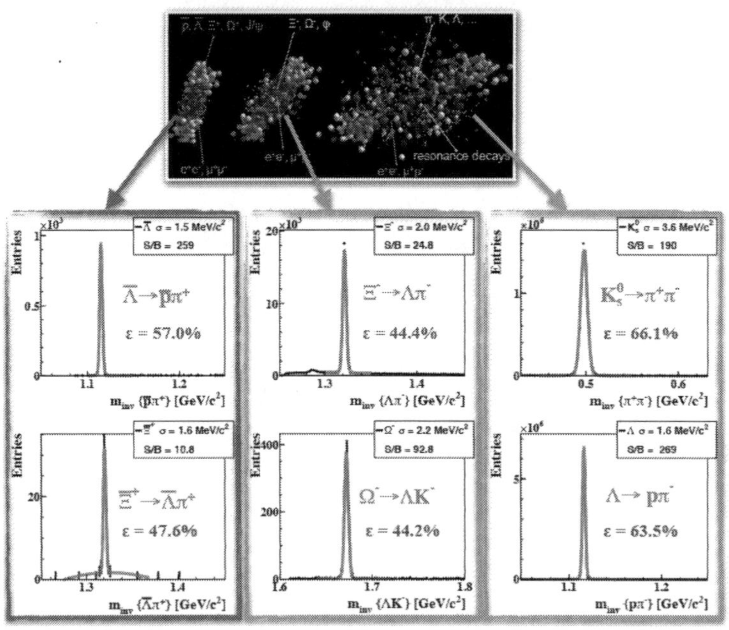

Figure 6. The FLES package provides clean probes of various stages of the collision.

The use of the Kalman filter at all stages of particle reconstruction allows in many cases to get rid almost completely of the combinatorial background and to obtain clean sets of particles, which can serve as probes of various stages of the collision (Fig. 6).

3. Discussion

Absence of a hardware trigger for selection of rare events in the CBM experiment requires the complete reconstruction of the event topology, including short-lived particles, in order to select events with rare decays, which carry important information about the process of heavy ion collision. The high interaction rate of up to 10^7 collisions/s and a relatively low data recording speed on the storage media lead to the necessity of the online event selection stage in the FLES package in order to reduce the data flow by 3-4 orders of magnitude. Such high reduction ratio, as well as the need to select events with rare decays online, require the offline quality of the data processing already in real time.

A low-quality online data processing, for example, due to poor detector alignment or the presence of uncontrolled errors in the reconstruction algorithms and in the operation of the detector system, will inevitably make all the selected and stored data useless. Therefore, use of the theoretical analysis of the reconstructed data already in real-time at the event selection stage will ensure the high quality of data processing, eliminate the possibility of irreparable loss of data and, finally, avoid the waste of the accelerator operation time.

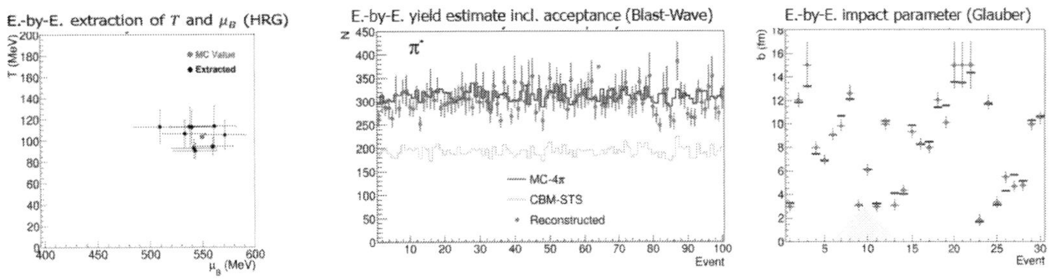

Figure 7. Online extraction of properties of medium created in heavy-ion collisions.

The offline quality of the online event processing, on the other hand, allows the use of theoretical models to estimate the collision parameters [8] immediately during the collection of experimental data on the accelerator with a delay of several milliseconds after the collision itself occurred.

As it is shown in Fig. 6, within the standard thermal analysis the fit to π^-/π^+, K^-/K^+, K^+/π^+ and p/π^-, reconstructed by the KF Particle Finder, gives the temperature and μ_B on the event-by-event basis. Integrating dN/dy using the blast-wave model gives the multiplicity estimate after correcting on the acceptance and the reconstruction efficiency. There are also several ways to correlate impact parameter to data: number of nucleon participants (wounded nucleons); number of spectators; number of binary collisions and charged multiplicity. At known number of participant protons the impact parameter can be estimated using the Galuber model with the accuracy up to 1 fm in the ideal 4π case.

4. Conclusion

The FLES package has been developed for the CBM experiment. It contains track finding, track fitting, short-lived particles finding and physics selection. The Cellular Automaton and the Kalman Filter algorithms are used for finding and fitting tracks, that allows to achieve a high track reconstruction efficiency up to 97% and the track parameters quality with 1% of the momentum resolution. Reconstruction of more than 100 decay channels of short-lived particles is currently implemented in the KF particle finder. The package shows a high reconstruction efficiency with an optimal signal to background ratio.

The FLES package is portable to different many-core CPU architectures. The package is vectorized using SIMD instructions and parallelized between CPU cores. All algorithms are optimized with respect to the memory usage and the speed. The FLES package shows a strong scalability on the many-core CPU systems and the processing and selection speed of 1700 events per second on a server with 80 CPU cores.

References

[1] FAIR — Facility for Antiproton and Ion Research. Green Paper. The Modularized Start Version. GSI. October 2009.

[2] CBM Collaboration, Compressed Baryonic Matter Experiment, Tech. Stat. Rep., GSI, Darmstadt, 2005; 2006 update.

[3] I. Kisel, Event reconstruction in the CBM experiment, Nucl. Instr. and Meth. A566 (2006) 85-88.

[4] I. Kisel, I. Kulakov and M. Zyzak, Standalone first level event selection package for the CBM experiment, IEEE Trans. Nucl. Sci. 60 (5) (2013) 3703-3708.

[5] V. Akishina and I. Kisel, Online 4-dimensional reconstruction of time-slices in the CBM experiment, IEEE Trans. Nucl. Sci. 62 (6) (2015) 3172-3176.

[6] V. Akishina, "4D Event Reconstruction in the CBM Experiment", Dissertation thesis, Goethe university, Frankfurt am Main (2017).

[7] M. Zyzak, "Online Selection of Short-Lived Particles on Many-Core Computer Architectures in the CBM Experiment at FAIR", Dissertation thesis, Goethe university, Frankfurt am Main (2016).

[8] V. Vovchenko, Theoretical description and on-line extraction of properties of medium created in heavy-ion collisions, Workshop on CBM-STAR Common Approaches in Data Reconstruction and Analysis, BNL, 10 December 2015.

IOP Conf. Series: Journal of Physics: Conf. Series **1070** (2018) 012016 doi:10.1088/1742-6596/1070/1/012016

Latest Results on Quarkonium Production from CMS

Dong Ho Moon on behalf of the CMS collaboration

Chonnam National University

E-mail: dhmoon@chonnam.ac.kr

Abstract. The understanding of quarkonium production in heavy-ion collisions requires the inclusion of many phenomena such as dissociation in the QGP, partonic energy loss, statistical recombination, on top of cold nuclear matter effects (modifications of nPDFs, initial-state energy loss, nuclear break-up). The Compact Muon Solenoid (CMS) collaboration has measured various quarkonium states via their decay into muon pairs in pp, pPb and PbPb collisions at 5.02 TeV, and can address some of these phenomena. In this talk, the most recent CMS results on quarkonium production will be presented.

1. Introduction

The measurement of quarkonium production in heavy ion collisions constitutes a powerful probe for studying properties of matter at high energy densities and temperatures. A strongly interacting medium of deconfined quarks and gluons, the quark-gluon-plasma (QGP), has been predicted in such an environment. Recent measurements of quarkonium production performed by the CMS Collaboration have been found to support theoretical models of sequential melting of quarkonia in QGP [1, 2, 3]. However, quarkonimum production in heavy-ion collisions is affected also by initial state effects, such as modification of the nuclear parton distribution functions (nPDFs) [4], initial state parton energy loss [6], and Cronin effect [5, 6]. These are in addition to final state effects such as Debye color screening and statistical recombination.

In this talk, a report was made on the measurements of the nuclear modification factors, R_{AA}, of prompt J/ψ, $\psi(2S)$, $\Upsilon(1S)$, $\Upsilon(2S)$, and $\Upsilon(3S)$ mesons, measured in their dimuon decay channel. The results use pp and PbPb data collected at the end of 2015 by the CMS Collaboration at $\sqrt{(s_{NN})} = 5.02$ TeV, corresponding to an integrated luminosities of 28 pb^{-1} and 351-368 μb^{-1}, respectively.The results are compared to those obtained at 2.76 TeV. The larger integrated luminosities allow for more precise and more differential measurements of R_{AA}, as functions of event centrality, rapidity (y), and transverse momentum (p_T).

2. Environmental setup and analysis procedure

The central feature of the CMS apparatus is a superconducting solenoid, providing an axial magnetic field of 3.8 T. Immersed in the magnetic field are the silicon pixel and strip tracker, the lead-tungstate crystal electromagnetic calorimeter, and the brass/scintillator hadron calorimeter. Muons are measured in gas ionization detectors embedded in the steel return yoke and in the pseudorapidity window $|\eta| < 2.4$, with detection planes made of three technologies: Drift Tubes, Cathode Strip Chambers, and Resistive Plate Chambers. Matching the muons to the tracks measured in the silicon tracker results in a transverse momentum resolution better

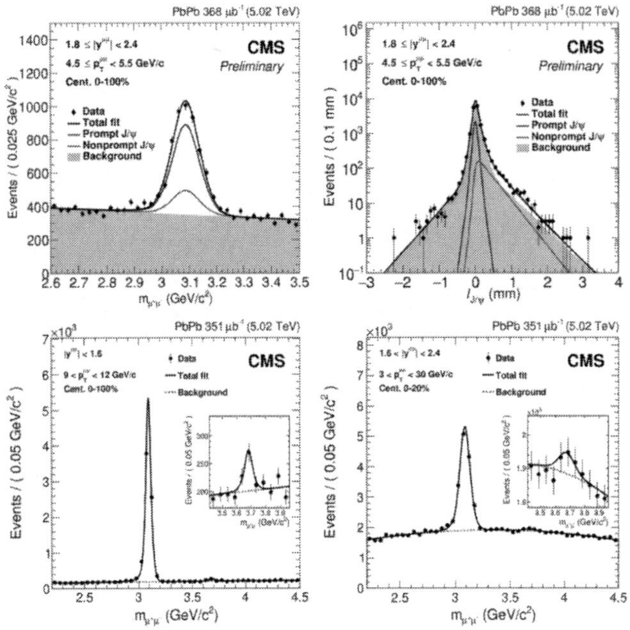

Figure 1. Top: Dimuon invariant-mass (left) and pseudo-proper decay length (right) distributions, for the J/ψ analysis [8]. The spectra are integrated over centrality, rapidity range $1.8 < |y| < 2.4$, and p_T range $4.5 < p_T < 5.5$ GeV/c. The projections of the two-dimensional fit onto the respective axes are overlaid as solid black lines. The dashed red lines show the fitted contribution of non-prompt J/ψ. The fitted background contributions are shown as dotted blue lines. **Bottom:** Dimuon invariant-mass distributions used in the ψ(2S) analysis [9], for $9 < p_T < 12$ GeV/c, $|y| < 1.6$, and 0–100% centrality (left), and $3 < p_T < 30$ GeV/c, the $1.6 < |y| < 2.4$ and 0–20% centrality (right).

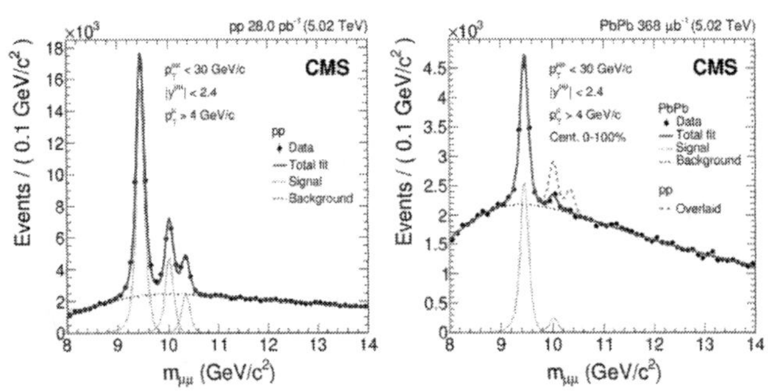

Figure 2. Dimuon invariant mass distributions in pp (left) and PbPb (right) collisions for the Υ analysis [12]. In both figures, the results of the fits to the data are shown as solid blue lines. The dashed red lines in the right panel are the result of the fits in PbPb (blue solid line) but with the fitted Υ yield for each state scaled by the inverse of its measured R_{AA}.

than 1.5% for p_T smaller than 100 GeV/c. A more detailed description of the CMS detector can be found in Ref. [7].

The invariant mass spectrum of all $\mu^+\mu^-$ pairs used in the J/ψ PbPb analysis [8] is shown in

IOP Conf. Series: Journal of Physics: Conf. Series **1070** (2018) 012016 doi:10.1088/1742-6596/1070/1/012016

the upper-left of Figure 1. The black curve is an extended unbinned maximum likelihood fit to the spectrum, using the sum of two Crystal Ball (CB) functions for signal, and an exponential for the background. To measure the fraction of non-prompt J/ψ (the so called b-fraction), the pseudo-proper decay length $l^{3D}_{J/\psi}$, shown in the uppwer-right of Figure 1, is computed as $L_{xyz} m_{J/\psi}/p_T$. Here, the L_{xyz} is defined as,

$$L_{xyz} = \frac{\hat{u}^T S^{-1} \vec{r}}{\hat{u}^T S^{-1} \hat{u}}, \tag{1}$$

where $\hat{u}$ is the unit vector in the direction of the $J/\psi p_T$ and S is the sum of the primary and secondary vertex covariance matrices. In the ψ(2S) analysis [9], the nonprompt component is removed by requesting the dimuon l^{3D}_{ψ} be smaller than a l_0 threshold. The threshold was chosen using MC simulations, such that it keeps 90% of the prompt ψs. Since the l^{3D}_{ψ} resolution improves with increasing dimuon p_T, from ≈100 μm to ≈20 μm, the l_0 cut is p_T dependent [9].

Figure 2 shows the invariant mass distribution for pp and PbPb data, in the mass range used in the Υ(nS) analysis [12]. The yields are extracted from an unbinned maximum likelihood fit. The signal shape of each Υ(nS) state is modeled by a CB function. The Υ(nS) mass ratios are fixed to the their world average values [11]. The background shape is modeled by an exponential function multiplied by an error function, and all its parameters are left free in the fit, as explained in Ref. [12].

3. Results

The hot and dense nuclear matter effects can be quantified by the nuclear modification factors R_{AA} computed as:

$$R_{\mathrm{AA}} = \frac{\mathcal{L}_{\mathrm{pp}}}{T_{\mathrm{AA}} N_{\mathrm{MB}}} \frac{N_{\mathrm{PbPb}}}{N_{\mathrm{pp}}} \cdot \frac{\varepsilon_{\mathrm{pp}}}{\varepsilon_{\mathrm{PbPb}}} \tag{2}$$

where $T_{\mathrm{AA}} = < N_{\mathrm{coll}} > / \sigma^{\mathrm{NN}}_{\mathrm{inel}}$ is calculated from a Glauber model, to account for the nuclear collision geometry [13], $\mathcal{L}_{\mathrm{pp}}$ is the integrated luminosity of pp collisions, and $N_{\mathrm{pp/PbPb}}$ are the raw yields measured in pp and PbPb collisions, N_{MB} is the count of minimum bias events in PbPb, and $\varepsilon_{\mathrm{pp}}/\varepsilon_{\mathrm{PbPb}}$ is the multiplicity dependent ratio of the efficiencies in pp and PbPb collisions for trigger and reconstruction, which is determined by a Monte-Carlo simulation based on embedding PYTHIA signal event [14] to a HYDJET background event [15].

3.1. Charmonia in pp and PbPb collisions

In Figure 3, the R_{AA} of prompt J/ψ mesons as function of N_{part}, p_T, and rapidity are shown, integrated in each case over the other two non-plotted variables. The results are compared to those obtained at $\sqrt{s_{NN}} = 2.76$ TeV [16], and found to be in good overall agreement: no strong dependence of the suppression is observed even for the most peripheral bin (70-100%), with the suppression slowly increasing with N_{part}. Although no clear dependence is observed in the R_{AA} as a function of p_T in the range of 5-20 GeV/c, an indication of less suppression at higher p_T is seen for the first time [8]. No significant rapidity dependence between the two energies is observed.

Figure 4 shows the double ratio as functions of p_T (left) and N_{part} (middle and right). The double ratios, $(N_{\psi(2S)}/N_{J/\psi})_{PbPb}/(N_{\psi(2S)}/N_{J/\psi})_{pp}$, measured at 5.02 TeV are below unity in all bins. Since J/ψ is suppressed [8], the results show that ψ(2S) is more suppressed than the J/ψ in PbPb collisions. No strong dependences are observed with centrality or transverse momentum. A reasonable agreement with the measurement made at $\sqrt{s_{NN}} = 2.76$ TeV can be seen in most of the bins. Systematic uncertainties are uncorrelated between the two datasets. In the range $1.6 < |y| < 2.4$ and $3 < p_T < 30$ GeV/c, the double ratios are consistently lower in the 5.02 TeV data, especially in the most central collisions. The difference is at the level of around 3 standard deviations in the centrality-integarted sample [9].

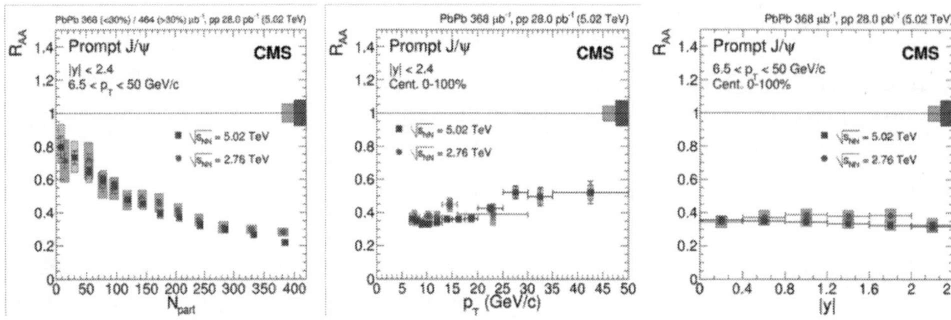

Figure 3. The nuclear modification factor of prompt J/ψ as a function of N_{part} (left), p_T (middle) and rapidity (right) at $\sqrt{s_{NN}}$ = 5.02 TeV [8] and 2.76 TeV [16]. For the results as a function of N_{part} the most central bin corresponds to 0-5%, and the most peripheral one to 70-100% centrality. The bars (boxes) represent statistical (systematic) point-by-point uncertainties. The boxes plotted at R_{AA} = 1 indicate the size of the global relative uncertainties.

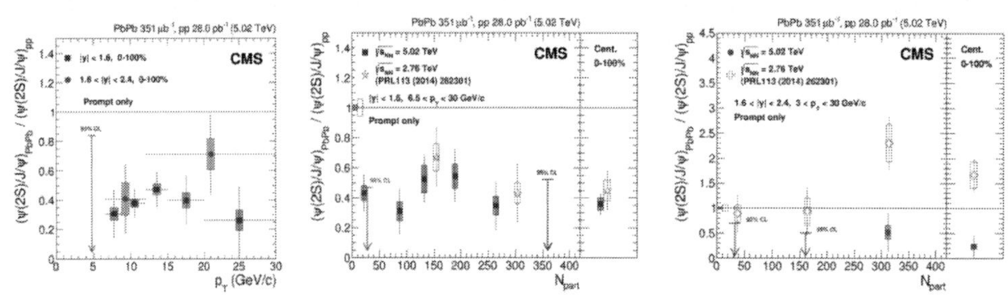

Figure 4. The double ratio $(N_{\psi(2S)}/N_{J/\psi})_{PbPb}/(N_{\psi(2S)}/N_{J/\psi})_{pp}$ versus p_T (left) and N_{part} at mid- (middle) and forward- (right) rapidity [9]. The closed blue squares and red circles are 5.02 TeV and, the open stars and crosses are 2.76 TeV. The arrow represents the 95% CL interval in the bin where the measurement is consistent with 0. Vertical lines (boxes) represent the statistical (systematic) uncertainties.

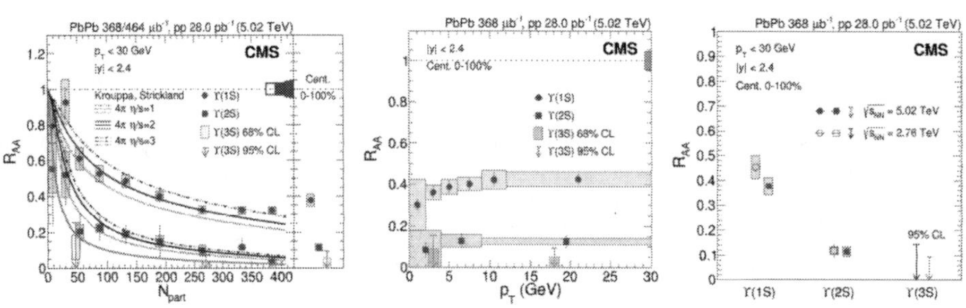

Figure 5. Nuclear modification factor of Υ(1S), Υ(2S), and Υ(3S) mesons as a function of $< N_{part} >$ (left), p_T (middle), and integrated over p_T, $< N_{part} >$ and y (right) [12]. For the Υ(3S) meson, the upper limits at 68% (green box) and 95% (green arrow) C.L are shown. The lines in the left panel represent calculations from Ref. [19].

3.2. Bottomonia in pp and PbPb collisions

Figure 5 shows the nuclear modification factor for the Υ(1S), Υ(2S), and Υ(3S) mesons as functions of N_{part} (right) and p_T (middle) [12]. Within the systematic uncertainties, the R_{AA}

IOP Conf. Series: Journal of Physics: Conf. Series **1070** (2018) 012016 doi:10.1088/1742-6596/1070/1/012016

values show no dependence on p_T. The excited Υ states are found to have larger suppression than the ground state, with $R_{AA} < 0.2$ over the full kinematic range examined here. The kinematic dependence of R_{AA} is useful to constrain models of Υ meson suppression in a deconfined medium [19]. The strong suppression of the Υ(3S) meson is observed in both centrality bins studied, 0-30% and 30-100%. The R_{AA} decreases with increasing centrality in the case of the Υ(1S) and Υ(2S) mesons. The theoretical calculations of Krouppa and Strickland, shown in the left panel of figure 5, are in agreement with the results. The model incorporates color-screening effects on the bottomonium family and feed-down contributions from decays of heavier quarkonia. In the right panel of Fig. 5 the centrality-integrated R_{AA} values at $\sqrt{s_{NN}} = 2.76$ TeV and 5.02 TeV are compared. The suppression at 5.02 TeV is larger by a factor of 1.2 ± 0.15, although the two R_{AA} values are compatible within the uncertainties.

4. Summary

Prompt J/ψ and ψ(2S) meson productions have been studied in pp and PbPb collisions at $\sqrt{s_{NN}}$ = 5.02TeV, as functions of rapidity, transverse momentum, and collision centrality. The R_{AA} results of prompt J/ψ show a strong centrality dependence, with an increasing suppression for increasing centrality. The double ratio, $(N_{\psi(2S)}/N_{J/\psi})_{PbPb}/(N_{\psi(2S)}/N_{J/\psi})_{pp}$, is below unity in all bins, suggesting that the ψ(2S) yield is more suppressed than the J/ψ yield in the kinematic range explored. A gradual decrease in R_{AA} with centrality for the Υ(1S) and Υ(2S) states is observed, while no significant dependence on p_T or y is found in the measured region. The R_{AA} of the Υ(3S) state is measured to be below 0.094 at 95% confidence level, making this the smallest R_{AA} value found for any hadron species in heavy ion collisions.

References

[1] T. Matsui and H. Satz, *Phys. Lett. B* **178** (1986) 416
[2] S. Digal, P. Petreczky, and H. Satz, *Phys. Rev. C* **64** (2001) 904015
[3] A. Mocsy and P. Petreczky, *Phys. Rev. Lett* **99** (2007) 211602
[4] R. Vogt, *Phys. Rev. C* **81** (2010) 044903
[5] R. Sharma and I. Vitev, *Phys. Rev. C* **87** (2013) 044905
[6] F. Arleo and S. Peigne, *JHEP* **03** (2013) 122
[7] CMS Collaboration, *JINST* **3** (2008) S08004
[8] CMS Collaboration, *Eur. Phys. J. C* **78** (2018) 059
[9] CMS Collaboration, *Phys. Rev. Lett.* **118** (2017) 162301
[10] CMS Collaboration, *Phys. Rev. Lett.* **113** (2014) 262301
[11] Particle Data Group Collaboration *Phys. Rev. D* **86** (2012) 010001
[12] CMS Collaboration, CMS-HIN-PAS-16-023, arXiv:1805.09215
[13] B. Alver et al., *Phys. Rev. C* **77** (2008) 014906
[14] T. Sjostrand, S. Mrenna and P. Z. Skands, *JHEP* **0605** (2006) 026
[15] I. P. Lokhtin and A. M. Snigirev, *Eur. Phys. J. C* **45** (2006) 211
[16] CMS Collaboration, *Eur. Phys. J. C* **77** (2017) 252
[17] ALICE Collaboration, *Phys. Lett. B* **732** (2014) 314
[18] ALICE Collaboration, arXiv:1606.08197
[19] B. Krouppa and M. Strickland, *Universe* **2** (2016) 16
[20] CMS Collaboration, *Phys. Lett. B* **770** (2017) 35

IOP Conf. Series: Journal of Physics: Conf. Series **1070** (2018) 012017 doi:10.1088/1742-6596/1070/1/012017

PHENIX measurements of bottom and charm quark production

Timothy Rinn (For the PHENIX Collaboration)
Iowa State University, Ames, IA 50010
E-mail: trinn@iastate.edu

Abstract.
Heavy quarks are produced primarily in the early stage of high-energy collisions, before the formation of any quark-gluon plasma (QGP). Therefore heavy quarks provide valuable insight to the properties of the QGP, especially by changing the mass of the heavy quark. PHENIX is able to measure the nuclear modification factor of electrons from bottom and charm composed hadrons. PHENIX has also made measurements of the momentum integrated nuclear modification factor of $B \to J/\psi$ and prompt J/ψ production, observing no nuclear modification for B mesons and a significant suppression of prompt J/ψ due to its breaking in the medium. Additionally, the $b\bar{b}$ cross-section was measured in p+p collisions where it shows a factor of two larger than the center value of the Fixed-Order-Next-to-Leading-Log (FONLL) prediction.

1. Introduction

Heavy Flavor quarks are a valuable probe of the Quark Gluon Plasma (QGP) at RHIC, this is due to the fact that at RHIC energies Heavy Flavor quarks, such as bottom and charm, are produced predominately in the initial hard scattering collision. Since heavy flavor quarks observe the full evolution of the collision system, modification of heavy flavor production in larger systems compared to p+p can provide valuable insight to effects of the medium [1].

In hard scattering p+p collisions it is possible to use perturbative Quantum Chromodynamics (pQCD) to calculate heavy flavor production. One such Monte Carlo which is used in PHENIX is PYTHIA, which uses leading order calculations to calculate particle generation. Using Fixed-order plus Next-to-Leading-Log (FONLL) [2] one can make more accurate predictions on open heavy flavor productions. Using the Color Evaporation Model (CEM) [3], which allows for the calculation of closed heavy flavor production such as J/ψ, there is a fairly comprehensive collection of models for comparison of heavy flavor quark production at PHENIX. These models have been probed by PHENIX through various avenues. FONLL/CEM have been studied by looking at J/ψ production for both prompt and through B decays. Additionally, FONLL was probed through its estimation of the $b\bar{b}$ cross section. Using pythia as a Monte Carlo model of p+p collisions, PHENIX was able to probe and understand the various production modes of heavy flavor at RHIC energies through the study of $q\bar{q}$ azimuthal correlations.

In heavy ion collisions there are two main effects whose contributions can be studied: initial state effects which are modeled to first order using parton distribution functions such as EPS09[4]; and final state interactions with the nuclear matter which can be modeled using transport calculations [5]. PHENIX has probed the effects of interactions with the nuclear medium by looking at heavy flavor production of separated charm and bottom in Au+Au collisions. By comparing measurements to calculations done using a Transport Matrix approach one is able to gane insights onto the strength of coupling within the QGP. By looking at the nuclear modification factor (R_{AA}) of separated bottom and charm quarks we can learn about the mass dependence of interactions within the nuclear medium. The nuclear modification factor of integrated momentum B mesons at forward/backward rapidity in Cu+Au collisions can tell us if initial state effects can affect the production of b-quarks.

2. Nuclear Modification of electrons from bottom and charm decays

PHENIX has measured electrons from the semi-leptonic decays of bottom and charm hadrons in 0-10% central Au+Au events at $\sqrt{s} = 200GeV$. In PHENIX electrons candidates are identified at mid rapidity using a combination of a ring imaging Cherenkov detector and an energy/momentum cut to remove charged hadrons, a schematic of the central arm of PHENIX is seen in Figure 1. This electron candidate sample has contributions from many sources including but not limited to heavy flavor decays, photonic, non-photonic (such as J/ψ), and hadron contamination. For each electron the distance of closest approach in the transverse direction (dca_T) was measured using the Silicon Vertex Tracker (VTX). Due to life time and decay kinematics the dca_T allows PHENIX to distinguish between electrons from bottom, charm, and various background sources as seen in Figure 1.

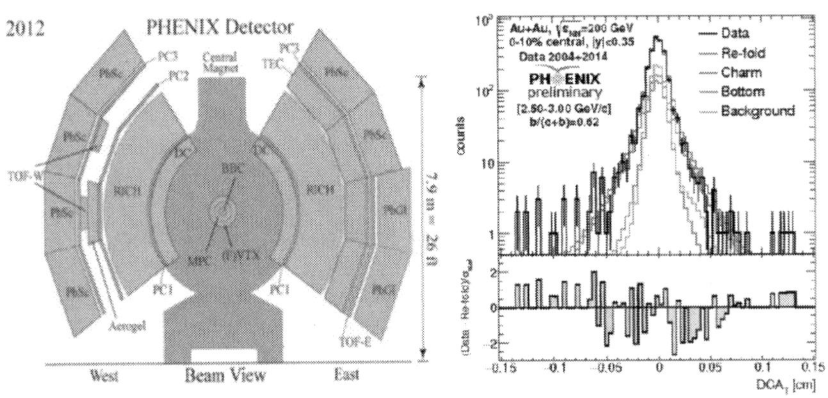

Figure 1. Left: Diagram of the Central arm of the PHENIX detector.
Right: Measured electron dca_T with separated contributions from bottom, charm, and background sources

The separated bottom and charm measurement was done using Bayesian Inference Techniques to simultaneously fit previously published inclusive heavy flavor electron invariant yields and measured electron dca_T distributions, the details of this technique are discussed in reference [6].The b-fraction, as defined as $b \to e/(b \to e + c \to e)$ and shown in shown in Figure 2, agrees with a model implying strong coupling in the QGP, and at $p_T > 5GeV/c$ it observes agreement with DGLV predictions implying both radial and collisional energy loss occurring withing the QGP.

Combining the measured heavy flavor electron invariant yields in central Au+Au with STAR measurements in p+p done using electron-hadron correlations PHENIX was able to extract an R_{AA} for both electrons from charm and bottom, as seen in Figure 2. We observe that electrons from bottom are less suppressed than those from charm at p_T between 3 and 4 GeV/c. This would imply that there is a mass dependence to the energy loss experienced due to interactions within the QGP. In PHENIX there is currently an effort underway to measure a continuous p+p baseline using a similar technique for the range of 1-9 GeV/c in combination with analyzing the full 2014 Au+Au statistics (4x) one can expect an updated R_{AA} result with reduced uncertainties soon.

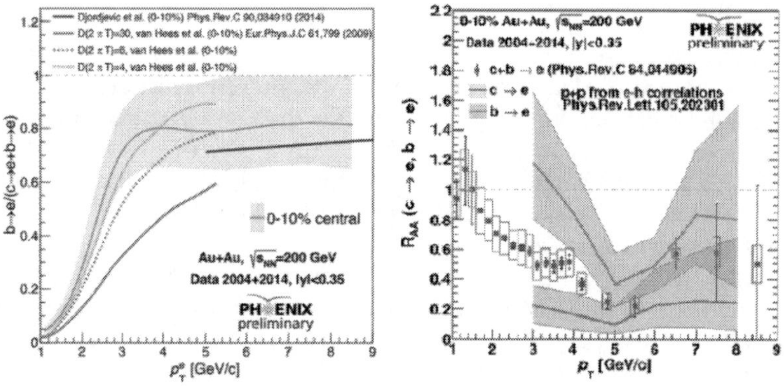

Figure 2. Left: Preliminary b-fraction result for 0-10% central Au+Au.
Right: Preliminary R_{AA} result for electrons from bottom and charm in 0-10% central Au+Au

3. Nuclear Modification of Prompt J/ψ and J/ψ from B decays

PHENIX has measured muons from J/ψ decays within a rapidity range of $1.2 < |\eta| < 2.2$. The muons are measured using a combination of a muon identification detector, which provides rejection of charged pions, and a muon tracking detector, used to measure the muon momentum from the trajectory in the magnetic field. J/ψ's were reconstructed as opposite charge muon pairs with $2.8 < mass[GeV/c^2] < 3.5$. For each muon from a di-muon pair within the J/ψ mass region the distance of closest approach along the radial distance (dca_R) was measured using the forward silicon vertex detector. This quantity allows one to distinguish muons from prompt J/ψ and those from $B \to J/\psi$ decays, as those come from a displaced vertex, as seen in Figure 3. By doing a fit to the measured dca_R distributions using templates for the shapes of prompt and non-prompt J/ψ's PHENIX was able to extract measurements of both prompt and $B \to J/\psi$ in both p+p and Cu+Au. Specific details on the analysis method can be found in reference [7]

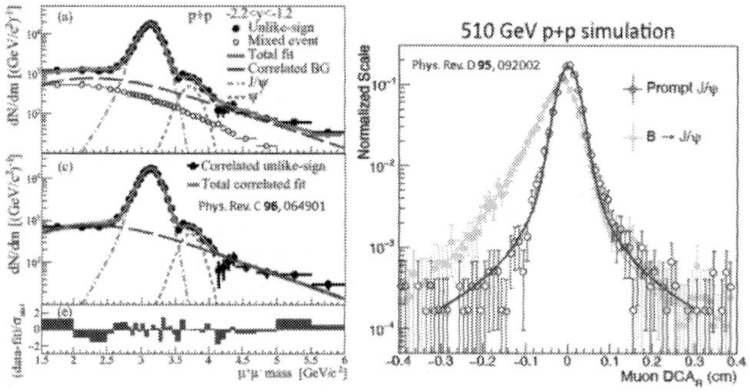

Figure 3. Left: di-muon mass spectrum
Right: dca_R distribution of simulated muons from prompt J/ψ and those from B to J/ψ decays

PHENIX measurements of the fraction of J/ψ's which come from B decays show agreement in p+p with FONLL/CEM predictions, while in Cu+Au we observe a significant increase in the $B \rightarrow J/\psi$ fraction. This indicates that $B \rightarrow J/\psi$ observe less suppression than prompt J/ψ. By combining the $B \rightarrow J/\psi$ fraction with an inclusive J/ψ R_{CuAu} an R_{CuAu} was extracted for both prompt and $B \rightarrow J\psi$. It is observed that the $B \rightarrow J/\psi$ is consistent within uncertainties with no nuclear modification, while prompt J/ψ observe significant suppression. This is consistent with the number of b-quarks being conserved or having small modification in nuclear collisions, whereas prompt J/ψ are breaking/melting within the nuclear medium.

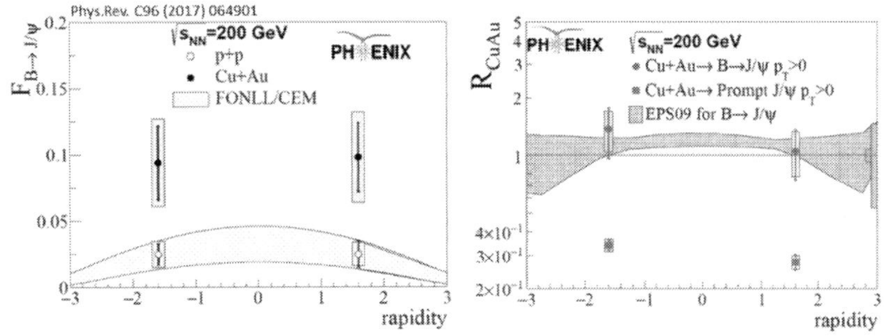

Figure 4. Left: Fraction of J/ψ which come from B decays
Right: Nuclear modification factors R_{CuAu} for both prompt J/ψ and $B \rightarrow J/\psi$

4. Heavy Flavor measurements through Di-Muons

PHENIX has made a measurement of separated bottom and charm at forward and backwards rapidity in p+p by studying di-muon pairs. This analysis, similar to the J/ψ analysis, uses the same detector systems to look at the di-muon mass spectra of both opposite sign and same sign di-muon pairs. This analysis takes advantage of the fact that although D mesons decay to $\mu^{\pm}\mu^{\mp}$ B mesons are able to decay into $\mu^{\pm}\mu^{\pm}$. This allows for the use of the like sign di-muon pairs to be used to measure $b\bar{b}$. As seen in Figure 5 we are able to get a very clean measurement of $b\bar{b}$ in the like sign high mass region as there are very few sources of like sign di-muon pairs in PHENIX. Additionally, by looking at the intermediate mass region in un-like sign pairs we are able to make a measurement of $c\bar{c}$. These measurements were made by combining the measured di-muon mass spectra with a di-muon mass cocktail to account for the various background sources and applying a fit to extract the contributions from $b\bar{b}$ and $c\bar{c}$.

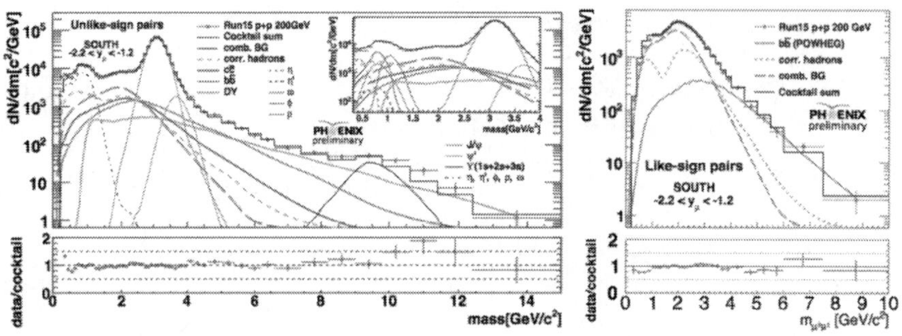

Figure 5. Left: Unlike sign di-muon mass spectrum with the di-muon mass cocktail
Right: Like sign di-muon mass spectrum with the di-muon mass cocktail

Using this measurement PHENIX was able to extract the $b\bar{b}$ cross-section at both forward and backwards rapidity. Combining this result with PHENIX measurements at mid rapidity we observe good agreement across the full rapidity coverage with MC@NLO and POWHEG, however there is an apparent factor of two larger than the center value of FONLL calculation for the $b\bar{b}$ cross-section as seen in Figure 6.

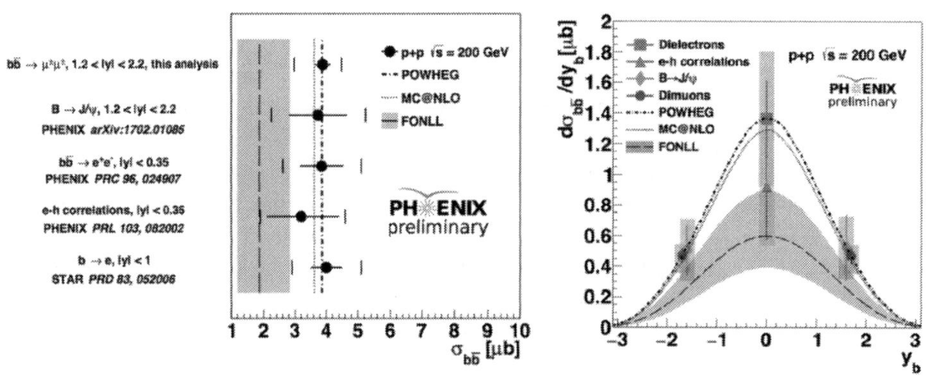

Figure 6. Left: $b\bar{b}$ cross section world data comparison
Right: $b\bar{b}$ cross section measurement over full PHENIX rapidity range

Looking at azimuthal correlations between the two muons from $b\bar{b}$ or $c\bar{c}$ decays it was observed that heavy flavor at PHENIX is well described by PYTHIA, this is seen in Figure 7. At RHIC at $\sqrt{s} = 200GeV$ pair creation and flavor excitation are the dominant modes of production, while the contribution due to gluon splitting is very small. This is in contrary to what is observed at the LHC, where gluon splitting is the dominant mode of heavy flavor production.

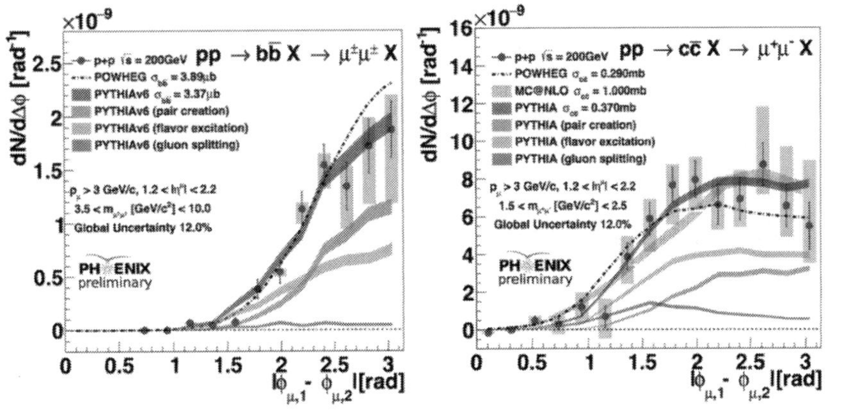

Figure 7. Left: $b\bar{b}$ cross section measurement over full PHENIX rapidity range
Right: Azimuthal correlation measurements comparison to PYTHIA and various production modes

5. Summary

PHENIX has measured bottom and charm quarks in p+p, Au+Au, and Cu+Au providing valuable insight into both production mechanisms and nuclear matter effects. It was observed in central Au+Au collisions that electrons from bottom between 3 and 4 GeV/c are less suppressed than electrons from charm, which provides some indication of mass dependence to interactions within the QGP. PHENIX showed that $B \to J/\psi$ measurements are consistent with no nuclear modification in Cu+Au collisions, while prompt J/ψ observe significant suppression consistent with J/ψ experiencing melting/breaking in the medium. In p+p $b\bar{b}$ cross section measurements are consistent with the global data, a factor of two larger than the central FONLL prediction and mainly produced by pair creation and flavor excitation processes.

References

[1] A. Adare *et all* (PHENIX) 2006 *Phys. Rev. Let.* **96** 032301
[2] M. Cacciari, M. Greco, and P. Nason 1998 *J. High Energy Phys.* **05** 007
[3] A.D. Frawley, T. Ullrich, and R. Vogt 2008 *Phys. Rept* **462** 125
[4] K.J. Eskola, H. Paukkunen, and C. A. Salgado 2009 *High Energy Phys.* **04** 065
[5] V. Heese *et all* 2009 *Eur. Phys. J. C* **61** 799
[6] A. Adare *et all* (PHENIX) 2016 *Phys. Rev. C* **93** 034904
[7] C. Aidala *et all* (PHENIX) 2016 *Phys. Rev. C* **96** 064901

IOP Conf. Series: Journal of Physics: Conf. Series **1070** (2018) 012018 doi:10.1088/1742-6596/1070/1/012018

Recent results on hard processes in p+Pb, Pb+Pb, and pp collisions from the ATLAS Experiment at the LHC

Laura Havener on behalf of the ATLAS collaboration
Physics Department, Columbia University, New York, NY 10027 USA
E-mail: lbh3133@columbia.edu

Abstract. In relativistic heavy–ion collisions, a hot medium with a high density of unscreened colour charges is produced. Hard processes from parton-parton scattering in the early stages of the collision can be used as probes of the medium since the initial state is mostly understood such that differences from pp collisions in the final state can be attributed to interactions with the medium. The initial and final state effects are separated by comparing p+Pb and pp collisions. Recent results from the ATLAS experiment at the LHC on the production of electroweak bosons, jets, and quarkonia are discussed in p+Pb and Pb+Pb collisions, compared to pp. Recent results on ultra–peripheral collisions are also shown.

1. Introduction

In relativistic heavy–ion (HI) collisions, a hot dense medium with a high density of unscreened colour charges is produced, called quark-gluon plasma (QGP) [1]. Hard processes from parton–parton scattering are produced in the early stages of the collision. The initial state of these processes are mostly understood when considering cold nuclear matter (CNM) effects. Thus products from hard interactions can be used as probes of the medium since differences from pp collisions in the final state can be attributed to interactions with the medium. The initial state effects, like CNM, can be separated out by using p+Pb collisions and comparing to pp collisions. Recent hard probes results using the ATLAS detector [2] at the LHC, including electroweak bosons, quarkonia, jets, and charged hadrons, in Pb+Pb, p+Pb, and pp at $\sqrt{s_{NN}} = 2.76$ TeV and 5.02 TeV are presented in this proceeding.

2. Electroweak bosons

Electroweak bosons, like photons, Z, and W bosons, are treated as controls in HI collisions. As they are colorless and do not interact with the medium, there should be no modification to their production rates. They are useful for verifying that we understand the nuclear geometry of the collision. The production of Z bosons in the di-muon decay channel is measured in 5.02 TeV Pb+Pb and pp data. Z boson candidates from pairs of oppositely charged muons are used to calculate the Z boson rate in Pb+Pb and pp. The rates are compared through the nuclear modification factor, $R_{\rm AA}$, defined below:

$$R_{\mathrm{AA}} = \frac{\left.\frac{1}{N_{\mathrm{evt}}^{\mathrm{tot}}}\frac{\mathrm{d}N_{\mathrm{Pb+Pb}}}{\mathrm{d}y}\right|_{\mathrm{cent}}}{\left.T_{\mathrm{AA}}\frac{\mathrm{d}\sigma_{pp}}{\mathrm{d}y}\right|_{pp}}, \tag{1}$$

where T_{AA} is the nuclear thickness function which accounts for the geometric enhancement of per-collision nucleon-nucleon luminosity to allow for a proper assessment of the quenching effects [3] and $N_{\mathrm{evt}}^{\mathrm{tot}}$ is the total number of Pb+Pb collisions within a chosen centrality interval.

The Z boson R_{AA} as a function of N_{part} is shown in the bottom panel of figure 1. It is consistent with unity and has no dependence on N_{part} which indicates that the T_{AA} factor is under control since no modification is expected in the medium [6]. The same analysis is performed in p+Pb collisions and the R_{pPb}, defined in equation 2, is evaluated as a function of the Z boson rapidity, y^{Z}, in the right-hand side of figure 1. This shows a slight forward-backward asymmetry that is consistent with nPDF (nuclear parton distribution function) effects [7].

$$R_{\mathrm{pPb}} = \frac{1}{A_{\mathrm{Pb}}}\frac{\mathrm{d}\sigma_{\mathrm{pPb}}/\mathrm{d}y}{\mathrm{d}\sigma_{pp}/\mathrm{d}y} \tag{2}$$

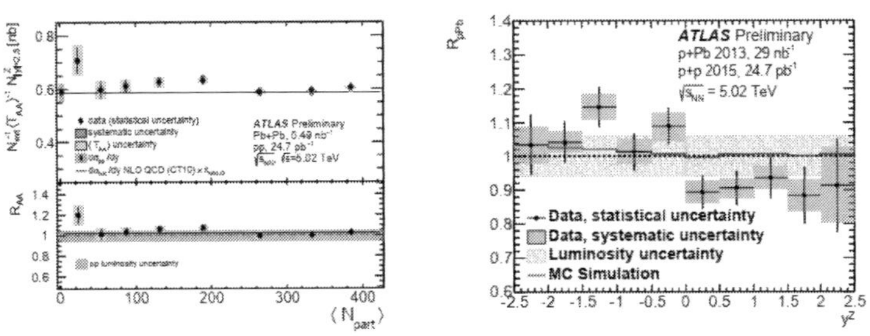

Figure 1. The left top panel shows the Z-boson yield in Pb+Pb collisions divided by T_{AA} (circles) as a function of N_{part} and the cross section for pp collisions at $N_{\mathrm{part}} = 2$ (diamond). The bottom panel shows the R_{AA} as a function of N_{part} [6]. The right panel shows the R_{pPb} as a function of Z boson rapidity. The blue line is simulation (see reference) [7].

The production of W bosons in the $W \to \mu\nu$ channel is measured in 5.02 TeV Pb+Pb data. The yields of positively and negatively charged muons are shown in the left panel in figure 2 as a function of N_{part} and scaled by T_{AA}. Both are independent of N_{part} which is excepted since W bosons should not interact with the medium. The W^+ yield is 10% higher than the W^- yield which is consistent with the MC simulation. The right panel shows the muon asymmetry as a function of $|\eta_\mu|$, which is described well by the MC simulation [8].

3. Jets

Jets originiate from partons that are expected to lose energy in interactions with the medium through a phenomena called 'jet quenching' [9, 10]. Recent results from ATLAS are more precise with better control over the systematic and the background subtraction due to the underlying event. They are unfolded for detector effects which enables direct theory comparisions. They have better statistics that allow for measurements at high jet p_{T} and differentially in rapidity and jet p_{T}, as well as measurements of boson+jet systems.

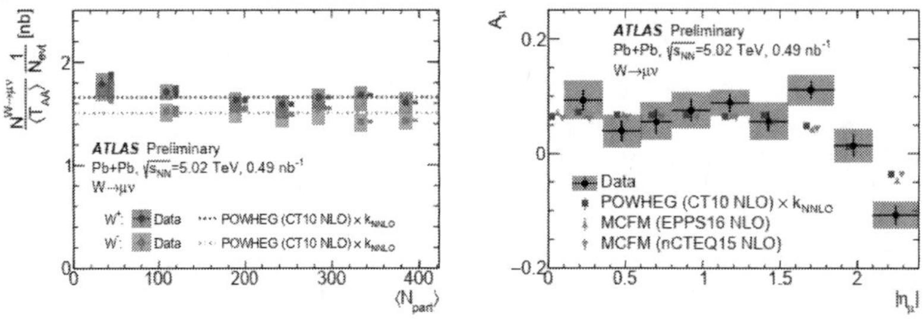

Figure 2. The left panel shows the W boson yields divided by $T_{\rm AA}$ as a function of $N_{\rm part}$ for W^+ (circles) and W (squares). The MC simulation (see reference) is in the dashed lines. The right panel shows the muon charge asymmetry in 0–80% centrality range as a function of η. Theory predictions (see reference) are indicated with the markers shifted along the x-axis [8].

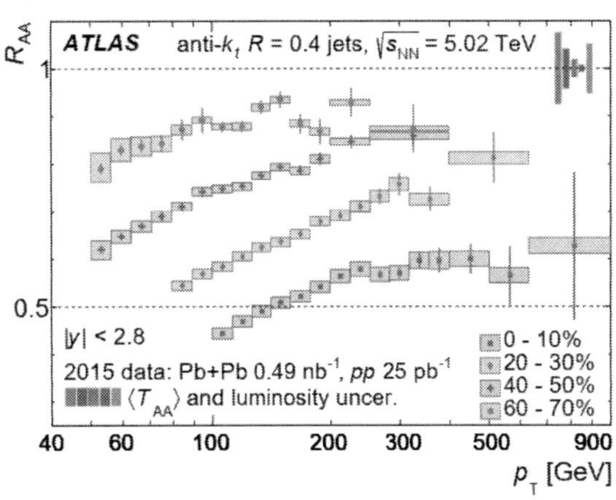

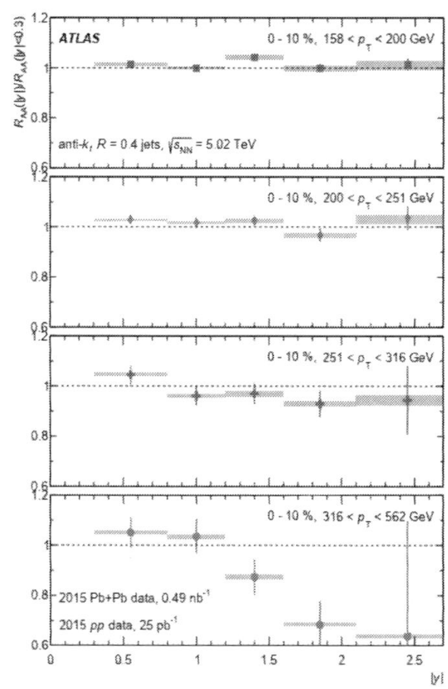

Figure 3. The left panel shows the $R_{\rm AA}$ as a function of jet $p_{\rm T}$ for jets for three centrality intervals. The right panel shows the ratio of the $R_{\rm AA}(|y|)$ to the $R_{\rm AA}(|y| < 0.3)$ for jets in 0-10% centrality in different $p_{\rm T}$ intervals [11].

3.1. Jet suppression

The jet yield is expected to be suppressed in central collisions relative to *pp* collisions at fixed $p_{\rm T}$ in HI collisions due to in-medium energy loss. This section describes new measurements of jet yields in Pb+Pb and *pp* data collected at 5.02 TeV. The individual jet spectra and cross sections are unfolded to correct for bin migration due to experimental effects [4, 5]. The $R_{\rm AA}$, defined in a similar way as in equation 1, as function of $p_{\rm T}$ is shown in figure 3 in different centrality intervals. The $R_{\rm AA}$ is less than 1 for all centralities, but it is lower in more central collision, where it is around 0.5, than in peripheral collisions. The rapidity dependence of the $R_{\rm AA}$ is shown in Figure 3 by evaluating the ratio of the $R_{\rm AA}(|y|)$ to the $R_{\rm AA}$ at $|y| < 0.3$ in intervals of $p_{\rm T}$. The ration is flat at lower $p_{\rm T}$, but as the $p_{\rm T}$ is increased the $R_{\rm AA}$ starts to decrease with rapidity. This decrease is the most significant in the highest $p_{\rm T}$ interval [11].

3.2. Jet imbalance

Dijets are an effective probe of jet quenching because jets originating from the same hard scattering can lose different amounts of energy as they traverse the medium whereas in *pp* collisions the jets are expected to be balanced. This section present a new measurement of the dijet asymmetry in Pb+Pb and *pp* collisions at 2.76 TeV using the observable $x_J = p_{T1}/p_{T2}$ that has been fully corrected for detector effects by unfolding in 2D in the two jet's p_T [4, 5]. The left panels in figure 4 show the centrality dependence of x_J, which demonstrates that as more central collisions are probed the dijets become more asymmetric. The most probable configuration for *pp* is 1, whereas for central Pb+Pb it is 0.5. The right panels show the leading jet p_{T1} dependence in 0–10% centrality, where a drastic p_{T1} dependence is observed in which the Pb+Pb become like the *pp* at higher p_{T1} [12].

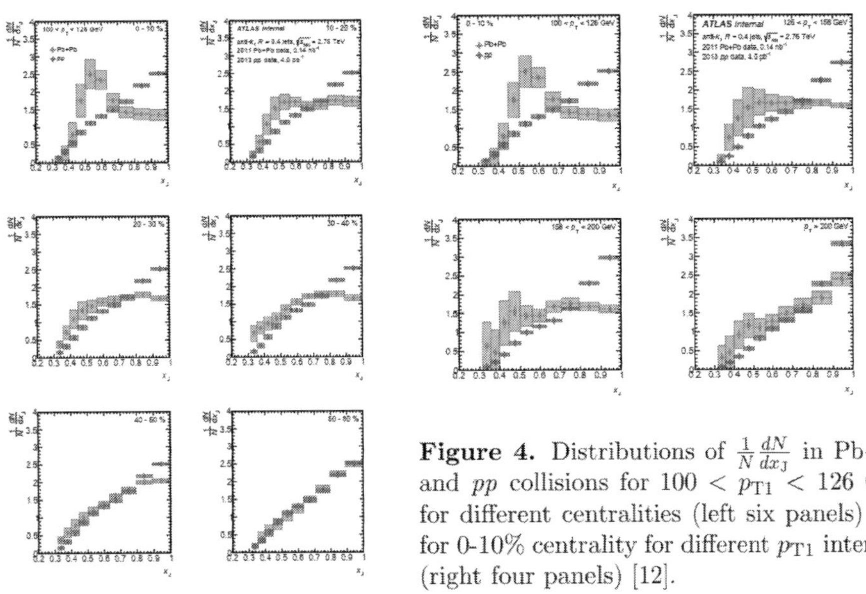

Figure 4. Distributions of $\frac{1}{N}\frac{dN}{dx_J}$ in Pb+Pb and *pp* collisions for $100 < p_{T1} < 126$ GeV for different centralities (left six panels) and for 0-10% centrality for different p_{T1} intervals (right four panels) [12].

Photons are not strongly interacting so are unmodified by the medium produced in HI collisions. Thus measuring γ+jets in HI collisions allow for an estimate of the energy of the original parton before it loses energy in the medium. Also, γ+jets are more likely to originate from quarks than dijets so can help constrain the flavor dependence of energy loss. This section presents a measurement of γ+jet correlations in Pb+Pb and *pp* collisions at 5.02 TeV using the observable $x_{J_\gamma} = p_T^{\mathrm{jet}}/p_T^{\gamma}$. The distributions for 0-10% Pb+Pb, *pp*, and MC simulations are show in figure 5. The Pb+Pb has more asymmetric pairs than in MC and *pp* [13].

3.3. Jet fragmentation

The internal structure of jets is expected to be modified in Pb+Pb collisions compared to *pp* collisions. This section describes a new measurement of the jet fragmentation functions (FF) in Pb+Pb, *p*+Pb, and *pp* collisions at 5.02 TeV. The *p*+Pb results help determine if there are modifications in the jet FF due to the presence of a large nucleus. The FF are defined as:

$$D(z) = \frac{1}{N_{\mathrm{jet}}}\frac{\mathrm{d}N_{\mathrm{ch}}}{\mathrm{d}z}, \tag{3}$$

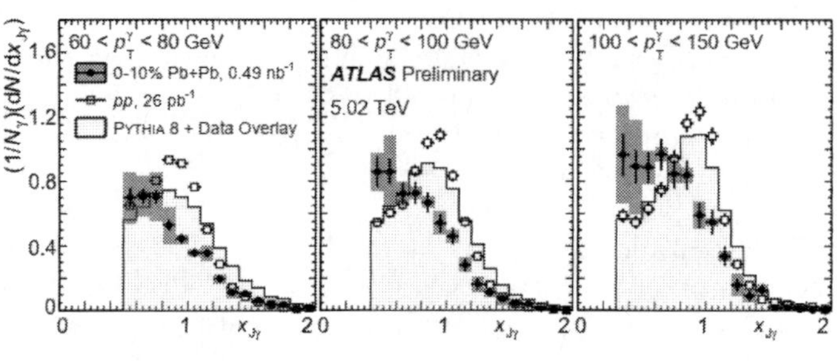

Figure 5. Distributions of $\frac{1}{N_\gamma}\frac{dN_\gamma}{dx_{J_\gamma}}$ for 0-10% Pb+Pb data (black circles), *pp* (white boxes), and MC (shaded yellow) in different p_T^γ intervals [13].

where $N_{\rm ch}$ is the number of charged particles associated with the jet, $N_{\rm jet}$ is the number of jets, and $z = p_T^{\rm trk}\cos\Delta R/p_T^{\rm jet}$, where $\Delta R = \sqrt{\Delta\eta^2 + \Delta\phi^2}$. The FF are then compared between Pb+Pb or p+Pb and *pp* collisions using the ratio $R_{D(z)} = D(z)_{\rm PbPb,pPb}/D(z)_{\rm pp}$.

The FF in Pb+Pb data are measured for a jet p_T range of 126-501 GeV and are corrected for detector effects using 2D unfolding in jet p_T and track z [4, 5]. The left panel of figure 6 shows that the $R_{D(z)}$ is consistent for three jet p_T intervals for $z \gtrsim 0.3$ which indicates that the enhancement of hard fragments is nearly independent of jet p_T. The right panel shows the $R_{D(z)}$ for a jet p_T range 126-158 GeV for this measurement and a previous measurement at 2.76 TeV. An enhancement at low z and high z, with a suppression in between is shown. The low and mid z behavior are indicative of jet internal structure modification where the energy loss is transferred into soft particles inside and near the jet. The high z enhancement could be explained by differences in expected energy loss for quark and gluon jets since there are more quark jets at higher z and quark jets are expected to lose less energy than gluon jets [14]. The modification of the jet internal structure is shown to be similar between the two center-of-mass energies [16].

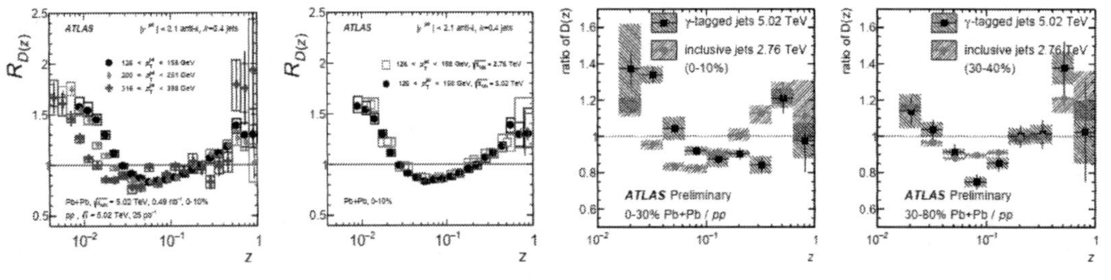

Figure 6. The left-most panel shows the $R_{D(z)}$ for different $p_T^{\rm jet}$ intervals in 0–10% centrality. The left center panel shows the $R_{D(z)}$ for $126 < p_T^{\rm jet} < 158$ GeV at 5.02 TeV (black circles) compared to 2.76 TeV (open squares) [16]. The right center panel shows the $R_{D(z)}$ ratios for 0–30% γ-tagged (red circles) and 0–10% inclusive jets (black squares). The right-most panel shows the $R_{D(z)}$ ratios for 30–80% γ-tagged (red circles) and 30–40% inclusive jets (black squares) [18].

The FF are also measured for p+Pb collisions for a jet p_T range is 45-260 GeV. The $R_{D(z)}$ are shown in figure 7 in six different jet p_T intervals. The ratios are consistent with unity which indicates no evidence for modification of the jet internal structure in p+Pb collisions [15].

The internal structure of γ-tagged jets should also be modified by the medium but in a potentially different way than inclusive jets for the reasons mentioned in section 3.2. Also, inclusive jets have already been quenched making them an indirect comparison to jets in *pp*

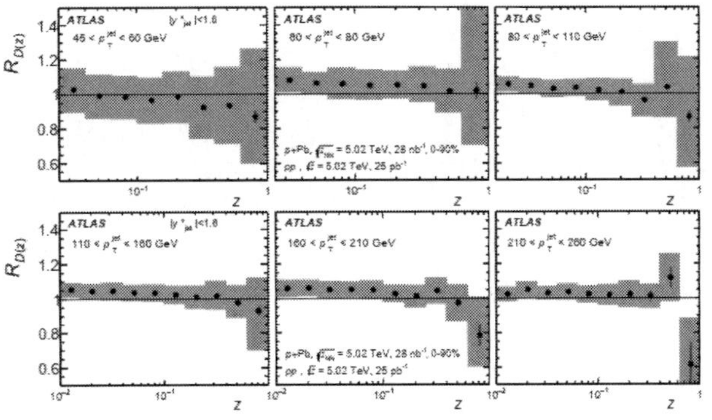

Figure 7. Distributions of $R_{D(z)}$ for different p_T^{jet} intervals [15].

collisions. They also preferentially select jets that have been less quenched [14, 17]. The photon is not expected to be quenched so comparing γ-tagged jets in pp and Pb+Pb collisions is a more direct comparison of the same jets. This section describes a measurement of γ-tagged jet FF in Pb+Pb collisions at 5.02 TeV. They are measured by selecting a photon with E_T^{γ} between 79.6-126 GeV that is azimuthally balanced to a jet with p_T between 63.1-144 GeV. The $R_{D(z)}$ are shown in the right panels of figure 6 compared between the γ-tagged and inclusive jets in central and peripheral collisions. The γ-tagged jets have a stronger modification than the inclusive jets in central collisions and become in better agreement in more peripheral collisions [18].

4. Quarkonia

Color charge screening is where bound quarkonia states screen themselves from the medium produced in HI collisions [19]. Eventually they will break apart, leading to a suppression of bound states. The more loosely bound the state, the easier it is to break up (melt) and the stronger the suppression at a fixed temperature. This is called sequential melting which can probe the temperature of the medium since different states will melt at different temperatures.

$J/\psi \rightarrow \mu^+\mu^-$ and $\psi(2S) \rightarrow \mu^+\mu^-$ charmonia production are measured in Pb+Pb and pp at 5.02 TeV. Both the "prompt J/ψ", from the immediate formation of the composite $c\bar{c}$ bound state, and the "non-prompt J/ψ", from the decay of b-quarks, are measured. The prompt can study sequential melting and the non-prompt can study b-quark suppression [20]. The left panels of figure 8 shows the R_{AA} as a function of p_T for prompt J/ψ on the left and non-prompt J/ψ on the right. The ratio of the R_{AA} for $\psi(2S)$ to J/ψ ($\rho_{\rm PbPb}^{\psi(2S)/J/\psi}$) is evaluated in the right panels of figure 8. For prompt the ratio is less than unity indicating that the $\psi(2S)$ are more suppressed than the J/ψ. For non-prompt the ratio is consistent with unity, which is expected since b-quark quenching should be the same regardless of which charmonia state forms [21].

$J/\psi \rightarrow \mu^+\mu^-$ and $\psi(2S) \rightarrow \mu^+\mu^-$ charmonia production are also measured in the p+Pb data at 5.02 TeV to look for CNM effects. The $R_{\rm pPb}$ as a function of p_T for prompt J/ψ, shown is the left panel of figure 9, is consistent with unity indicating no CNM effects. The right hand panel of figure 9 shows the ratio of the $R_{\rm pPb}$ for $\psi(2S)$ and J/ψ ($\rho_{\rm pPb}^{\psi(2S)/J/\psi}$) as function of y* which descrease slightly with rapidity [22].

5. Ultra-peripheral collisions

HI collisions are an intense source of photons in ultra-peripheral collisions (UPC) allowing both photo-nuclear dijet production and light-by-light (LbyL) scattering. In HI UPC events, where strong interactions do not play a role, the large electromagnetic field strengths, which scale with Z^2, allows for LbyL scattering ($\gamma\gamma \rightarrow \gamma\gamma$) [23, 24, 25]. A measurement of the cross section for LbyL scattering in UPC events in Pb+Pb data at 5.02 TeV is shown in the left panel of figure 10.

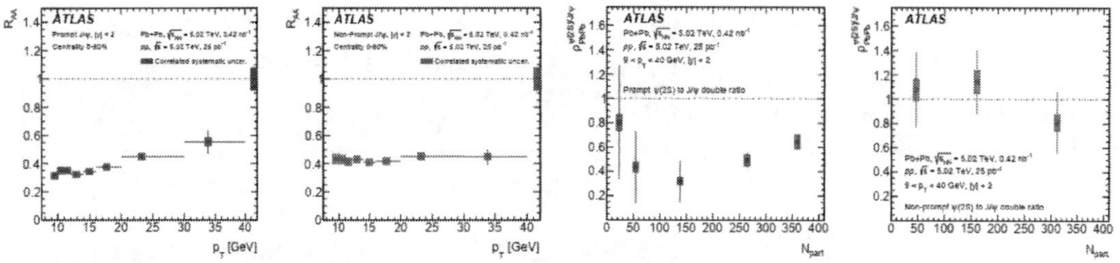

Figure 8. The R_{AA} as a function of p_T for the prompt J/ψ (left-most) and non-prompt J/ψ (left center) for 0–80% centrality. The R_{AA} ratio for $\psi(2S)$ to J/ψ, as a function of N_{part}, for prompt meson production (right center) and non-prompt meson production (right-most) [21].

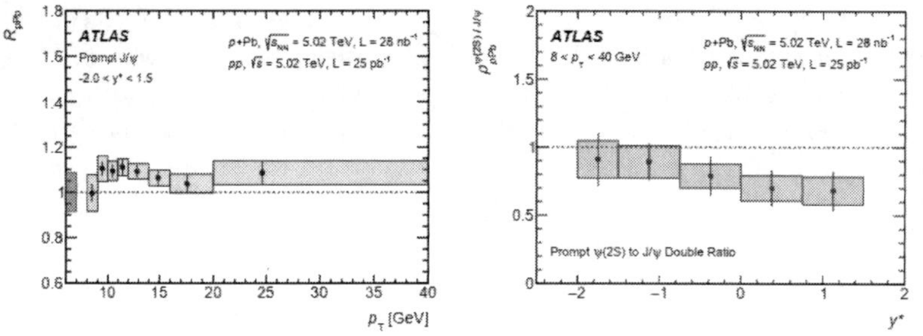

Figure 9. The left panel shows R_{pPb} as a function of p_T for prompt J/ψ in 0–80% centrality. The right panel shows the R_{pPb} ratio for prompt $\psi(2S)$ to J/ψ, as a function of y* [22].

The excess in the data is consistent with MC predictions of LbyL scattering. The cross section is measured to be 70±24 (stat.)±17 (syst.) nb, which is consistent with predicted values [26].

Photo-nuclear dijets in UPC events can study the nPDF at low x [27]. These events are found by looking for a rapidity gap in the main detector since no significant particle production in the rapidity region between the dijet system and the nucleus is expected. Also, zero neutrons in the ATLAS zero degree calorimeters (ZDC) on one side and one or more neutrons on the other side is required because the photon breaks up one nucleus while the other stays intact. A measurement of the cross section for photo-nuclear jet production in UPC events in Pb+Pb data at 5.02 TeV is shown in the right panel of figure 10 as a function of $x_A = (m_{jets}/\sqrt{s})e^{-y_{jets}}$, where m_{jets} and y_{jets} are the combined mass and rapidity of all the jets in the event. It is shown in $H_T = \Sigma_i p_{Ti}$ intervals, where i runs over all the measured jets in an event. It is compared to Pythia+STARlight [28] and is shown to be mostly consistent [29].

6. Conclusion

A variety of new measurements from ATLAS at the LHC in Pb+Pb, p+Pb, and pp collisions have been shown. Measurements of EW bosons show no effect from the medium in Pb+Pb collisions thus indicating an understanding of the nuclear geometry. Measurements of jets show jet suppression up to a TeV in energy, jet imbalance in both photon+jet and dijet systems, and modifications of the jet internal structure in both photon+jets and inclusive jets. J/ψ and $\psi(2S)$ production show evidence of sequential melting. Finally, new data from UPC collisions allow for direct photon and photo-nuclear production measurements.

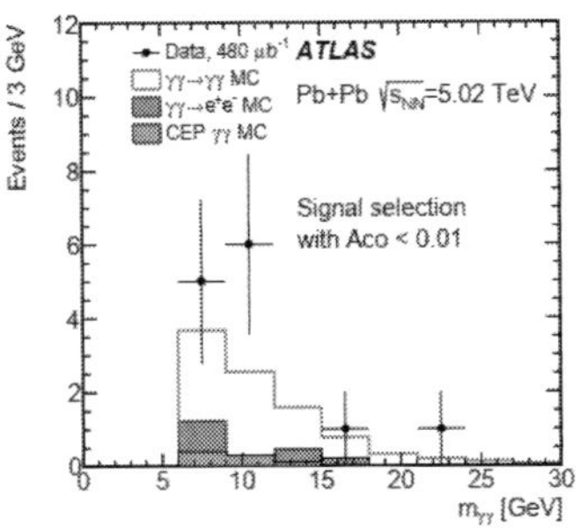

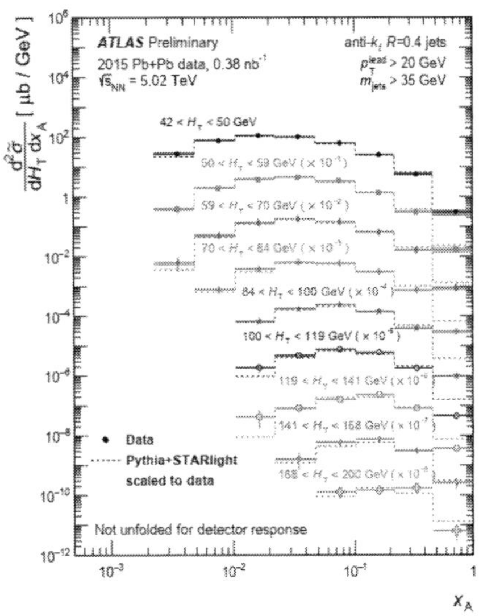

Figure 10. The left panel shows the diphoton invariant mass distribution for the data (points) and MC predictions (histograms, see reference) [26]. The right panel shows the photo-nuclear jet double–differential cross-section for as a function of x_{A} for different intervals of H_{T} compared to PYTHIA+STARLIGHT [28] (dashed lines) [29].

References

[1] Andronic A and Braun-Munzinger P 2004 *Lect. Notes Phys.* **652** 35
[2] ATLAS Collaboration 2008 *JINST* **3** S08003
[3] Miller M, Reygers K, Sanders S and Steinberg P 2007 *Ann. Rev. Nucl. Part. Sci.* **57** 205
[4] DAgostini G 1995 *Nucl. Instrum. Meth.* A **362** 487
[5] Adye T 2011 arXiv: 1105.1160 [physics.data-an]
[6] ATLAS Collaboration 2017 ATLAS-CONF-2017-010, http://cdsweb.cern.ch/record/2244821
[7] ATLAS Collaboration 2016 ATLAS-CONF-2016-107, https://cds.cern.ch/record/2220375
[8] ATLAS Collaboration 2017 ATLAS-CONF-2017-067, http://cdsweb.cern.ch/record/2285571
[9] Mehtar-Tani Y, Milhano J and Tywoniuk K 2013 *Int. J. Mod. Phys.* A**28** 1340013
[10] Qin G and Wang X 2015 *Int. J. Mod. Phys.* E**24** 1530014
[11] ATLAS Collaboration 2018 arXiv: 1805.05635 [nucl-ex]
[12] ATLAS Collaboration 2017 *Phys. Lett.* B**774** 379
[13] ATLAS Collaboration 2016 ATLAS-CONF-2016-110, http://cdsweb.cern.ch/record/2220772
[14] Spousta M and Cole B 2016 *Eur. Phys. J.* C**76** 50
[15] ATLAS Collaboration 2017 arXiv: 1706.02859 [hep-ex]
[16] ATLAS Collaboration 2018 arXiv: 1805.05424 [nucl-ex]
[17] Casalderrey-Solana J, Gulhan D, Milhano G, Pablos D and Rajagopal K 2017 *JHEP* C**03** 135
[18] ATLAS Collaboration 2017 ATLAS-CONF-2017-074, https://cds.cern.ch/record/2285812
[19] Matsui T and Satz H 1986 *Phys. Lett.* B**178** 416
[20] Brambilla N et al. 2011 *Eur. Phys. J.* C**71** 1534
[21] ATLAS Collaboration 2018 arXiv: 1805.04077 [nucl-ex]
[22] ATLAS Collaboration 2018 *Eur. Phys. J.* B**78** 171
[23] Bertulani A and Baur G 1988 *Phys. Rep.* **163** 299
[24] Natale A Roldao C and Carneiro J 2002 *Phys. Rev.* C**65** 014902
[25] Sengul M et al. 2016 *Eur. Phys. J.* C**76** 428
[26] ATLAS Collaboration 2017 *Nature Physics* **13** 852
[27] Strikman M, Vogt R and White S 2006 *Phys. Rev. Lett.* **96** 082001
[28] Klein S, Nystrand J, Seger J, Gorbunov Y and Butterworth J 2017 *Comput. Phys. Commun.* **212** 258
[29] ATLAS Collaboration 2017 ATLAS-CONF-2017-011, http://cdsweb.cern.ch/record/2244822

IOP Conf. Series: Journal of Physics: Conf. Series **1070** (2018) 012019 doi:10.1088/1742-6596/1070/1/012019

Studying jet formation in the medium with the help of correlation and jet shape observables

Eliane Epple for the ALICE Collaboration
Yale University, New Haven, Connecticut, USA
E-mail: eliane.epple@yale.edu

Abstract. The ALICE collaboration at CERN has dedicated a significant part of its physics program to study jet formation and modification in the hot and dense environment formed in heavy-ion collisions. These proceedings will summarize the latest insights we have gained about this process using data collected during Pb–Pb collisions at $\sqrt{s_{\rm NN}} = 2.76\,{\rm TeV}$. In particular π^0-hadron correlations, jet mass measurements, and jet shapes of small-radius jets show a significant influence of the hot and dense medium on the measured jet fragmentation process.

1. Introduction

In recent years the qualitative investigation of jet modification in terms of jet suppression, such as $R_{\rm AA}$ measurements of individual high-$p_{\rm T}$ particles (with $p_{\rm T}$ being their transverse momentum) or reconstructed jets, has been succeeded by more differential observables that dissect the jet structure itself [1, 2]. Examples of such observables are measurements of jet mass [3, 4], various jet shapes [5], jet fragmentation [6, 7, 8, 9], and jet splitting functions [10]. Through such measurements it was observed that the jets reconstructed in heavy-ion collisions have a modified structure as compared to their counterparts in pp collisions. In particular, observation of the distribution of energy inside of the jet cone showed that soft fragments appear at large angles with respect to the jet axis [11, 12]. Such a redistribution of energy among the jet fragments was confirmed by complementary observables of jet-fragment distributions as a function of the radial distance to the jet axis (r) [12, 13, 14] and the parton momentum fraction carried by the final state hadron (z) [6, 7, 8, 9].

These initial observations motivated researchers to intensify their investigation of jet substructures and the interplay between the pure jet fragmentation modification and the medium response to the presence of the jet. Measurements of the modified jet fragmentation function in Pb–Pb collisions [6] led to the discussion of "survivor bias" caused by a different quenching of quark vs. gluon jets [15].

Future measurements and their theoretical interpretation will have to disentangle the pure impact of a modified jet fragmentation due to the medium from modifications originating from medium recoils. Further, the impact of the underlying jet spectrum (increased number of quark vs. gluon jets in heavy-ion collisions) must be disentangled in the observables in order to understand the influence of the medium to the many effects that modify the measured jet profile.

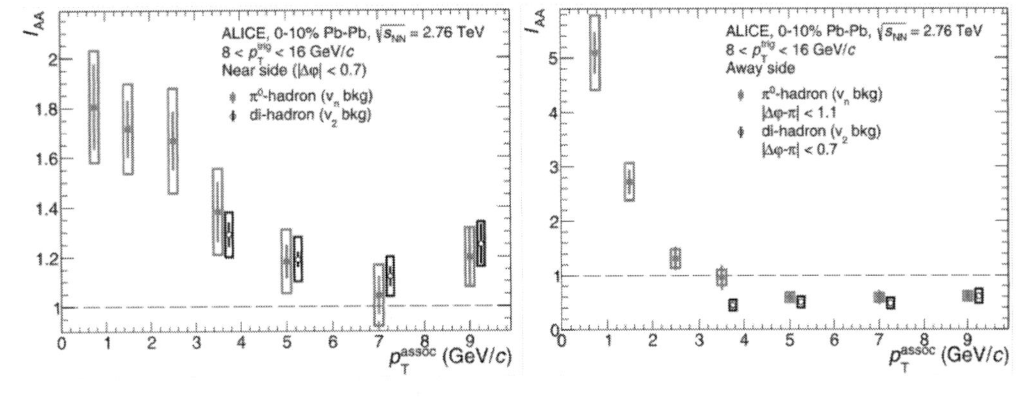

Figure 1. Ratio of associated hadron per trigger yields in Pb–Pb compared to pp collisions, I_{AA} as a function of p_T of the associated hadrons. The left panel presents the ratio for the near-side yield and the right panel for the away-side yield. Both panels show data selected with a high-p_T π^0-trigger with p_T = 8–16 GeV/c and a 0–10% centrality selection for the Pb–Pb data set [16].

2. Correlations of π^0s and hadrons

Jet fragmentation can be studied without necessarily reconstructing the jet as an object itself. Instead, one can rely on a high-p_T trigger particle and study the pseudo rapidity (η) and azimuthal angle (φ) difference ($\Delta\eta$, $\Delta\varphi$) of associated hadrons with respect to the trigger particle. This combination of trigger and associated hadrons forms typical 2D correlation functions that can be studied to collect information about the associated hadron per trigger yield or the width of associated hadron distributions in $\Delta\eta$ or $\Delta\varphi$.

One example of such correlation measurements with high-p_T trigger particles is a measurement by ALICE of π^0-hadron correlations in Pb–Pb collisions at $\sqrt{s_{NN}}$ = 2.76 TeV [16]. In this measurement π^0s of p_T = 8–16 GeV/c were correlated with hadrons of p_T = 0.5–10 GeV/c for the 0–10% most central collisions. The correlation function was corrected for acceptance effects and background. The associated hadron yield was integrated on the near ($|\Delta\varphi| < 0.7$) and away ($|\Delta\varphi - \pi| < 1.1$) side. These integrated associated yields were evaluated for different p_T intervals. To study medium effects on the measured associated yield, which is connected to the modification of the jet fragmentation function, the associated yield in pp collisions was compared to Pb–Pb collisions, at the same collision energy. Figure 1 shows the ratio of these two yields as a function of p_T of the associated hadrons. The ratio is shown for the near side (left) and the away side (right). While there is a general enhancement of associated hadrons on the near side (the jet in which the π^0-trigger is produced) there is a more differential modification on the away side (the jet balancing the jet with the π^0-trigger). This enhancement of the near-side jet yield is not yet fully understood. It could be due to a modified fragmentation function, the medium response, or a different underlying parton sample that fragments. The away side, on the other hand, shows a depletion of high momentum particles and an enhancement of low momentum particles in the balancing jet. This picture is consistent with other measurements that show a transition of depletion to enhancement at a p_T of the associated hadrons of about 3.5 GeV/c [17].

More properties of such correlation observables are currently studied in Pb–Pb and p–Pb collisions at $\sqrt{s_{NN}}$ = 5.02 TeV and will be available in the future for comparisons to the previously published results at the lower collision energy.

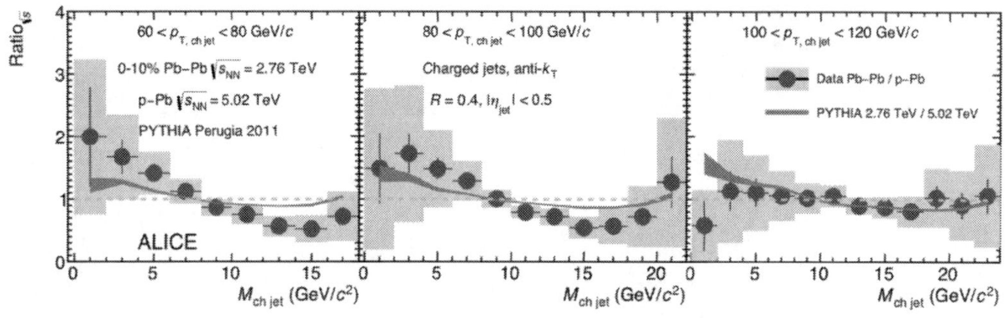

Figure 2. Ratio of the reconstructed jet mass for two collision systems: 0–10% most central Pb–Pb collisions at $\sqrt{s_{NN}} = 2.76$ TeV, and minimum bias p–Pb collisions at $\sqrt{s_{NN}} = 5.02$ TeV. The results are shown for three jet-p_T ranges. The measured ratios are compared to ratios of PYTHIA simulations at the two collision energies to illustrate the effect induced by the energy difference [3].

3. Jet mass measurements

ALICE has published the first jet mass measurement of charged jets produced in Pb–Pb and p–Pb collisions [3]. The jet mass is related to the virtuality of the parton that originates the jet shower [18]. There are several effects that may compete in modifying the virtuality of the parton traversing the hot and dense phase. The interaction of propagating partons with the hot nuclear matter might increase their virtuality, which in turn could increase the measured mass; energy loss experienced by the partons could decrease the mass; significant jet profile broadening could shift constituents out of the jet cone radius and thus decrease the mass; and the medium response to the jet could add particles to the jet which would increase the reconstructed jet mass [3].

The measurement was carried out for charged jets with resolution parameter of $R = 0.4$ and jet momenta of 60–120 GeV/c. The jets were constructed from particles with a track p_T = 0.15–100 GeV/c and required at least one constituent with $p_T >$4 GeV/c. The jets measured in the 0–10% most central Pb–Pb collisions were corrected for background contributions on a jet-by-jet basis and then unfolded back to the particle level. Jets measured in minimum bias p–Pb collisions were corrected for their small background contributions in the unfolding procedure.

The unfolded results of the jet mass in the two collision systems were compared to each other in three jet-p_T intervals. Figure 2 shows the ratio of Pb–Pb to p–Pb data. Since the measurement is performed at two different collision energies this ratio is compared to the respective PYTHIA reference ratio to illustrate the effect induced by the energy difference. The difference in collision energies is expected to cause a slight shift of the ratio to lower jet masses; the observed ratio indicates a shift to lower masses beyond that expectation. In order to gain insight into which of the above-mentioned effects dominates the jet mass shift, a comparison to theoretical models is necessary. Figure 3 summarizes a comparison of the measured jet mass in Pb–Pb collisions to several theoretical predictions. It appears that the inclusion of the medium response in the models (JEWEL w. recoils and Q-PYTHIA) causes a significant shift of the jet mass to higher values, which is not confirmed by the experimental data. Surprisingly, PYTHIA, with its vacuum mass, describes the measurement best. It is obvious that more experimental results on the reconstructed jet mass at different jet radii and momenta are necessary in order to clarify the origin of this discrepancy.

Preliminary results from ATLAS on a related observable also show no significant medium

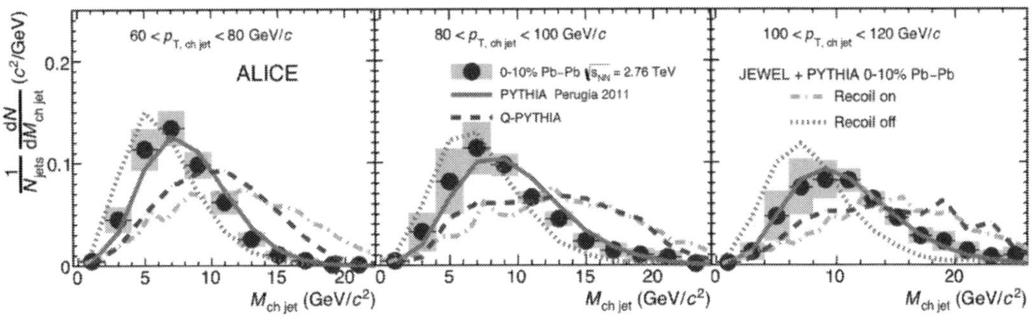

Figure 3. Distributions of the fully-corrected jet mass in 0–10% most central Pb–Pb collisions compared to predictions of PYTHIA simulations with the Perugia 2011 tune and predictions from JEWEL and Q-PYTHIA [3].

effect on the measured jet mass [4]. ATLAS has measured the nuclear modification factor R_{AA} in different jet p_T bins as a function of m/p_T (with m being the jet mass). There is no obvious dependence of R_{AA} on the jet mass in a given p_T bin that would indicate a suppression of higher or lower masses in respect to pp collisions. For higher jet p_T there might be an increased suppression for higher jet masses visible, which would also shift the average mass to lower values, as indicated by ALICE.

4. Jet substructure measurements of small-radius jets

To further study the modification of the jet fragmentation function and the medium recoil influence on reconstructed jet shapes, the ALICE collaboration has measured three additional jet shape observables [5]: the difference between leading and sub-leading track $p_T(LeSub)$, the momentum dispersion ($p_T D$), and the first radial moment or girth (g). They are defined as:

$$LeSub = p_{T,track}^{lead} - p_{T,track}^{sublead}, \quad p_T D = \frac{\sqrt{\sum_{i \in jet} p_{T,i}^2}}{\sum_{i \in jet} p_{T,i}}, \quad g = \sum_{i \in jet} \frac{p_{T,i}}{p_{T,jet}} \Delta R_{jet,i}. \tag{1}$$

$LeSub$, which measures a momentum difference between the two leading tracks, is sensitive to the hardness of the fragmentation process. The momentum dispersion is a measure for the quantity of jet fragments to which the original parton p_T is distributed. In case the original parton p_T is only distributed into very few jet constituents $p_T D$ will approximate 1; in case it is distributed to many jet constituents $p_T D$ will approach 0. The girth of very collimated jets or jets where one only finds very soft fragments at larger radii will have small values around 0 and jets with a very broad shower or hard fragments far away from the center will exhibit larger values of g.

These three jet shape variables were reconstructed in Pb–Pb collisions at $\sqrt{s_{NN}} = 2.76$ TeV and in pp collisions at $\sqrt{s} = 7$ TeV. The jet samples that were used to measure the jet shapes contained charged jets with $R = 0.2$ and $p_{T,jet}^{ch}$ of 40–60 GeV/c. The jets in the Pb–Pb data sample were corrected jet-by-jet for the contribution of the underlying event. The jets in both data sets were unfolded back to the particle level in a two-dimensional unfolding procedure. The jet shape measurement for the pp collision system is presented in Fig. 4. The data were compared to PYTHIA simulations which shows a good agreement to the vacuum jet shape distributions. Assuming no major collision energy dependence on the quality of description, this validates to use PYTHIA as a reference for the shapes measured in Pb–Pb collisions. Any major difference of these measurements from PYTHIA simulations can thus be attributed to medium effects. Figure 5 presents the comparison of the fully corrected small-radius jet shapes, measured in

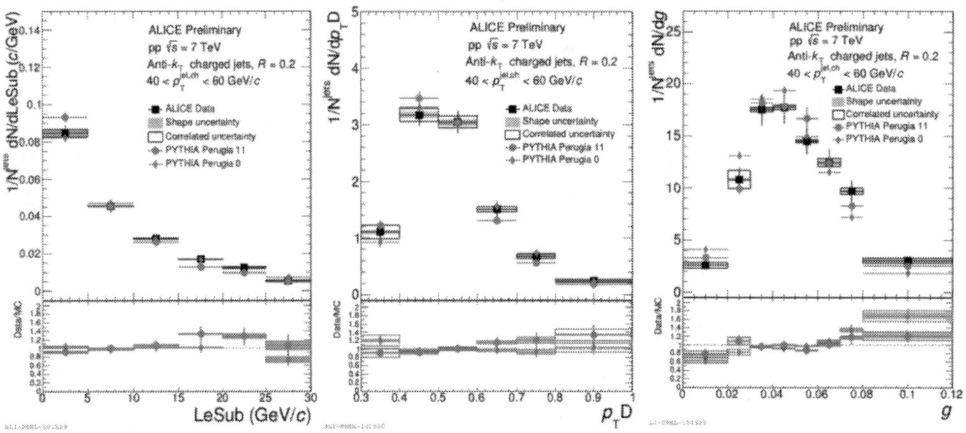

Figure 4. The jet shape distributions for small-radius jets of $R = 0.2$ and $p_{\mathrm{T}}^{\mathrm{ch\,jet}} = 40$–$60$ GeV/c in minimum bias pp collisions with $LeSub$ (left), $p_{\mathrm{T}}D$ (middle), and g (right). The lower panel shows the ratio to PYTHIA simulations with two different tunes [5].

Pb–Pb collisions, to PYTHIA simulations at the same collision energy. A significant difference of the jet shapes between data and simulation is observed. $LeSub$ is overall well described by the vacuum distribution of PYTHIA. The $p_{\mathrm{T}}D$ distribution is shifted to higher values as compared to the vacuum distribution indicating a harder fragmentation with fewer constituents in the jet. The g distribution is shifted to lower values indicating a more collimated jet shower of jet cores in Pb–Pb collisions as compared to pp collisions.

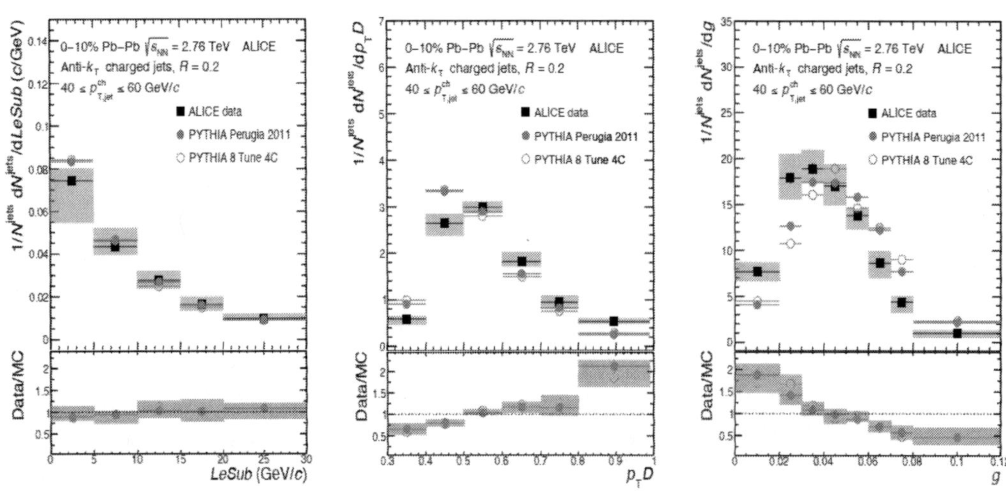

Figure 5. Jet shape distributions for small-radius jets of $R = 0.2$ and $p_{\mathrm{T}}^{\mathrm{ch\,jet}} = 40$–$60$ GeV/c in 0–10% most central Pb–Pb collisions with $LeSub$ (left), $p_{\mathrm{T}}D$ (middle), and g (right). The data are compared to PYTHIA simulations made with the Perugia 11 tune [5].

Overall, the fragmentation of jets reconstructed in Pb–Pb collisions appears to be more quark-like (harder, more collimated fragmentation) than gluon-like (wider, softer fragmenting jet) which could point to a survivor bias of quark jets being measured in Pb–Pb collisions.

References

[1] M. Dasgupta, A. Fregoso, S. Marzani and G. P. Salam, JHEP **1309**, 029 (2013) doi:10.1007/JHEP09(2013)029 [arXiv:1307.0007 [hep-ph]].
[2] L. Asquith *et al.*, arXiv:1803.06991 [hep-ex].
[3] S. Acharya *et al.* [ALICE Collaboration], Phys. Lett. B **776**, 249 (2018) [arXiv:1702.00804 [nucl-ex]].
[4] The ATLAS collaboration [ATLAS Collaboration], ATLAS-CONF-2018-014.
[5] S. Acharya *et al.* [ALICE Collaboration], arXiv:1807.06854 [nucl-ex].
[6] G. Aad *et al.* [ATLAS Collaboration], Phys. Lett. B **739**, 320 (2014) doi:10.1016/j.physletb.2014.10.065 [arXiv:1406.2979 [hep-ex]].
[7] A. M. Sirunyan *et al.* [CMS Collaboration], arXiv:1801.04895 [hep-ex].
[8] M. Aaboud *et al.* [ATLAS Collaboration], arXiv:1805.05424 [nucl-ex].
[9] The ATLAS collaboration [ATLAS Collaboration], ATLAS-CONF-2017-074.
[10] A. M. Sirunyan *et al.* [CMS Collaboration], Phys. Rev. Lett. **120**, no. 14, 142302 (2018) doi:10.1103/PhysRevLett.120.142302 [arXiv:1708.09429 [nucl-ex]].
[11] V. Khachatryan *et al.* [CMS Collaboration], JHEP **1601**, 006 (2016) doi:10.1007/JHEP01(2016)006 [arXiv:1509.09029 [nucl-ex]].
[12] A. M. Sirunyan *et al.* [CMS Collaboration], JHEP **1805**, 006 (2018) doi:10.1007/JHEP05(2018)006 [arXiv:1803.00042 [nucl-ex]].
[13] S. Chatrchyan *et al.* [CMS Collaboration], Phys. Lett. B **730** (2014) 243 doi:10.1016/j.physletb.2014.01.042 [arXiv:1310.0878 [nucl-ex]].
[14] The ATLAS collaboration [ATLAS Collaboration], ATLAS-CONF-2018-010.
[15] M. Spousta and B. Cole, Eur. Phys. J. C **76**, no. 2, 50 (2016) doi:10.1140/epjc/s10052-016-3896-0 [arXiv:1504.05169 [hep-ph]].
[16] J. Adam *et al.* [ALICE Collaboration], Phys. Lett. B **763**, 238 (2016) doi:10.1016/j.physletb.2016.10.048 [arXiv:1608.07201 [nucl-ex]].
[17] J. D. Osborn [PHENIX Collaboration], Nucl. Phys. A **967**, 476 (2017) doi:10.1016/j.nuclphysa.2017.05.045 [arXiv:1704.04457 [nucl-ex]].
[18] A. Majumder and J. Putschke, Phys. Rev. C **93**, no. 5, 054909 (2016) [arXiv:1408.3403 [nucl-th]].

IOP Conf. Series: Journal of Physics: Conf. Series **1070** (2018) 012020 doi:10.1088/1742-6596/1070/1/012020

Inverting the mass hierarchy of jet quenching effects with prompt b-jet substructure

Ivan Vitev

Los Alamos National Laboratory, Theoretical Division, Los Alamos, NM 87545, USA

E-mail: ivitev@lanl.gov

Abstract. In these proceedings, we discuss the role of heavy quark mass on the formation of parton showers. Mass effects are not well understood when parton branching occurs in nuclear matter, such as the quark-gluon plasma. Recently, a theoretically consistent picture of open heavy flavor production in ultra relativistic nuclear collisions has begun to emerge based on effective theories of QCD, such as soft collinear effective theory with Glauber gluons. We show that implementation of in-medium splitting processes containing heavy quarks into next-to-leading order calculations of heavy flavor production leads to larger cross section suppression when compared to traditional energy loss phenomenology. To better constrain the important mass dependence of in-medium splitting functions, we propose a new measurement in relativistic heavy ion collisions, based on a two-prong structure inside a reconstructed heavy flavor jet. In the region of jet transverse momenta where parton mass effects are leading, we predict a unique reversal of the mass hierarchy of jet quenching effects in heavy ion relative to proton collisions. We find that the momentum sharing distribution of prompt b-tagged jets is more strongly modified in comparison to the one for light jets. The work summarized here opens new directions of research on the substructure of heavy flavor jets.

1. Introduction

Understanding the production of heavy flavor and hadronic jets is crucial to test perturbative Quantum Chromodynamics (QCD) and make full use of the data from the world's high energy collider experiments. These questions are even more pertinent today, as the center-of-mass energies $\sqrt{s_{\rm NN}}$ of hadronic and heavy ion collisions continue to increase and heavy quarks, such as charm (c) and bottom (b), are copiously produced in parton showers. The fraction of jets initiated by prompt heavy quarks is also becoming sizable, underscoring the importance of more precise theoretical control on the effects of parton mass. In heavy ion collisions, it was suggested more than a decade ago that these mass effects should be readily observable as reduced energy loss of charm and bottom quarks relative to light quarks [1].

Experimental measurements of D-mesons, B-mesons, and b-quark jets relative to light hadrons and light quark jets in ultrarelativistic nuclear collision have not clearly established the argued "dead cone" effect. This puzzle has stimulated extensive theoretical work, focusing on collisonal energy loss, heavy meson dissociation, and heavy flavor transport. A comprehensive review that covers theory and experimental measurements as of a few years ago is available [2]. More recently, an important effort has been made to quantify the uncertainties in the theoretical model description of open heavy flavor and the extraction of the transport properties of the QGP [3].

IOP Conf. Series: Journal of Physics: Conf. Series **1070** (2018) 012020 doi:10.1088/1742-6596/1070/1/012020

The purpose of these proceedings is to summarize insights on heavy meson production in heavy ion reactions that come from high energy physics and from effective theories of QCD for heavy quarks. A few years ago, while studying the quenching of b-jets, we noticed that a very large fraction of them originate from prompt hard gluons and to some extent from light quarks [4]. Global analysis of D-meson and B-meson has reliably identified gluon fragmentation a major contributor to open heavy flavor production [5, 6]. Soft collinear effective theory has been generalized to include heavy quark masses [7, 8], yielding $\mathrm{SCET_M}$. In what follows, we discuss the formal development of such ideas for heavy ion physics and their applications to phenomenology.

2. Heavy quark splitting functions

An important step toward a unified picture of particle and jet production in hadronic and nuclear collisions was the formulation of an effective theory of QCD in the background of strongly-interacting matter. It allowed for the derivation of a full set of in-medium splitting functions for light quarks and gluons [9], the exploration of higher-order branchings [10] and applications ranging from an evolution approach to the quenching of light hadrons, to the suppression of light jets and their substructure modification [11, 12, 13]. See figure 1 for a schematic of in-medium branching processes. Recently, we performed the next logical step in this line of work by including finite mass effects in the $\mathrm{SCET_G}$ Lagrangian. This enables the effective theory study the of interactions of heavy quarks with the QCD medium [14]. The SCET Lagrangian in the vacuum with quark masses was first derived in [7, 8]. The corresponding theory in the vacuum is commonly referred to as $\mathrm{SCET_M}$. We labelled the new effective field theory $\mathrm{SCET_{M,G}}$.

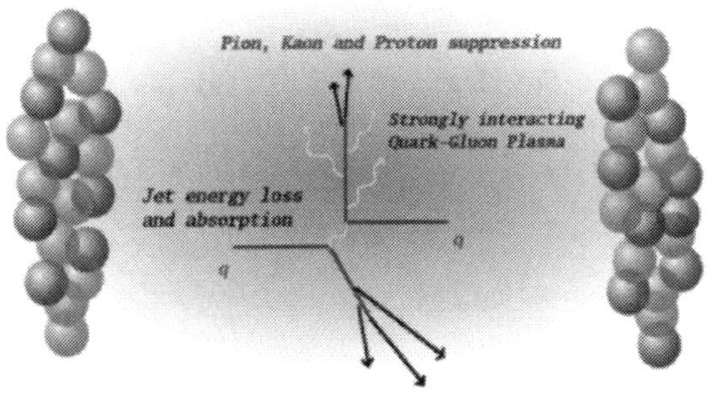

Figure 1. Illustration of parton shower formation in the QGP produced in ultrarelativistic nuclear collisions.

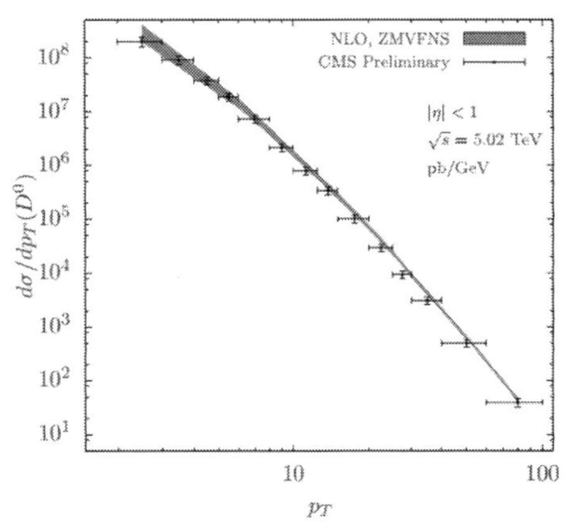

Figure 2. Calculation of D^0 production at $\sqrt{s}$ = 7 TeV within the ZM-VFNS scheme is compared to CMS data [15] from p+p collisions at the LHC.

The recent advances in generalizing $\mathrm{SCET_G}$ to include finite heavy quark masses is the stepping stone for the first calculation of the heavy flavor splitting functions in the QGP. We are interested in the limit $0 < m < p^+$ [7, 8], where m is the heavy quark mass and p^+ is the large lightcone momentum. Starting with an off-shell parton of momentum $[p^+, p^-, \mathbf{0}_\perp]$ that splits into two daughter partons $[zp^+, \mathbf{k}_\perp^2/zp^+, \mathbf{k}_\perp]$ and $[(1-z)p^+, \mathbf{k}_\perp^2/(1-z)p^+, -\mathbf{k}_\perp]$, in the absence of a QCD medium we derive the massive vacuum splitting kernels $Q \to Qg$, $Q \to gQ$, and $g \to Q\bar{Q}$

$$\left(\frac{dN^{\rm vac}}{dzd^2\mathbf{k}_\perp}\right)_{Q\to Qg} = \frac{\alpha_s}{2\pi^2}\frac{C_F}{\mathbf{k}_\perp^2+z^2m^2}\left(\frac{1+(1-z)^2}{z}-\frac{2z(1-z)m^2}{\mathbf{k}_\perp^2+z^2m^2}\right), \quad (1)$$

$$\left(\frac{dN^{\rm vac}}{dzd^2\mathbf{k}_\perp}\right)_{g\to Q\bar{Q}} = \frac{\alpha_s}{2\pi^2}\frac{T_R}{\mathbf{k}_\perp^2+m^2}\left(z^2+(1-z)^2+\frac{2z(1-z)m^2}{\mathbf{k}_\perp^2+m^2}\right), \quad (2)$$

$$\left(\frac{dN^{\rm vac}}{dzd^2\mathbf{k}_\perp}\right)_{Q\to gQ} = \left(\frac{dN^{\rm vac}}{dzd^2\mathbf{k}_\perp}\right)_{Q\to Qg}(z\to 1-z)\,. \quad (3)$$

Here, C_F and C_A are Casimir operators of the fundamental and the adjoint representation of SU(3) and $T_R = 1/2$. It is important to note that the longitudinal momentum fraction z and the transverse momentum $\mathbf{k}_\perp$ dependencies do not factorize. Furthermore, the mass term that regulates the collinear singularity is different for the different splitting kernels in Eqs. (1) - (3). Results for in-medium splitting functions have been obtained to first order in opacity and share similar kinematic structure to the vacuum ones, but include the non-Abelian Landau-Pomeranchuk-Migdal interference effects [14].

Heavy quark splitting functions enter parton showers and the framework of next-to-leading order (NLO) calculations. In the absence of a medium, we use the zero mass variable flavor number scheme (ZM-VFNS) to calculate $D-$meson and $B-$meson production. The NLO cross sections [16] reads

$$\frac{d\sigma_{pp}^H}{dp_Td\eta} = \frac{2p_T}{s}\sum_{a,b,c}\int_{x_a^{\rm min}}^1\frac{dx_a}{x_a}f_a(x_a,\mu)\int_{x_b^{\rm min}}^1\frac{dx_b}{x_b}f_b(x_b,\mu)\int_{z_c^{\rm min}}^1\frac{dz_c}{z_c^2}\frac{d\hat{\sigma}_{ab}^c(\hat{s},\hat{p}_T,\hat{\eta},\mu)}{dvdz}D_c^H(z_c,\mu), \quad (4)$$

where $\sum_{a,b,c}$ stands for a sum over all the parton flavors including light and heavy quarks and gluons. Results for D^0 meson production at the LHC are shown in figure 2 and compared to CMS measurements [15] at the LHC. It is important to note that $\sim 50\%$ of open heavy flavor mesons at these energies are produced by gluon fragmentation. This leads to significantly larger than naively expected quenching of intermediate p_T mesons in the QGP from purely radiative processes.

2.1. Global analysis of heavy flavor fragmentation functions and its implications for heavy ion phenomenology

The critical importance of understanding the mechanism of open heavy flavor production has lead to novel global QCD analysis of charged D^*-meson fragmentation functions (FFs) at NLO accuracy [17]. The key advance in this work is that, in addition to making use of the available data for single-inclusive D^*-meson production in electron-positron annihilation and hadron-hadron collisions, for the first time in-jet fragmentation to D^* data [18] is included. We denote this fit as AKSRV16. We compare this fit to fragmentation functions into D^{*+} mesons to the one obtained in the KKKS08 fit [5] in figure 3. It can be seen that one of the main differences is that our fit returns a significantly larger gluon contribution, when compared to KKKS08 at intermediate values of z. One also notices, that the valence charm FFs are somewhat shifted in z with respect to each other and the height of the peak is different. The larger gluon fragmentation contribution to open charm production, however, is the most important from heavy ion physics perspective. It implies even larger quenching of open heavy flavor.

We investigate the implications of AKSRV16 for the D^* cross section in Pb+Pb collisions by adding the one-loop medium correction to the vacuum NLO result. In other words, we have

$$d\sigma_{\rm PbPb}^H = d\sigma_{pp}^{H,\rm NLO} + d\sigma_{\rm PbPb}^{H,\rm med}, \quad d\sigma_{\rm PbPb}^{H,\rm med} = \hat{\sigma}_i^{(0)}\otimes D_i^{H,\rm med}. \quad (5)$$

Here, $d\sigma_{pp}^{H,\mathrm{NLO}}$ is the NLO cross section in the vacuum, and $d\sigma_{\mathrm{PbPb}}^{H,\mathrm{med}}$ is the one-loop medium correction. We emphasize that $d\sigma_{\mathrm{PbPb}}^{H,\mathrm{med}}$ is negative, which leads to the quenching of the inclusive hadron cross section in heavy-ion collisions. Phenomenological results for 0-7.5% central Pb+Pb reactions at the LHC are shown in figure 4 and compared to ALICE D-meson measurements at $\sqrt{s_{NN}} = 2.76$ TeV [19]. The NLO calculation with the new AKSRV16 fragmentation functions gives noticeably larger suppression than the one with the KKKS08 FFs at intermediate and low p_T, and leaves little room for additional suppression effects.

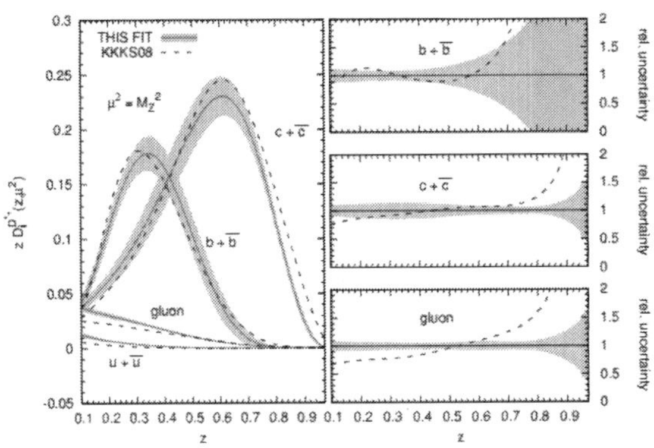

Figure 3. The fragmentation functions $zD_i^{D^{*+}}(z, \mu^2)$ at scale $\mu^2 = M_Z^2$ (solid red lines) along with the obtained uncertainty estimates (shaded bands). The right panels give the ratios of our uncertainty estimates (shaded bands) and the KKKS08 fit relative to our best fit for the $b + \bar{b}$, $c + \bar{c}$ and the gluon FF, respectively.

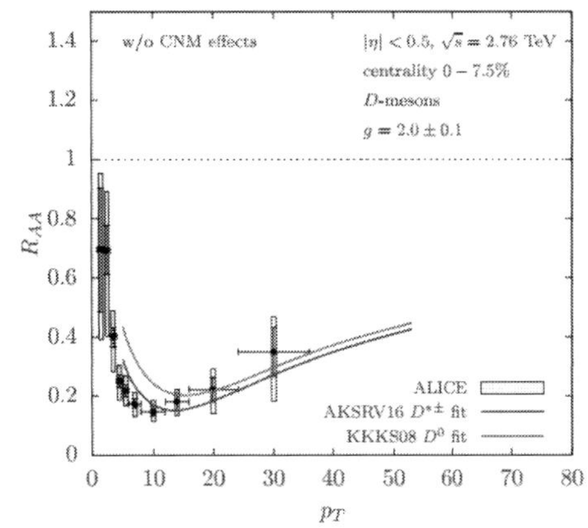

Figure 4. The effect of the new AKSRV16 fit on D-meson quenching in heavy ion collisions at the LHC. Comparison is shown to ALICE D-meson suppression in 0-7.5% Pb+Pb collisions at $\sqrt{s_{NN}} = 2.76$ TeV [19].

3. Heavy jet substructure

To place stringent constraints on mass effects in parton showers, it is necessary to identify new experimental observables [20]. The distribution of the two leading subjets inside a reconstructed jet [21] is particularly suitable, since it is related to the $1 \to 2$ QCD splitting functions in the vacuum [22] and in QCD matter [23]. The jet momentum sharing variable is defined as follows

$$z_g = \frac{\min(p_{T1}, p_{T2})}{p_{T1} + p_{T2}}, \quad z_g > z_{\mathrm{cut}} \left(\frac{\Delta R_{12}}{R}\right)^\beta . \tag{6}$$

In Eq. (6), p_{T1} and p_{T2} are the transverse momenta of the subjets and soft bremsstrahlung is eliminated through the minimum z_g requirement. ΔR_{12} is the distance between two subjets and R is the radius of the original jet. In Ref. [24] and simulations of heavy flavor splitting functions were performed using a Monte Carlo event generator.

We are interested a in different aspect of heavy quark splitting functions – the precise effect of mass on parton shower formation. To prepare for the jet splitting function calculation in heavy ion collisions, we start with the vacuum case, denote by $j \to i\bar{i}$ the parton branchings and define $r_g = \theta_g R$. The θ_g and z_g distribution for parton j, after soft-drop grooming at the lowest non-trivial order is

$$\left(\frac{dN^{\mathrm{vac}}}{dz_g d\theta_g}\right)_j = \frac{\alpha_s}{\pi} \frac{1}{\theta_g} \sum_i P_{j\to i\bar{i}}^{\mathrm{vac}}(z_g) . \tag{7}$$

When the splitting probability becomes large, resummation can be performed to modified leading-logarithmic (MLL) accuracy and the resummed distribution for a j-type jet, initiated by a massless quark or a gluon, is given by

$$\frac{dN_j^{\text{vac,MLL}}}{dz_g d\theta_g} = \sum_i \left(\frac{dN^{\text{vac}}}{dz_g d\theta_g}\right)_{j\to i\bar{i}} \underbrace{\exp\left[-\int_{\theta_g}^{1} d\theta \int_{z_{\text{cut}}}^{1/2} dz \sum_i \left(\frac{dN^{\text{vac}}}{dz d\theta}\right)_{j\to i\bar{i}}\right]}_{\text{Sudakov Factor}} . \tag{8}$$

The normalized joint probability distribution in the domain $z_{\text{cut}} \le z_g \le 1/2$, $0 \le \theta \le 1$ can be straightforwardly obtained.

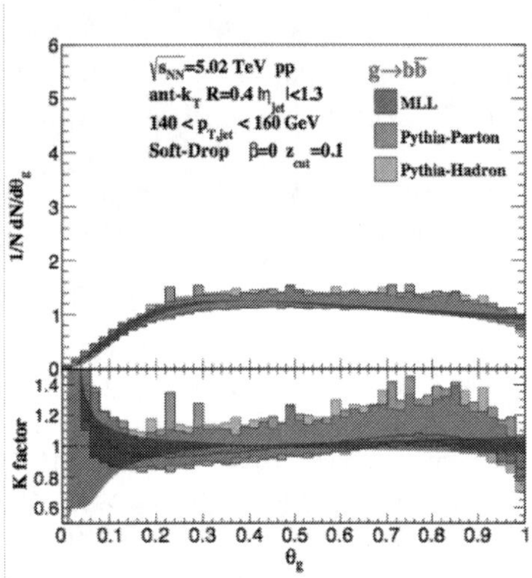

Figure 5. The MLL calculation of the angular distribution of subjets is shown versus Pythia parton level simulations, as well as simulations that include non-perturbative hadronization effects.

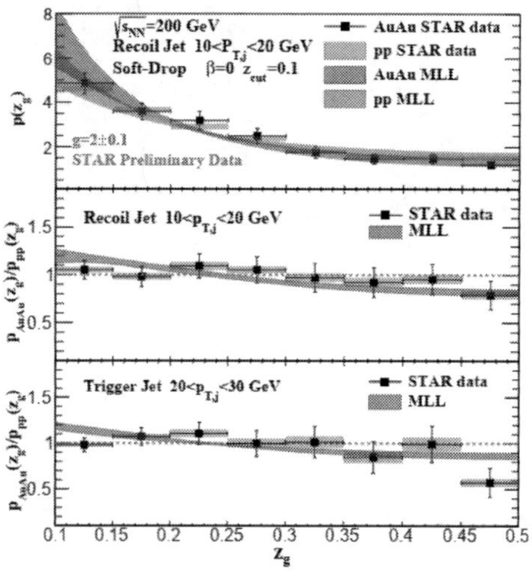

Figure 6. Upper panel: the theoretical predictions for the z_g distribution of recoil jets to the preliminary STAR data [25]. The middle and bottom panels show the predictions and measurements for the modification of recoil and trigger jets, respectively.

Eq. (8) can be extended to the case of prompt heavy flavor jet splitting, such as $b \to bg$ or $c \to cg$, in a straight forward way. For gluon splitting into heavy quark pairs greater care must be taken for the probability function and it is defined as

$$p(\theta_g, z_g)|_{g\to Q\bar{Q}} = \frac{\left(\frac{dN^{\text{vac}}}{dz_g d\theta_g}\right)_{g\to Q\bar{Q}} \Sigma_g(\theta_g)}{\int_0^1 d\theta \int_{z_{\text{cut}}}^{1/2} dz \left(\frac{dN^{\text{vac}}}{dz d\theta}\right)_{g\to Q\bar{Q}} \Sigma_g(\theta)} . \tag{9}$$

Here, $\Sigma_g(\theta_g)$ is the Sudakov factor for gluon evolution and it exponentiates all the possible contributions from gluon splitting, such as $g \to gg$ and $g \to q\bar{q}$. Thus, MLL resummation can change significantly the predictions for the $g \to Q\bar{Q}$ channel relative to the leading order (LO) results. Finally, the distribution needs to be convolved with the quark and gluon jet production cross section, for which we use MADGRAPH5_AMC@NLO. Our results for the $g \to b\bar{b}$ channel, are shown in figure 5. The agreement between the theory framework and the Pythia simulations is at the 5% level, as seen from the bottom panel of the figure, and non-perturbative effects are small. In contrast, for such channels the lowest order calculation shows $\mathcal{O}(100\%)$ deviations.

In the presence of a QCD medium, we replace the vacuum splitting functions by the full splitting kernels, as rigorously derived in [9, 14]

$$\left(\frac{dN^{\rm full}}{dzd^2\mathbf{k}_\perp}\right)_j = \left(\frac{dN^{\rm vec}}{dzd^2\mathbf{k}_\perp}\right)_j + \left(\frac{dN^{\rm med}}{dzd^2\mathbf{k}_\perp}\right)_j . \tag{10}$$

To reduce any model dependence, we evaluate the QGP-induced component in the hydrodynamic background used to describe quarkonium suppression [26]. Since resummed calculations of jet splitting functions have not been performed in heavy ion collisions prior to our work [20], we start with comparison to light jets at RHIC. Results are shown in figure 6 for both the absolute momentum sharing distributions and their modification in Au+Au relative to p+p collisions. We have compared theory to STAR preliminary measurements [25] for both tagged and recoil jets and observe good agreement.

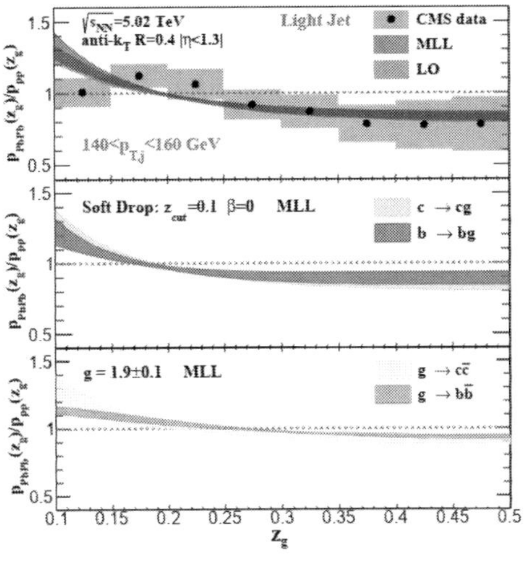

Figure 7. The modification of the jet splitting functions in 0-10% central Pb+Pb collisions at $\sqrt{s_{\rm NN}} = 5.02$ TeV for the p_T bin $140 < p_{T,j} < 160$ GeV. The upper panels compare the LO and MLL predictions to CMS light jet substructure measurements [27]. The middle and lower panels present the MLL modifications for heavy flavor tagged jet - the $Q \to Qg$ and $\to Q\bar{Q}$, respectively.

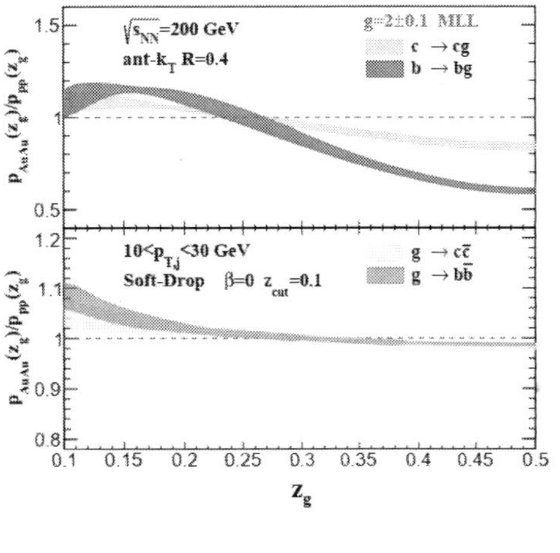

Figure 8. The modifications of the splitting functions for heavy flavor tagged jet is shown for $\sqrt{s_{\rm NN}} = 200$ GeV Au+Au collisions. An important feature is the strong quenching effects for prompt b-jets contrasted by the lack of QGP-induced modification for the $g \to Q\bar{Q}$ splitting.

We now move to the substructure of heavy flavor jets at the LHC, with first results shown in figure 7 for an illustrative kinematic range $140 < p_{T,j} < 160$ GeV in 0-10% central Pb+Pb collisions at $\sqrt{s_{\rm NN}} = 5.02$ TeV. The top panel confirms the good description of the CMS splitting function modification measurements [27], in this case anti-k_T jets with $R = 0.4$ and $|\eta| < 1.3$ with the two subjets separated by $\Delta R_{12} > 0.1$. Predictions for the modification of prompt b-jet and c-jet substructure are given in the middle panel, which show that jet quenching effects for $p(z_g)$ are comparable to that of light jets in this high $p_T \sim 100$ GeV region. This is not

IOP Conf. Series: Journal of Physics: Conf. Series **1070** (2018) 012020 doi:10.1088/1742-6596/1070/1/012020

unexpected, since mass effect slowly vanishes with increasing jet energy. The bottom panel of figure 7 illustrates QGP effects for the $g \to Q\bar{Q}$ channel, where they are somewhat smaller for $g \to b\bar{b}$ in comparison to the other splitting functions.

The most important finding from our recent work [20] is that by going to lower jet transverse momenta, we can enter a regime where the leading mass dependence comes from the propagator in the splitting kernels. The relevant condition reads $\mathbf{k}_\perp^2 \ll z_g^2 m^2$, $\mathbf{k}_\perp^2 \ll m^2$, $\mathbf{k}_\perp^2 \ll (1-z_g)^2 m^2$ and in the medium induced splitting functions we typically have 2 such propagators. This leads to a unique dependence of the jet momentum sharing distributions in heavy ion collisions, which is different for light jets, and the different heavy flavor jet channels. Despite the tremendous importance of open heavy flavor production for heavy ion collisions, our theoretical expectations so far have been limited to the energy loss hierarchy $\Delta E_b^{\rm rad} < \Delta E_c^{\rm rad} < \Delta E_{u,d}^{\rm rad} < \Delta E_g^{\rm rad}$ from radiatve processes in the QGP. In contrast, we can predict analytically that the substructure modification will be the largest for prompt heavy flavor subjets, followed by the one for light jets. If both subjets are heavy, corresponding to $g \to Q\bar{Q}$, we expect no modification. This unique inversion of the mass hierarchy of jet quenching effects can be tested by experiments in the near future. Numerical results for RHIC energies and momentum sharing distribution ratios for heavy flavor tagged jets in Au+Au to p+p collisions at $\sqrt{s_{\rm NN}} = 200$ GeV are presented in figure 8 for $10 < p_{T,j} < 30$ GeV jets. These corroborate the analytic expectations, show that the magnitude of the effects is large, and provide new and promising ways to pin down the mass dependence of in-medium parton showers.

4. Conclusions

To summarize, we focused here on the effects of heavy quark mass on parton shower formation in the QGP, and the production of open heavy flavor and heavy flavor jets in general. The recent results that we report are important steps toward the much needed development of a unified theory that puts the description of jets and heavy flavor production in hadronic and heavy ion collisions on the same footing. These advances are linked to development of modern effective theories of QCD in the environment of strongly interacting matter, such as soft collinear effective theory with Gluaber gluons. SCET$_{\rm G}$ has been extended to open heavy flavor, leading to the first consistent NLO calculation to of $D-$meson and $B-$meson suppression in heavy ion collisions. The fact that half of open heavy flavor production, especially at LHC energies, comes from gluon fragmentation, motivated a new global analysis of D^* FFs that includes for the first time novel results on hadron production inside reconstructed jets. Our results show that the fraction of open heavy flavor mesons that come from the decay of a prompt gluon is even larger than previously anticipated and their suppression through purely radiative processes in the QGP exhausts the observed magnitude of jet quenching.

We also looked for the most promising observables to accurately constrain the role of parton mass in the formation of parton showers. To this end, we presented the first resummed calculations of the soft-dropped momentum sharing distributions in heavy ion collisions. For light jets, the MLL result for this observable in Au+Au and Pb+Pb reactions agrees well with the recent experimental measurements over a wide range of center-of-mass energies. For heavy flavor tagged jets, we demonstrated that jet splitting functions can be used to constrain the still not well understood dead cone effect in the QGP. We also identified the kinematic domain where those effects are important and predicted a unique inversion of the mass hierarchy of jet quenching effects, with the modification of the momentum sharing distribution for prompt b-jets being the largest. The work reported here sets new direction of research on heavy flavor jet substructure in ultrarelativistic nuclear collisions and underscores the importance of future studies of flavor-separated jets [28].

Acknowledgments
This work is supported on US Department of Energy, Office of Science under Contract No. DE-AC52-06NA25396, the DOE Early Career Program, and the LDRD program at LANL.

References
[1] Dokshitzer Y L and Kharzeev D E 2001 *Phys. Lett.* **B519** 199–206 (*Preprint* hep-ph/0106202)
[2] Andronic A *et al.* 2016 *Eur. Phys. J.* **C76** 107 (*Preprint* 1506.03981)
[3] Beraudo A *et al.* 2018 (*Preprint* 1803.03824)
[4] Huang J, Kang Z B and Vitev I 2013 *Phys. Lett.* **B726** 251–256 (*Preprint* 1306.0909)
[5] Kneesch T, Kniehl B A, Kramer G and Schienbein I 2008 *Nucl. Phys.* **B799** 34–59 (*Preprint* 0712.0481)
[6] Kniehl B A, Kramer G, Schienbein I and Spiesberger H 2008 *Phys. Rev.* **D77** 014011 (*Preprint* 0705.4392)
[7] Rothstein I Z 2004 *Phys. Rev.* **D70** 054024 (*Preprint* hep-ph/0301240)
[8] Leibovich A K, Ligeti Z and Wise M B 2003 *Phys. Lett.* **B564** 231–234 (*Preprint* hep-ph/0303099)
[9] Ovanesyan G and Vitev I 2012 *Phys. Lett.* **B706** 371–378 (*Preprint* 1109.5619)
[10] Fickinger M, Ovanesyan G and Vitev I 2013 *JHEP* **07** 059 (*Preprint* 1304.3497)
[11] Kang Z B, Lashof-Regas R, Ovanesyan G, Saad P and Vitev I 2015 *Phys. Rev. Lett.* **114** 092002 (*Preprint* 1405.2612)
[12] Chien Y T and Vitev I 2016 *JHEP* **05** 023 (*Preprint* 1509.07257)
[13] Kang Z B, Ringer F and Vitev I 2017 *Phys. Lett.* **B769** 242–248 (*Preprint* 1701.05839)
[14] Kang Z B, Ringer F and Vitev I 2017 *JHEP* **03** 146 (*Preprint* 1610.02043)
[15] Collaboration C (CMS) 2016
[16] Jager B, Schafer A, Stratmann M and Vogelsang W 2003 *Phys. Rev.* **D67** 054005 (*Preprint* hep-ph/0211007)
[17] Anderle D P, Kaufmann T, Stratmann M, Ringer F and Vitev I 2017 *Phys. Rev.* **D96** 034028 (*Preprint* 1706.09857)
[18] Aad G *et al.* (ATLAS) 2012 *Phys. Rev.* **D85** 052005 (*Preprint* 1112.4432)
[19] Grelli A (ALICE) 2013 *Nucl. Phys.* **A904-905** 635c–638c (*Preprint* 1210.7332)
[20] Li H T and Vitev I 2017 (*Preprint* 1801.00008)
[21] Larkoski A J, Marzani S and Thaler J 2015 *Phys. Rev.* **D91** 111501 (*Preprint* 1502.01719)
[22] Larkoski A, Marzani S, Thaler J, Tripathee A and Xue W 2017 *Phys. Rev. Lett.* **119** 132003 (*Preprint* 1704.05066)
[23] Chien Y T and Vitev I 2017 *Phys. Rev. Lett.* **119** 112301 (*Preprint* 1608.07283)
[24] Ilten P, Rodd N L, Thaler J and Williams M 2017 *Phys. Rev.* **D96** 054019 (*Preprint* 1702.02947)
[25] Kauder K (STAR) 2017 *Nucl. Part. Phys. Proc.* **289-290** 137–140 (*Preprint* 1703.10933)
[26] Aronson S, Borras E, Odegard B, Sharma R and Vitev I 2017 (*Preprint* 1709.02372)
[27] Sirunyan A M *et al.* (CMS) 2017 (*Preprint* 1708.09429)
[28] Chien Y T and Kunnawalkam Elayavalli R 2018 (*Preprint* 1803.03589)

IOP Conf. Series: Journal of Physics: Conf. Series **1070** (2018) 012021 doi:10.1088/1742-6596/1070/1/012021

Azimuthal correlations of the longitudinal structure of the mid-rapidity charged-particle multiplicity in Pb-Pb collisions at $\sqrt{s_{\rm NN}} = 2.76$ TeV with ALICE

Saehanseul Oh (for the ALICE Collaboration)
Wright Laboratory, Yale University
266 Whitney Avenue, New Haven, CT 06511, USA
E-mail: saehanseul.oh@yale.edu

Abstract. Studies of longitudinal correlations of the charged-particle multiplicity in heavy-ion collisions have provided insights into the asymmetry and fluctuations of the initial-state collision geometry. In addition to the expansion of the medium in the transverse direction, commonly quantified using Fourier coefficients (v_n), the initial geometry and resulting longitudinal expansion as a function of azimuthal angle enable us to better understand the full 3-dimensional picture of heavy-ion collisions. In these proceedings, azimuthal correlations of the longitudinal structure of charged-particle multiplicity are reported for Pb-Pb collisions at a nucleon-nucleon center-of-mass energy of 2.76 TeV. The azimuthal angle distribution is divided into regions of in-plane and out-of-plane with respect to the second-order event plane, and the coefficients of Legendre polynomials are estimated from a decomposition of the longitudinal structure of the charged-particle multiplicity at midrapidity ($|\eta| < 0.8$) on an event-by-event basis in each azimuthal region for different centralities. Correlations between the coefficients of various orders in different azimuthal regions are studied and exhibit collective features of longitudinal structure in the azimuthal direction. The results are compared with HIJING and AMPT simulations.

1. Introduction

Collective flow is one of the key features in relativistic heavy ion physics. The initial state anisotropy and its fluctuations are translated into the anisotropy in the final state particle distribution. This has been investigated extensively in the transverse direction via the Fourier decomposition of particle distributions and its coefficients, v_n [1]. Similar studies in the longitudinal direction were proposed recently [2, 3], where results from the ATLAS Collaboration indicate different levels of contributions for short-range and long-range correlations in particle distributions in the longitudinal direction for various collision systems [4]. While these approaches are limited either to the transverse or longitudinal direction, there are studies aimed to extend the analysis to both azimuthal and longitudinal directions, such as measurements of pseudorapidity (η) dependent v_n [5, 6] and the decorrelation of the transverse flow in the longitudinal direction [7]. Using the decomposition in the transverse direction as a basis, they reveal hints of an interplay between transverse and longitudinal dynamics. Such measurements are essential to understand the full 3-dimensional expansion of the medium created in heavy ion collisions.

The ALICE Collaboration has recently devised new observables in order to further investigate

IOP Conf. Series: Journal of Physics: Conf. Series **1070** (2018) 012021 doi:10.1088/1742-6596/1070/1/012021

3-dimensional features using particle correlations. The decomposition in the longitudinal direction is used as a basis, and correlations among longitudinal structures in different azimuthal regions are studied. In the present proceedings, these new observables will be introduced, and the corresponding results in Pb-Pb collisions at a nucleon-nucleon center-of-mass energy of 2.76 TeV at the LHC will be shown.

2. Definition of observables

In a similar manner to that of the decomposition of particle distributions in azimuthal angle (φ) using Fourier harmonics, particle distributions in the longitudinal direction can be decomposed using Legendre polynomials [4]. Adding azimuthal information in the definition, the single particle density ratio within a limited azimuthal (φ) range, $R_{\varphi_i}(\eta)$, is defined by

$$R_{\varphi_i}(\eta) \equiv \frac{1}{\int_{\varphi_i} \mathrm{d}\varphi} \int_{\varphi_i} \frac{N(\varphi,\eta)}{\langle N(\varphi,\eta)\rangle} \mathrm{d}\varphi \,, \tag{1}$$

where the integral range denoted by φ_i is equal to the range of the corresponding φ bin, $N(\varphi,\eta)$ is the multiplicity distribution as a function of (φ,η) in a single event, and $\langle\rangle$ corresponds to the average within the given centrality class. The advantage of using $R_{\varphi_i}(\eta)$ instead of $\mathrm{d}N/\mathrm{d}\eta$ within a limited φ range includes larger sensitivity to the event-by-event variations, a reduced influence from finite tracking efficiencies and related systematic uncertainties. For the range of φ_i in Eq. 1, the azimuthal angle is divided into four regions with respect to the second-order event plane (Ψ_2) in each event, i.e. two in-plane regions (φ_{in}) and two out-of-plane regions (φ_{out}). Two φ_{in} bins cover the φ ranges $\Psi_2 - \frac{\pi}{4} < \varphi < \Psi_2 + \frac{\pi}{4}$ and $\Psi_2 + \frac{3\pi}{4} < \varphi < \Psi_2 + \frac{5\pi}{4}$, and φ_{out} bins cover $\Psi_2 + \frac{\pi}{4} < \varphi < \Psi_2 + \frac{3\pi}{4}$ and $\Psi_2 + \frac{5\pi}{4} < \varphi < \Psi_2 + \frac{7\pi}{4}$.

Within the considered η range $[-Y, Y]$, $R_{\varphi_i}(\eta)$ is then decomposed as

$$R_{\varphi_i}(\eta) \propto 1 + \sum_{n=1} a_n T_n(\eta) \tag{2}$$

with

$$T_n(\eta) = \sqrt{n + \frac{1}{2}} P_n(\eta/Y) \,, \tag{3}$$

where P_n is the n-th order Legendre polynomial and $\sqrt{n+\frac{1}{2}}$ is a normalization factor, such that

$$\int_{-Y}^{Y} T_n(\eta) T_n(\eta)\, \mathrm{d}\eta = 1 \,. \tag{4}$$

The first few terms of Legendre polynomials are $P_0(x) = 1$, $P_1(x) = x$, $P_2(x) = (3x^2 - 1)/2$. The coefficients $a_n(\varphi_i)$ of the decomposition represent the longitudinal structure within a given φ_i range and centrality class.

Once the $a_n(\varphi_i)$ coefficients are extracted in each event, correlations among the coefficients can be used to describe the correlations among the longitudinal structures in different azimuthal regions. These correlations are studied through the conditional a_n, denoted by $a_n(\varphi_i|a_m(\varphi_j))$. $a_n(\varphi_i|a_m(\varphi_j))$ is defined as a mean $a_n(\varphi_i)$ value for given events with $a_m(\varphi_j)$. In other words, once events with a certain a_m value in φ_j bin are selected, the mean a_n in the φ_i bin is estimated for the value of $a_n(\varphi_i|a_m(\varphi_j))$. Various combinations of n, m, (n, m =1, 2, 3, ...) and i, j (i, j =in, out) can be investigated to reveal features of azimuthal correlations of the longitudinal structures, but only $n = m$ cases up to $n = 3$ are discussed in these proceedings. For (φ_i, φ_j) combinations, four groups of conditional a_n are studied:

- $a_n(\varphi_{\mathrm{opp.in}}|a_n(\varphi_{\mathrm{in}}))$
- $a_n(\varphi_{\mathrm{opp.out}}|a_n(\varphi_{\mathrm{out}}))$
- $a_n(\varphi_{\mathrm{out}}|a_n(\varphi_{\mathrm{in}}))$
- $a_n(\varphi_{\mathrm{in}}|a_n(\varphi_{\mathrm{out}}))$

where "opp.in" and "opp.out" correspond to the opposite in-plane and opposite out-of-plane φ bins, respectively.

3. Experimental setup

For the current analysis, ALICE data from the Pb–Pb run at the LHC in 2011 are used. A full description of the ALICE detector and its performance during this period can be found in [8], while a few details with direct relation to this analysis will be explained in this section.

The online event selection for the analysis combined the minimum-bias (MB) trigger and two centrality triggers, namely, central and semi-central triggers. Requiring signals in two scintillator arrays (V0) and Zero Degree Calorimeters (ZDC) in the MB trigger suppressed events from only electromagnetic interactions between the lead ions. Two centrality triggers required additional V0 signals, and selected large numbers of events particularly in central collisions. Events were then classified into centrality classes based on the signal in V0 detectors and 0–5%, 5–10%, 10–20%, 20–30%, 30–40%, 40–50% centrality classes were used for the analysis.

As explained in Sec. 2, the second-order event plane, Ψ_2, should first be determined for conditional a_n measurements. Ψ_2 is determined by

$$Q_{2,x} = \sum_{i=1}^{n} w_i \cos(2\varphi_i)\ ,\quad Q_{2,y} = \sum_{i=1}^{n} w_i \sin(2\varphi_i)\ , \tag{5}$$

and

$$\Psi_2 = \frac{1}{n}\tan^{-1}\left(\frac{Q_{2,y}}{Q_{2,x}}\right)\ , \tag{6}$$

using signals from the V0 detectors, which cover $2.8 < \eta < 5.1$ and $-3.7 < \eta < -1.7$. In Eq. 5, index i runs over sectors of V0 detectors, and w_i is the measured signal in sector i. Due to the non-uniform gain of V0 detectors, the calibration for the Ψ_2 estimation was performed using the procedures described in [9–11], such as a gain equalization of the V0 detector signal, a flattening of the Ψ_2 distribution through Q-vector re-centering, twisting, and rescaling. Estimating Ψ_2 with the V0 detectors ensures a sufficient η gap between ranges of Ψ_2 and conditional a_n measurements, as mid-rapidity tracks are used for the conditional a_n. Such a separation allows to avoid auto-correlation biases between Ψ_2 and tracks in the mid-rapidity.

The mid-rapidity charged tracks are reconstructed using clusters of the inner tracking system (ITS) and the time projection chamber (TPC) in order to achieve uniform tracking efficiency [12]. The default track selection required at least 70 space points in the TPC, the ratio of the number of recorded points to the findable points larger than 80%, a distance of closest approach (DCA) to the reconstructed vertex smaller than 2.4 cm in the transverse direction and 3.2 cm in the longitudinal direction, and the maximum fraction of shared clusters in the TPC lower than 40%. In the process of matching TPC tracks with ITS clusters, two approaches were followed: tracks reconstructed with at least one hit in the SPD, and tracks without associated SPD hit but the reconstructed vertex position in the fit. The p_{T} and η ranges of tracks are $0.2 < p_{\mathrm{T}} < 5.0\,\mathrm{GeV}/c$ and $-0.8 < \eta < 0.8$, respectively.

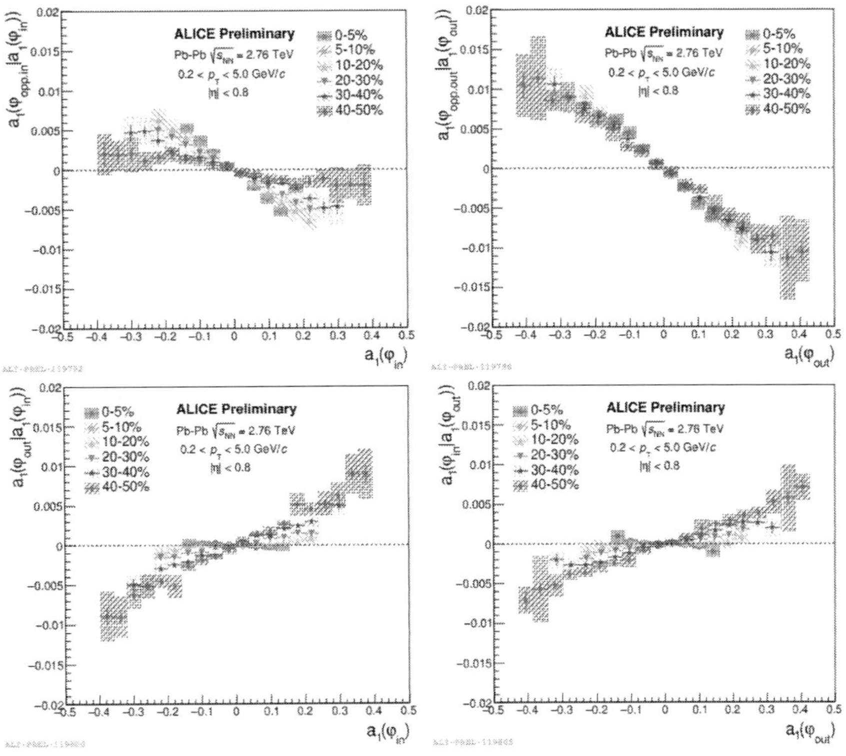

Figure 1. The $a_1(\varphi_{\rm opp.in}|a_1(\varphi_{\rm in}))$ (top left), $a_1(\varphi_{\rm opp.out}|a_1(\varphi_{\rm out}))$ (top right), $a_1(\varphi_{\rm out}|a_1(\varphi_{\rm in}))$ (bottom left), and $a_1(\varphi_{\rm in}|a_1(\varphi_{\rm out}))$ (bottom right) measured with charged particles with $0.2 < p_{\rm T} < 5.0\,{\rm GeV}/c$ and $|\eta| < 0.8$ in different centrality classes.

4. Results

In the measuremnts of the conditional a_n, two correction procedures are applied to take into account detector inefficiencies. First, an event-plane resolution correction accounts for the difference between the true second-order event plane and the measured second-order event plane. The corresponding correction procedure in the flow measurements in the transverse direction is well-established [9]. Another correction accounts for the multiplicity effects due to the limited tracking efficiency in the ITS and TPC. For a given true a_n value, the distribution of measured a_n depends on the charged-particle multiplicity used in the a_n measurement. This dependence is investigated with the Monte Carlo method, which simulates the response in the measured a_n from an input a_n with various multiplicity distributions. Deviations attributed to the difference in the measured multiplicity from the true multiplicity in observables are corrected, accordingly.

For each conditional a_n result, systematic uncertainties are assigned by varying the measurement conditions and correction details. These include centrality estimation methods, detector ranges for the event-plane determination, selected ranges of reconstructed primary vertex coordinate along the beam direction, track selection, functional forms in the event-plane resolution correction, and variationss in the correction for the multiplicity effects. The total systematic uncertainty is evaluated by combining deviations to the values from the nominal measurement in each variation quadratically.

The conditional a_1 of four azimuthal combinations for different centrality classes are shown in Fig. 1. Statistical uncertainties are marked with error bars and systematic uncertainties are

IOP Conf. Series: Journal of Physics: Conf. Series **1070** (2018) 012021 doi:10.1088/1742-6596/1070/1/012021

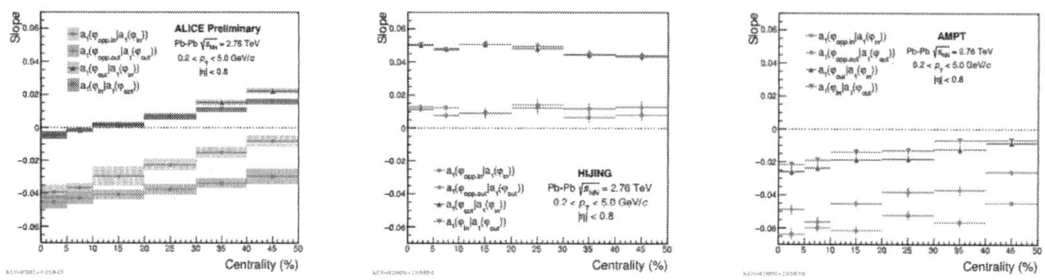

Figure 2. Slopes of conditional a_1 as a function of centrality from ALICE measurements at the LHC (left), HIJING (center), and AMPT (right) simulations.

marked with shaded areas in every figure. These conditional a_1 results are fitted with a linear function, and the extracted slopes are plotted in Fig. 2 along with the quantities obtained with HIJING [13] and AMPT [14] simulations.

The a_1 is a parameter quantifying the forward-backward asymmetry in particle production, and the plots in Fig. 1 and Fig. 2 show a distinctive level of contributions to the longitudinal structure from collective and non-collective correlations depending on the centrality. In general, physics mechanisms studied via geometrical correlations can be classified into two groups, collective and non-collective correlations. Particles originating from non-collective correlations, such as jets and resonances, are localized in a small range of (φ, η) space or opposite φ regions, while particles from collective correlations, i.e. flow, are distributed over a wide range in (φ, η) space. Due to the limited range of non-collective correlations in (φ, η) space, non-collective correlations can contribute to either one or two adjacent φ bins in the current analysis for most cases, as the size of each φ bin is $\pi/2$. For example, fragmented particles from a jet can be included in either a single φ bin or two adjacent φ bins. In the case that the jet axis is located in the positive η region, the corresponding fragmented particles can contribute to the positive a_1 in a given φ bin or two adjacent φ bins simultaneously. More positive values of slopes in $a_1(\varphi_{\rm out}|a_1(\varphi_{\rm in}))$ and $a_1(\varphi_{\rm in}|a_1(\varphi_{\rm out}))$ than those in $a_1(\varphi_{\rm opp.in}|a_1(\varphi_{\rm in}))$ and $a_1(\varphi_{\rm opp.out}|a_1(\varphi_{\rm out}))$ are mostly attributed to the involvement of the non-collective correlations in the observables, and indicate larger contributions in more peripheral classes. Comparison with HIJING results on Fig. 2, where no collective correlations are expected, indicates larger contributions from collective correlations in more central classes.

Collective correlations contribute to the conditional a_1 for all (φ_i, φ_j) combinations. Within the framework of collective correlations, various physics scenarios can result in the negative slopes observed in $a_1(\varphi_{\rm opp.in}|a_1(\varphi_{\rm in}))$ and $a_1(\varphi_{\rm opp.out}|a_1(\varphi_{\rm out}))$. The transverse distribution of the asymmetric number of participating nucleons from two incoming nuclei fluctuates event-to-event, and can be considered as a seed for the asymmetric expansion of the medium in the longitudinal direction. Then the negative slope in conditional a_1 with two opposite φ bins may originate solely from the initial bias in the transverse distribution of the asymmetric number of participating nucleons, although this initial bias may be diluted through the expansion. In the same sense, the difference between $a_1(\varphi_{\rm opp.in}|a_1(\varphi_{\rm in}))$ and $a_1(\varphi_{\rm opp.out}|a_1(\varphi_{\rm out}))$ may attribute to the different expansion rate in in-plane and out-of-plane directions, as the difference is the smallest in the most central class. Also, the momentum conservation of particles created from the expansion of the medium may impact these results. Slopes of conditional a_1 in AMPT are observed to have similar centrality dependence (Fig. 2), but generally shifted in the negative direction. This can be interpreted as a different level of contributions from collective and non-

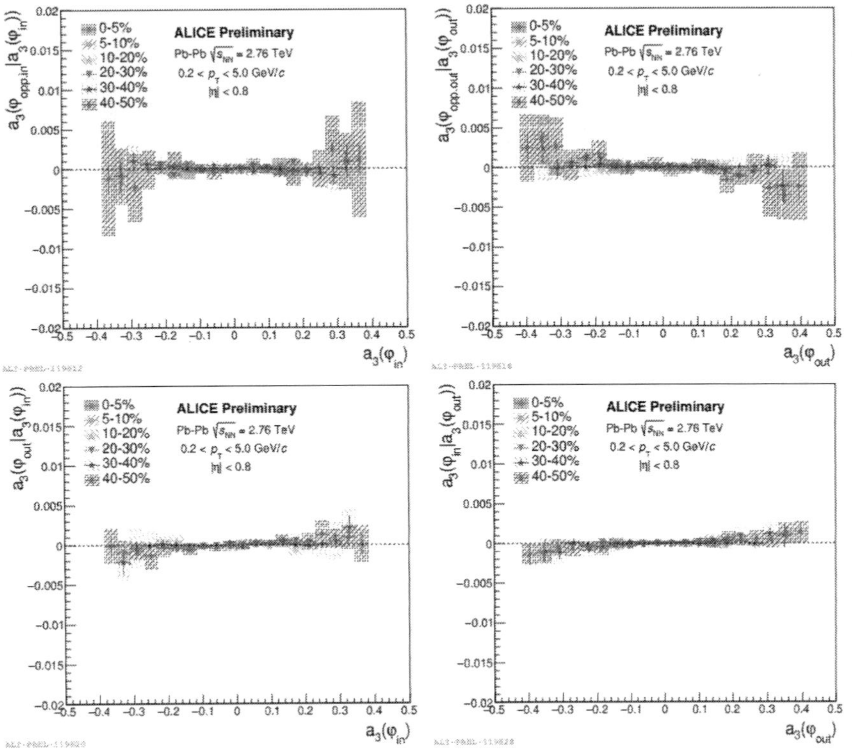

Figure 3. The $a_3(\varphi_{\mathrm{opp.in}}|a_3(\varphi_{\mathrm{in}}))$ (top left), $a_3(\varphi_{\mathrm{opp.out}}|a_3(\varphi_{\mathrm{out}}))$ (top right), $a_3(\varphi_{\mathrm{out}}|a_3(\varphi_{\mathrm{in}}))$ (bottom left), and $a_3(\varphi_{\mathrm{in}}|a_3(\varphi_{\mathrm{out}}))$ (bottom right) measured with charged particles with $0.2 < p_{\mathrm{T}} < 5.0\,\mathrm{GeV}/c$ and $|\eta| < 0.8$ in different centrality classes.

collective correlations to the data.

In [2, 4], a_3 and higher order coefficients are argued to mostly attribute to the non-collective or short-range correlations. The conditional a_3 results for different centrality classes are shown in Fig. 3. The conditional a_3 values remain near 0 in all (φ_i, φ_j) combinations, which indicates independence of a_3 in different φ regions. This is a genuine feature of non-collective correlations, as non-collective correlations can generally contribute to the local region in (φ, η) space. Similar features were found in HIJING and AMPT simulations although they are now shown in these proceedings.

Lastly, the conditional a_2 measurements for different centrality classes are shown in Fig. 4. The a_2 is a parameter quantifying the mid-peripheral asymmetry, and related to the amount of nuclear stopping power or shift of the effective center-of-mass of the collisions [4]. A positive a_2 represents more particles in peripheral regions than the central region, and a negative a_2 represents more particles in the central region than peripheral regions. Figure 4 shows a very distinctive feature of conditional a_2 relative to conditional a_1 or conditional a_3. For all (φ_i, φ_j) combinations, the value of $a_2(\varphi_i|a_2(\varphi_j))$ significantly increases for positive $a_2(\varphi_j)$ in central classes, while it remains near 0 for other $a_2(\varphi_j)$. This indicates that when a positive value of a_2 is observed in a given φ bin in central collisions, values of a_2 in other φ bins tend to be positive. Such collective features disappear in more peripheral classes, and are not observed in HIJING and AMPT simulations.

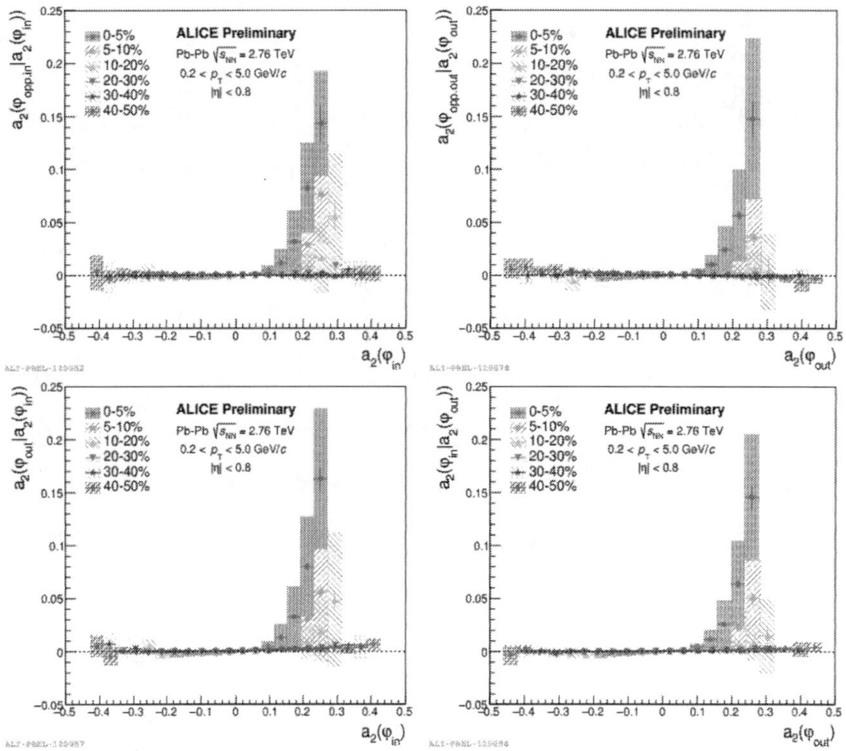

Figure 4. The $a_2(\varphi_{\rm opp.in}|a_2(\varphi_{\rm in}))$ (top left), $a_2(\varphi_{\rm opp.out}|a_2(\varphi_{\rm out}))$ (top right), $a_2(\varphi_{\rm out}|a_2(\varphi_{\rm in}))$ (bottom left), and $a_2(\varphi_{\rm in}|a_2(\varphi_{\rm out}))$ (bottom right) measured with charged particles with $0.2 < p_{\rm T} < 5.0\,{\rm GeV}/c$ and $|\eta| < 0.8$ in different centrality classes.

5. Summary

The ALICE Collaboration reports the first results of azimuthal correlations of longitudinal structure of charged-particle multiplicity in Pb-Pb collisions with newly devised observables, conditional a_n. While previous measurements in the longitudinal direction are limited to exclusively dealing with the pseudorapidity of produced particles, these new measurements extend the scope of longitudinal correlations into a 3-dimensional space. The conditional a_1 results show a different level of contributions from collective and non-collective correlations into a forward-backward asymmetry depending on the centrality. The conditional a_3 results confirm the origin of a_3, which is expected to be due to non-collective correlations. A unique collective feature in central classes is observed in conditional a_2, which indicates that a positive a_2 is preserved over the azimuthal angle in more central classes. Both HIJING and AMPT are not able to reproduce such features. These results represent new constraints for the 3-dimensional heavy-ion collision models, and require further theoretical investigation.

Acknowledgements

This work is supported by the U.S. Department of Energy under grant number DE-SC0004168.

References

[1] Aamodt K *et al.* (ALICE) 2010 *Phys. Rev. Lett.* **105** 252302 (*Preprint* 1011.3914)
[2] Bzdak A and Teaney D 2013 *Phys. Rev.* **C87** 024906 (*Preprint* 1210.1965)

IOP Conf. Series: Journal of Physics: Conf. Series **1070** (2018) 012021 doi:10.1088/1742-6596/1070/1/012021

[3] Bzdak A and Bozek P 2016 *Phys. Rev. C* **93** 024903 (*Preprint* 1509.02967)
[4] Aaboud M *et al.* (ATLAS) 2017 *Phys. Rev.* **C95** 064914 (*Preprint* 1606.08170)
[5] Sirunyan A M *et al.* (CMS) 2017 (*Preprint* 1710.07864)
[6] Aad G *et al.* (ATLAS) 2012 *Phys. Lett* **B707** 330–348 (*Preprint* 1108.6018)
[7] Aaboud M *et al.* (ATLAS) 2018 *Eur. Phys. J.* **C78** 142 (*Preprint* 1709.02301)
[8] Abelev B *et al.* (ALICE) 2014 *Int. J. Mod. Phys.* **A29** 1430044 (*Preprint* 1402.4476)
[9] Poskanzer A M and Voloshin S A 1998 *Phys. Rev.* **C58** 1671–1678 (*Preprint* nucl-ex/9805001)
[10] Selyuzhenkov I and Voloshin S 2008 *Phys. Rev.* **C77** 034904 (*Preprint* 0707.4672)
[11] Massacrier L 2013 *Nuclear Physics* **A910-911** 235–238 (*Preprint* 1208.5401)
[12] Alme J *et al.* 2010 *Nucl. Instrum. Meth.* **A622** 316 – 367 (*Preprint* 1001.1950)
[13] Wang X N and Gyulassy M 1991 *Phys. Rev.* **D44** 3501–3516
[14] Lin Z W, Ko C M, Li B A, Zhang B and Pal S 2005 *Phys. Rev.* **C72** 064901 (*Preprint* nucl-th/0411110)

IOP Conf. Series: Journal of Physics: Conf. Series **1070** (2018) 012022 doi:10.1088/1742-6596/1070/1/012022

Latest Minimum Bias and Underlying Event measurements with the ATLAS Detector

Valentina Maria Martina Cairo – on behalf of the ATLAS Collaboration
State University of New York at Albany
E-mail: valentina.maria.cairo@cern.ch

Abstract. The modelling of Minimum Bias and Underlying Event is a crucial component in the description of soft Quantum Chromodynamics processes. They are both described by multi-parton interaction models, the result of proton collisions containing more than one partonic interaction due to collective and beam remnant effects. Recent studies by the ATLAS Experiment at the LHC aiming at measuring Charged-Particle distributions and the properties of the Underlying Event are presented.

1. Introduction

Hard Quantum Chromodynamics (QCD) events constitute only a tiny fraction of the total proton-proton (*pp*) cross-section, which is dominated by soft QCD events (peripheral processes). While hard QCD processes can be studied by means of perturbative approaches, this is not possible for the soft QCD events. The development of Monte Carlo (MC) event generators began shortly after the discovery of the partonic structure of hadrons and the formalisation of QCD as the theory of strong interactions. Models have to be developed with a set of tunable parameters to describe the hadron-level properties of final states dominated by soft QCD. Inclusive charged-particle and underlying event (UE) measurements in *pp* collisions are the ideal test bed to provide insight into the soft QCD region: they are crucial for the tuning of the MC event generator, essential to understand and correctly simulate any other more complex phenomena and ideal to study tracking performance in the early stage of a new data taking.

2. Charged-particle multiplicity

The measurements of inclusive charged-particle spectra provide insight into the low energy non-perturbative region of QCD. A description of low-energy processes within a perturbative framework is not possible in this regime, thus charged-particle interactions are typically described by QCD-inspired models implemented in MC event generators. Measurements are used to constrain the free parameters of these models. Furthermore, soft processes, arising from additional *pp* interactions per beam crossing (pile-up) at high luminosity may also affect the topologies of events triggered by a specific hard-scattering interaction. An understanding of soft QCD processes is therefore important both in its own right and as a means of reducing systematic uncertainties in measurements of high transverse momentum phenomena. Charged-particle distributions have been measured previously in hadronic collisions at various centre-of-mass energies, see Refs. [1–7] and references therein. This note describes the most recent charged-particle spectra measured by using data collected with the ATLAS detector [8] at the centre-of-mass energy of 13 TeV [9, 10], with a particular emphasis on the tracking-related aspects. Some highlights from the high charged-particle multiplicity regions studied at the 8 TeV [11] centre-of-mass energy are also given. The average primary charged-particle densities at central pseudorapidity are compared to measurements at lower centre-of-mass energies.

2.1. Methodology

The methodology used in the 8 and 13 TeV analyses is similar to that used at lower centre-of-mass energies in ATLAS [1]. The events collected correspond to minimum-bias datasets based on inelastic pp interactions. The term *minimum bias* is taken to refer to trigger and event selections which are as unrestrictive as possible for the pp-induced final state. The data were recorded during special fills with low beam currents and reduced focusing to give a mean number of interactions per bunch crossing below 0.005. This procedure guarantees that the contribution from pile-up in these analyses is negligible. The measurements use tracks from primary charged-particles, corrected for detector effects to the particle level, and presented as inclusive distributions in a fiducial phase space region. Primary charged-particles are defined as charged-particles with a mean lifetime $\tau > 300$ ps, either directly produced in pp interactions or from subsequent decays of directly produced particles with $\tau < 30$ ps. Particles produced from decays of particles with $\tau > 30$ ps, called secondary particles, are excluded. This definition differs from earlier analyses in which charged-particles with a mean lifetime $30 < \tau < 300$ ps were included. Most of these are charged strange baryons and they have been removed due to the low reconstruction efficiency of their decay products and to large variations in the predicted rates which would lead to a significant model dependence of the results presented here.

The following distributions are presented for data and compared to MC predictions:

$$\frac{1}{N_{\rm ev}} \cdot \frac{{\rm d}N_{\rm ch}}{{\rm d}\eta}, \; \frac{1}{N_{\rm ev}} \cdot \frac{1}{2\pi p_{\rm T}} \cdot \frac{{\rm d}^2 N_{\rm ch}}{{\rm d}\eta {\rm d}p_{\rm T}}, \; \frac{1}{N_{\rm ev}} \cdot \frac{{\rm d}N_{\rm ev}}{{\rm d}n_{\rm ch}}, \; \langle p_{\rm T} \rangle \text{ vs } n_{\rm ch},$$

where $p_{\rm T}$ is the track momentum component that is transverse to the beam direction[1], η is the track pseudorapidity, $n_{\rm ch}$ is the number of primary charged particles in an event, $N_{\rm ev}$ is the number of selected minimum bias events, $N_{\rm ch}$ is the total number of primary charged particles in the data sample and $\langle p_{\rm T} \rangle$ is the average $p_{\rm T}$ for a given number of charged-particles[2]. In order to make a more complete study of particle properties in minimum-bias events, results are given for different multiplicity and kinematic selections (referred to as *phase spaces*). In the most inclusive phase spaces, a minimum $n_{\rm ch} \geq 2$ or 1 is required and the primary charged-particle must have $\eta < 2.5$ and $p_{\rm T} > 100$ MeV (referred to as *extended phase space*) or 500 MeV (referred to as *nominal phase space*) , respectively. In the 13 TeV case, the spectra are also measured in a phase space that is common to the ATLAS, CMS [12] and ALICE [13] detectors in order to ease comparison between experiments. For this purpose an additional requirement of $\eta < 0.8$ (referred to as *reduced phase space*) is made for all primary charged-particles with $p_{\rm T} > 500$ MeV and the results can be found in [9].

The PYTHIA 8 [14] (used as a baseline), EPOS [15] and QGSJET-II [16] MC models of inclusive hadron–hadron interactions were used to generate event samples and compare their distributions to data. Different parameter settings in the models are used in the simulation

[1] The ATLAS reference system is a Cartesian right-handed coordinate system, with the nominal collision point at the origin. The anti-clockwise beam direction defines the positive z-axis, while the positive x-axis is defined as pointing from the collision point to the center of the LHC ring and the positive y-axis points upwards. The azimuth angle ϕ is measured around the beam axis, and the polar angle θ is measured with respect to the z-axis. The pseudorapidity is defined as $\eta = -\ln\tan(\theta/2)$.

[2] The factor $2\pi p_{\rm T}$ in the $p_{\rm T}$ spectrum comes from the Lorentz invariant definition of the cross-section in terms of d^3p. Furthermore, the mass-less approximation is used: $y \approx \eta$.

IOP Conf. Series: Journal of Physics: Conf. Series **1070** (2018) 012022 doi:10.1088/1742-6596/1070/1/012022

to reproduce existing experimental data and are referred to as *tunes*. For PYTHIA 8, the A2 [17] tune is based on the MSTW2008LO PDF [18] while the Monash [19] UE tune uses the NNPDF2.3LO PDF [20] and incorporates updated fragmentation parameters, as well as SPS and Tevatron data to constrain the scaling with energy. For EPOS, the LHC [21] tune is used, while for QGSJET-II the default settings of the generator are applied. Detector effects are simulated using the GEANT4-based [22] ATLAS simulation framework [23]. The simulation also takes into account inactive and inefficient regions of the ATLAS detector. The resulting datasets were used to derive corrections for detector effects, evaluate systematic uncertainties and compare to the data corrected to particle level.

2.2. Charged-particle measurements at 13 TeV

2.2.1. Event Selection Collision events were selected using a trigger which required one or more minimum-bias trigger scintillators counters (MBTS) above threshold on either side of the detector. Each event is required to contain a primary vertex, reconstructed from at least two tracks with a minimum p_{T} of 100 MeV, as described in [24]. A veto is applied on additional primary vertices arising from split vertices or secondary interactions. A special configuration of the track reconstruction algorithms was used for this analysis to reconstruct low-momentum tracks with good efficiency and purity. Similar configurations were already used in Run 1, but a more robust and efficient low-p_{T} track reconstruction program is available in Run 2 thanks to the installation of an insertable B-layer, IBL [25], which provides a fourth measurement point in the pixel detector. In the *nominal phase space*, events are required to contain at least one selected track, passing the following criteria: $p_{\mathrm{T}} > 500$ MeV and $|\eta| < 2.5$; at least one pixel hit and at least six SCT hits (two, four or six SCT hits for $p_{\mathrm{T}} < 300$ MeV, $p_{\mathrm{T}} < 400$ MeV or $p_{\mathrm{T}} > 400$ MeV, respectively, in the case of the *extended phase space*), with the additional requirement of an innermost-pixel-layer hit if expected[3] (if a hit in the innermost layer is not expected, the next-to-innermost hit is required if expected); $|d_0^{BL}| < 1.5$ mm, where the transverse impact parameter, d_0^{BL}, is calculated with respect to the measured beam line position; and $|z_0^{BL} \cdot sin\theta| < 1.5$ mm, where $z_0^{BL}$is the difference between the longitudinal position of the track along the beam line at the point where d_0^{BL} is measured and the longitudinal position of the primary vertex, and θ is the polar angle of the track. Finally, in order to remove tracks with mismeasured p_{T} due to interactions with the material or other effects, the track-fit χ^2 probability is required to be greater than 0.01 for tracks with $p_{\mathrm{T}} < 10$ GeV.

Approximately 9 million events are selected, containing a total of $\sim$ 100 million reconstructed tracks. While the overall number of particles in the kinematic acceptance of the *extended phase space* is nearly double that in the *nominal phase space*, the measurements are more difficult for $p_{\mathrm{T}} < 500$ MeV, due to multiple scattering and imprecise knowledge of the material in the detector. These systematic uncertainties at low p_{T} need therefore to be carefully evaluated. The performance of the Inner Detector (ID) track reconstruction in the 13 TeV data and its simulation is described in Ref. [26]. Overall, good agreement between data and simulation is observed.

2.2.2. Analysis strategy The main steps of the analysis are related to the trigger, vertex and track reconstruction efficiencies, which need to be evaluated together with their uncertainties. The background contributions to the tracks from primary particles, which include fake tracks (those formed by a random combination of hits), strange baryons and secondary particles, need to be estimated as well. Observables of interest can be evaluated and, by means of an unfolding procedure, can be corrected to account for detector effects. The details can be found in Refs. [9,10], while, in the next section, a few insights will be given on the track reconstruction

[3] A hit is expected if the extrapolated track crosses a known active region of a pixel module.

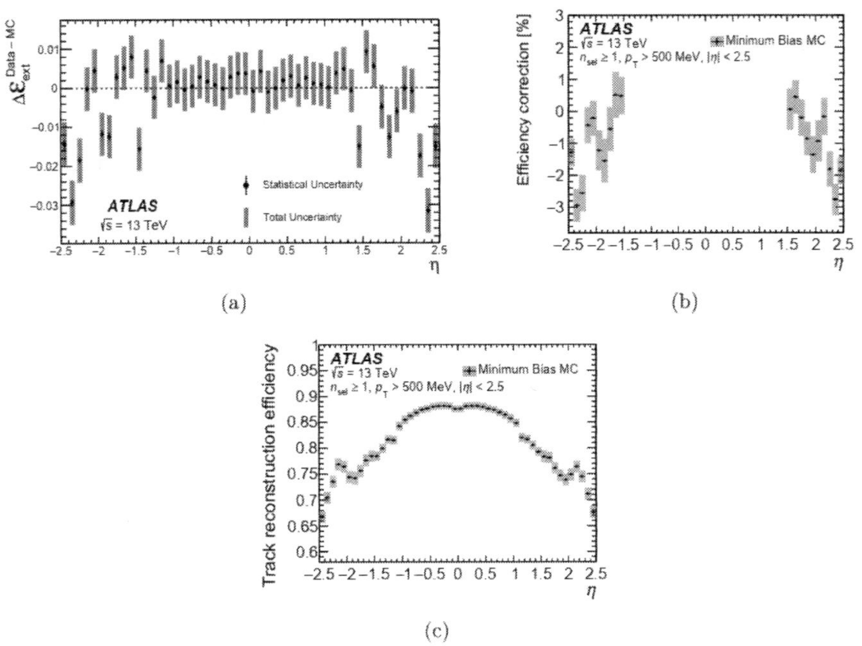

Figure 1. (a) The difference between the track-extension efficiency measured in data and in simulation as a function of η, $\Delta\mathcal{E}_{\rm ext}^{\rm Data-MC}(\eta)$, is shown together with the statistical and total uncertainties [27]. (b) Data-driven correction to the track reconstruction efficiency as a function of pseudorapidity η [9]. (c) The track reconstruction efficiency after this correction as a function of η [9]. $n_{\rm sel}$ is defined as the number of tracks passing all of the track selection requirements.

efficiency by highlighting the importance of a precise evaluation of the amount of material in the ATLAS ID, which represents the main source of systematic uncertainty for this analysis.

2.2.3. Track reconstruction efficiency The analysis is a track-based analysis and the evaluation of the track reconstruction efficiency and of the related systematics is crucial. The dominant uncertainty in the track reconstruction efficiency arises from imprecise knowledge of the amount of material in the ID. The primary track reconstruction efficiency $\varepsilon_{\rm trk}$ is determined from simulation. The efficiency is parameterised in two-dimensional bins of $p_{\rm T}$ and η, and is defined as:

$$\varepsilon_{\rm trk}(p_{\rm T},\eta) = \frac{N_{\rm rec}^{\rm matched}(p_{\rm T},\eta)}{N_{\rm gen}(p_{\rm T},\eta)}, \qquad (1)$$

where $p_{\rm T}$ and η are defined at generator level, $N_{\rm rec}^{\rm matched}(p_{\rm T},\eta)$ is the number of reconstructed tracks matched to a generated primary charged-particle and $N_{\rm gen}(p_{\rm T},\eta)$ is the number of generated primary charged-particles in the kinematic region of interest. A track is matched to a generated particle if the weighted fraction of track hits originating from that particle exceeds 50%. In the analysis performed in the *nominal phase space*, a data-driven correction to the efficiency was applied in order to account for material effects in the $|\eta| \geq 1.5$ region. The track reconstruction efficiency depends on the amount of material in the detector, due to particle interactions that lead to efficiency losses. The relatively large amount of material between the pixel and SCT detectors in the region $|\eta| \geq 1.5$ has changed between Run 1 and Run 2 due to the replacement of some pixel services, which are difficult to simulate accurately.

IOP Conf. Series: Journal of Physics: Conf. Series **1070** (2018) 012022 doi:10.1088/1742-6596/1070/1/012022

The track reconstruction efficiency in this region is corrected using a method, referred to as track-extension efficiency [27], that compares data and simulation for the efficiency to extend a track reconstructed in the pixel detector (referred to as pixel track segment) into the SCT. Differences in track-extension efficiency are quite sensitive to differences in the amount of material in this region, as can be seen in Figure 1(a). The correction, together with the systematic uncertainty, coming predominantly from the uncertainty of the particle composition in the simulation used to make the measurement, is shown in Figure 1(b). The uncertainty in the track reconstruction efficiency resulting from this correction is ±0.4% in the region $|\eta| > 1.5$. The resulting reconstruction efficiency as a function of η integrated over p_T is shown in Figure 1(c). The data-driven correction allows for a large reduction of the systematic uncertainty in the measurement with respect to previous studies but it cannot be applied in the *reduced phase space* due to the large uncertainties of this method for low-momentum tracks. In this case, the total uncertainty on the track reconstruction efficiency due to the amount of material is calculated as the linear sum of the contributions of 5% additional material in the entire ID, 10% additional material in the IBL and 50% additional material in the pixel services region for $|\eta| > 1.5$, as described in detail in [28].

The Track-extension efficiency only probes the material between the pixel and SCT detectors, but a good understanding of the material in the other regions of the ID is needed for good description of the track reconstruction efficiency. The material in the ID was studied extensively during Run 1 [29, 30], where the amount of material was known to ±5%. This gives rise to a systematic uncertainty in the track reconstruction efficiency of ±0.6% (±1.2%) in the most central (forward) region. Between Run 1 and Run 2, the IBL was installed, and its simulation was optimised with the Run 2 data. Two data-driven methods were used [27]: a study of secondary vertices from photon conversions and a study of secondary vertices from hadronic interactions, where the radial position of the vertex and the invariant mass of the outgoing particles are measured. Comparisons between data and simulation indicate that the material in the IBL is constrained to within ±10%. This leads to an uncertainty in the track reconstruction efficiency of ±0.1% (±0.2%) in the central (forward) region. This uncertainty is added linearly to the uncertainty from constraints from Run 1, to cover the possibility of missing material in the simulation in both cases. The resulting uncertainty is added in quadrature to the uncertainty from the data-driven correction.

The total uncertainty on the track reconstruction efficiency due to the imperfect knowledge of the detector material is ±0.7% in the most central region and it grows to ±1.5% in the most forward region. There is a small difference in efficiency, between data and simulation, due to the detector hit requirements described in Section 2.2.1. This difference is assigned as a further systematic uncertainty, amounting to ±0.5% for $p_T < 10$ GeV and ±0.7% for $p_T > 10$ GeV. The total uncertainty due to the track reconstruction efficiency determination, shown in Figure 1(c), is obtained by adding all effects in quadrature and is dominated by the uncertainty from the material description.

2.2.4. Corrections and final results To produce unfolded distributions at particle level, all distributions are first corrected for the loss of events due to the trigger and vertex requirements. The η and p_T distributions of selected tracks are then corrected using a track-by-track weight, as described in Refs. [9–11]. No additional corrections are needed for the η distribution because the resolution is smaller than the bin width. For the p_T distribution, an iterative Bayesian unfolding [31] is applied to correct the measured track p_T distribution to that for primary particles. After applying the event weight, the Bayesian unfolding is also applied to the multiplicity distribution. The total number of events, N_{ev}, used to normalise the distributions, is defined as the integral of the n_{ch} distribution, after all corrections are applied. The dependence of $\langle p_T \rangle$ on n_{ch} is obtained by first separately correcting the total number of tracks and $\sum_i p_T(i)$

IOP Conf. Series: Journal of Physics: Conf. Series **1070** (2018) 012022 doi:10.1088/1742-6596/1070/1/012022

(summing over the p_{T} of all tracks and all events), both versus the number of primary charged-particles. After applying the correction to all events using the event and track weights, both distributions are unfolded separately. The bin-by-bin ratio of the two unfolded distributions gives the dependence of $\langle p_{\mathrm{T}} \rangle$ on n_{ch}.

The corrected distributions for primary charged-particles in events with $n_{\mathrm{ch}} \geq 1$ in the kinematic range of the *nominal phase space* are shown in Figure 2, while, for the *extended phase space*, they can be seen in Figure 3.

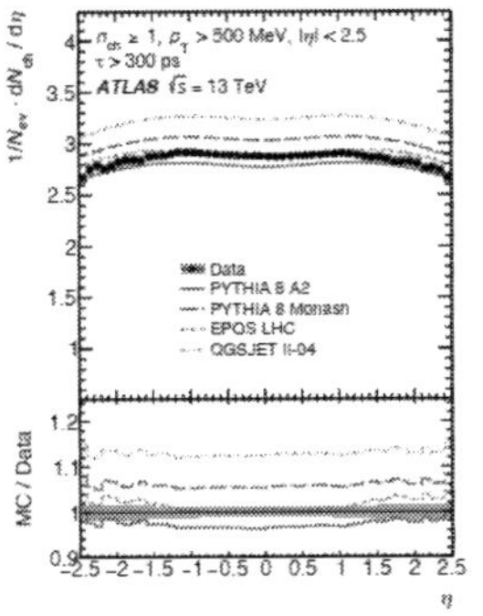

(a)

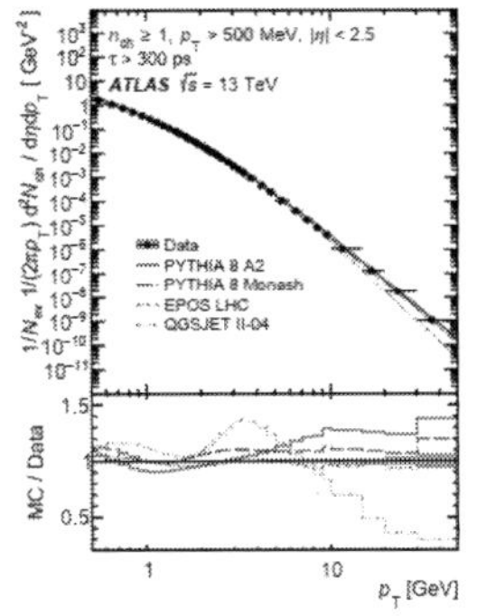

(b)

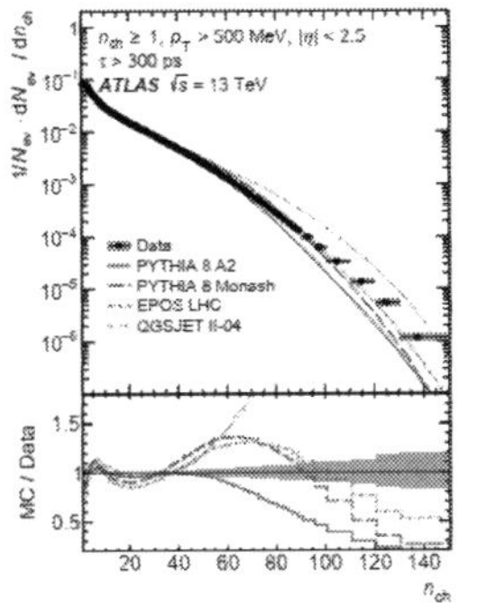

(c)

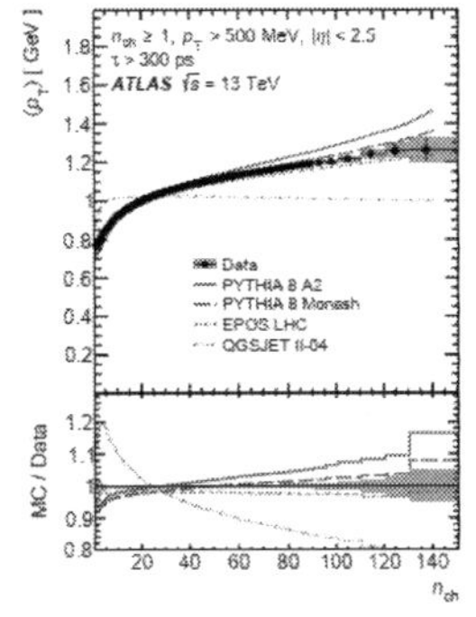

(d)

Figure 2. 13 TeV data [9]: Primary charged-particle multiplicities as a function of (a) pseudorapidity η and (b) transverse momentum p_{T}, (c) the primary charged-particle multiplicity n_{ch} and (d) the mean transverse momentum $\langle p_{\mathrm{T}} \rangle$ versus n_{ch} for events with at least one primary charged-particles with $p_{\mathrm{T}} > 500$ MeV, with $|\eta| < 2.5$, and with a lifetime $\tau > 300$ ps.

Figures 2(a) and 3(a) show the multiplicity of charged-particles as a function of η. In the *nominal phase space*, the mean particle density is roughly constant at 2.9 for $|\eta| < 1.0$ and decreases at higher values of $|\eta|$. EPOS describes the data well for $|\eta| < 1.0$, and predicts a slightly larger multiplicity at large $|\eta|$ values. QGSJET-II and PYTHIA 8 - Monash predict multiplicities that are too large by approximately 15% and 5% respectively. PYTHIA 8 - A2 predicts a multiplicity that is 3% too low in the central region, but describes the data well in the forward region. The total systematic uncertainty, dominated by the uncertainty on the track reconstruction efficiency, is below 1.5% in the entire η range. When moving to lower track-p_{T}, the situation changes and PYTHIA 8 - Monash, EPOS and QGSJET-II give a good description

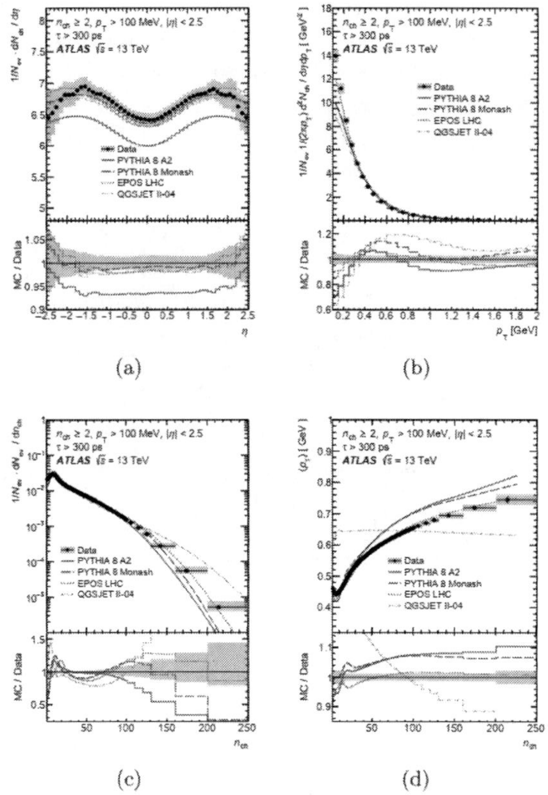

Figure 3. 13 TeV data [10]: Primary charged-particle multiplicities as a function of (a) pseudorapidity η and (b) transverse momentum p_T, (c) the primary charged-particle multiplicity n_{ch} and (d) the mean transverse momentum $\langle p_T \rangle$ versus n_{ch} for events with at least two primary charged-particles with $p_T > 100$ MeV, with $|\eta| < 2.5$, and with a lifetime $\tau > 300$ ps.

for $|\eta| < 1.5$. The prediction from PYTHIA 8 - A2 has the same shape as the predictions from the other generators, but lies below the data. It can be immediately noticed that much bigger systematic uncertainties affect the distribution in the high η region in the *extended phase space* (up to $\sim 7\%$ with respect to $\sim 1.5\%$ in the *nominal phase space*). They mainly come from the uncertainty in the amount of material in the ID, which was differently treated in the two phase spaces, as described above.

Figures 2(b) and 3(b) show the charged-particle transverse momentum distributions. EPOS describes the data well for $p_T > 300$ MeV. The PYTHIA 8 tunes describe the data reasonably well, but they are below the data in the low-p_T region. QGSJET-II gives a poor prediction over the entire spectrum, overshooting the data in the low-p_T region.

Figures 2(c) and 3(c) show the charged-particle multiplicity. PYTHIA 8 - A2 describes the data reasonably well in the low-n_{ch} region, but predicts too few events at larger n_{ch} values. PYTHIA 8 - Monash, EPOS and QGSJET-II describe the data reasonably well in the low-n_{ch} region but predict too many events in the mid-n_{ch} region, with PYTHIA 8 - Monash and EPOS predicting too few events in the high-n_{ch} region while QGSJET-II, which implements a model without colour-reconnection, describes the data poorly in this case also.

Figures 2(d) and 3(d) show how the mean transverse momentum, $\langle p_T \rangle$, rises versus the charged-particle multiplicity. This rise is expected because of colour coherence effects in dense parton environments and is modelled by a colour reconnection mechanism in PYTHIA 8 or by the hydrodynamical evolution model used in EPOS. EPOS describes the data better than PYTHIA 8, which predicts a steeper rise of $\langle p_T \rangle$ with n_{ch} than the data. If the high-n_{ch} region is assumed to be dominated by events with a large number of parton interactions within the same pp collision (MPI), without colour coherence effects, the $\langle p_T \rangle$ should be independent of n_{ch}, as predicted by QGSJET-II.

2.3. *Highlights from 8 TeV measurements*

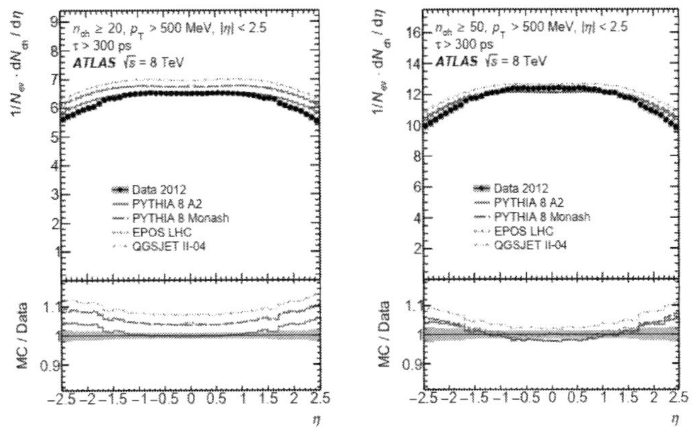

(a) (b)

Figure 4. 8 TeV data [11]: Primary charged-particle multiplicities as a function of pseudorapidity η for events with at least (a) 20 or (b) 50 primary charged-particles with $p_T >$ 500 MeV, with $|\eta| < 2.5$, and with a lifetime $\tau > 300$ ps.

In the context of the 8 TeV analysis, high multiplicity phase spaces with $n_{ch} \geq 20$ and 50 were studied for the first time. Figure 4 shows the multiplicity of charged-particles as a function of η for multiplicity phase spaces with $n_{ch} \geq 20$ or 50. For $n_{ch} \geq 20$, the generator which describes data best is PYTHIA 8 - A2. In the $n_{ch} \geq 50$ case, PYTHIA and EPOS give similar descriptions for $|\eta| > 1.5$ and overestimate the data, while for $|\eta| < 1.5$ the best description is given by EPOS.

The spectra for the other *phase spaces* studied at $\sqrt{s} = 8$ TeV can be found in [11].

2.4. *Charged-particle measurements at different $\sqrt{s}$*

Figure 5 shows the mean number of primary charged-particles in the central region of the detector as a function of $\sqrt{s}$. It is obtained by averaging over $|\eta| < 0.2$ and by correcting the 8 and 13 TeV data for the contribution from strange baryons in order to compare the results with lower centre-of-mass energies at which these particles were included. The mean number of primary charged-particles increases by a factor of 2.2 when $\sqrt{s}$ increases by a factor of about 14 from 0.9 TeV to 13 TeV. EPOS describes the dependence on $\sqrt{s}$ very well in several phase spaces, while PYTHIA 8 - A2 and PYTHIA 8 - Monash give a reasonable description, respectively for $p_T >$ 500 MeV and $p_T > 100$ MeV. QGSJET-II predicts a steeper rise in multiplicity with $\sqrt{s}$ than that shown by the data.

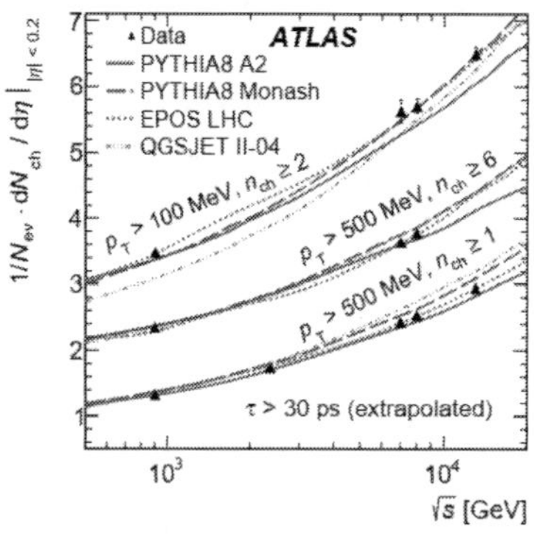

Figure 5. The average primary charged-particle multiplicity in pp interactions per unit of pseudorapidity η for $|\eta| < 0.2$ as a function of the centre-of-mass energy $\sqrt{s}$ [10]. The results at 8 and 13 TeV have been extrapolated to include charged strange baryons in order to compare the values with previous studies.

3. Pythia 8 - A3 tune

These results have been already used by the ATLAS Collaboration in order to produce a new PYTHIA 8 tune, referred to as PYTHIA 8 - A3 [33]. The A2 tune of PYTHIA 8 gives reasonable results in the prediction of the charged-particle multiplicity at 13 TeV, but an overestimation of the fiducial inelastic cross-section [32] was found compared to 13 TeV data. Since both measurements are important in order to define an optimal setup for the pile-up simulation, the PYTHIA 8 - A3 tune was developed with the aim of improving the prediction of the visible inelastic cross-section and, in the mean time, keeping a good prediction of the charged-particle multiplicity. All the details can be found in [33], where it can be seen that the results are promising, because as wanted, the prediction of the visible inelastic cross-section has improved and the charged-particle multiplicity is still reasonably described, even though discrepancies in the very low and very high $n_{\rm ch}$region remain.

4. Track-based Underlying Event

The activity accompanying any hard scattering in a collision event is generally referred to as the Underlying Event. This can include partons not participating in a hard-scattering process (beam remnants), multiple parton interactions, initial and final state gluon radiation. It is impossible to uniquely separate the UE from the hard scattering process on an event-by-event basis, but observables can be defined which are particularly sensitive to the properties of the UE. The most recent results released by the ATLAS collaboration are charged-particle distributions sensitive to the UE [34], measured in pp collisions at a centre-of-mass energy of 13 TeV on the same data set described earlier. These distributions characterise the angular distribution of energy and particle flows with respect to the charged particle with highest transverse momentum, as a function of both that momentum and of charged-particle multiplicity. The analysis strategy follows closely the one described above in terms of event selection, data-driven correction to the tracking efficiency, strange baryons treatment and unfolding of detector effects. The selected events were additionally required to contain at least one track with a transverse momentum above 1 GeV.

This measurement uses the established form of UE observables in which the azimuthal plane of the event is segmented into several distinct regions with differing sensitivities to the UE, as shown in Figure 6: the towards ($|\Delta\phi| < 60°$), away ($|\Delta\phi| > 120°$) and transverse regions

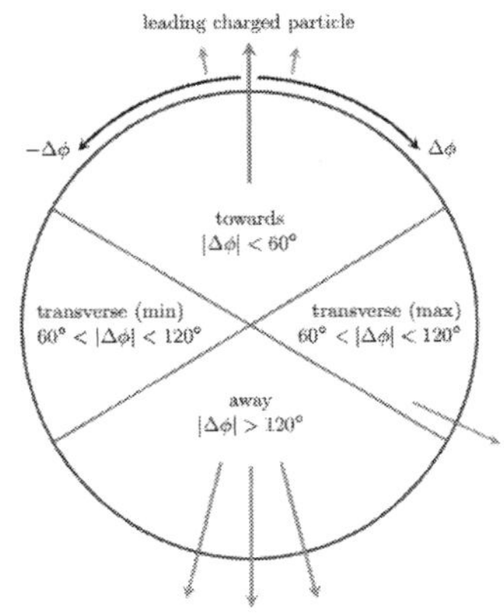

Figure 6. Definition of regions in the azimuthal angle with respect to the leading (highest-p_T) charged particle, with arrows representing particles associated with the hard scattering process and the leading charged particle highlighted in red. Conceptually, the presence of a hard-scatter particle on the right-hand side of the transverse region, increasing its $\sum p_T$, typically leads to that side being identified as the trans-max and hence the left-hand side as the trans-min, with maximum sensitivity to the UE [34].

($60° < |\Delta\phi| < 120°$). As the scale of the hard scattering increases, the leading charged particle acts as a convenient indicator of the main flow of hard-process energy. The towards and away regions are dominated by particle production from the hard process and are hence relatively insensitive to the softer UE. In contrast, the transverse region is more sensitive to the UE, and observables defined inside it are the primary focus of UE measurements.

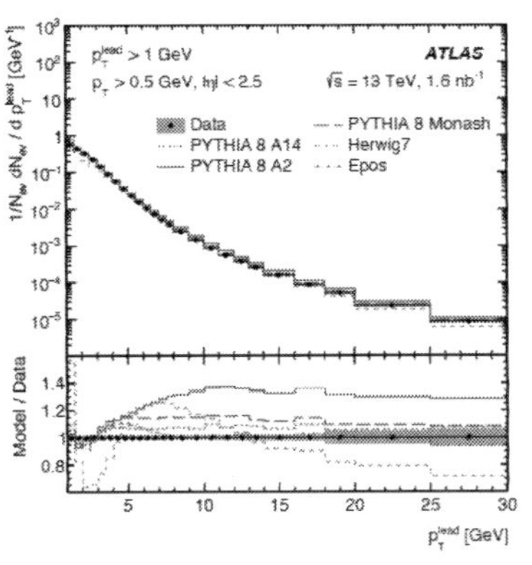

Figure 7. Unit-normalised distribution of the transverse momentum of the leading charged particle, $p_T^{lead} > 1$ GeV, compared to various generator models. The error bars on data points represent statistical uncertainty and the blue band the total combined statistical and systematic uncertainty [34].

Figure 7 shows the unit-normalised distribution of events with respect to the transverse momentum of the leading charged particle, p_T^{lead}. This is a steeply falling distribution, with a

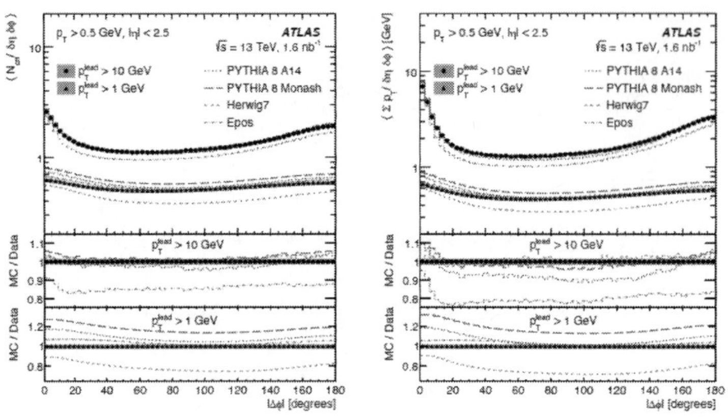

(a) (b)

Figure 8. Distributions of mean densities of charged-particle multiplicity $N_{\rm ch}$(a) and $p_{\rm T}$ (b) as a function of $|\Delta\phi|$ (with respect to the leading charged particle) for $p_{\rm T}^{lead} > 1$ GeV and $p_{\rm T}^{lead} > 10$ GeV separately, with comparisons to MC generator models. combined statistical and systematic uncertainty. The error bars on data points represent statistical uncertainty and the blue band the total combined statistical and systematic uncertainty [34].

change of slope for $p_{\rm T}^{lead} > 5$ GeV: a form which is broadly modelled by all generators. The Pythia 8 - A14 and Monash tunes, as well as EPOS, model the distribution within 15% out to plead = 30 GeV, while the Pythia 8 - A2 minimum-bias tune predicts too hard a spectrum in the high plead tail. Herwig 7 peaks strongly at the lowest plead and alternates between under- and over-shooting the data at higher scales, finally producing a softer tail than seen in data.

Figure 8 shows the mean multiplicity and $p_{\rm T}$ distributions as a function of azimuthal angle with respect to the leading particle for different $p_{\rm T}^{lead}$ requirements. Two event selections are shown here: the $p_{\rm T}^{lead} > 1$ GeV cut common to all observables, and a harder $p_{\rm T}^{lead} > 10$ GeV requirement. The difference between these two selections illustrates the transition from relatively isotropic minimum-bias scattering to the emergence of hard partonic scattering structure and hence a dominant axis of energy flow. This event structure with least activity perpendicular to the leading-object axis, i.e. away from $\Delta\phi = 0°$ and $\Delta\phi = 180°$, is seen for both selections and both observables but is much stronger for the event subset with the higher $p_{\rm T}^{lead} > 10$ GeV requirement: this demonstrates the evolution of event shape as a hard scattering component develops. There is no clear best MC model for these observables. EPOS performs best in the more inclusive $p_{\rm T}^{lead} > 1$ GeV selection, followed by Pythia 8 - A14; Herwig 7 significantly undershoots while Pythia 8 - Monash is everywhere above the data. But in the hard-scattering $p_{\rm T}^{lead} > 10$ GeV event selection, Herwig 7 and Monash perform best, with a slight undershoot from Pythia 8 - A14 and a large one from EPOS. These orderings apply to both distributions, although to different extents.

Many other distributions at the centre-of-mass energy of 13 TeV can be found in [34], while previous results can be found in the references therein.

5. Conclusion

The ATLAS collaboration published a number of soft QCD results in *pp* collisions at the centre-of-mass energy of 13 TeV at the LHC. Among them, charged-particle distributions, including those sensitive to the UE, were presented and the detector performance related to track

IOP Conf. Series: Journal of Physics: Conf. Series **1070** (2018) 012022 doi:10.1088/1742-6596/1070/1/012022

reconstruction were highlighted. A multitude of MC event generators and tunes was investigated. For the charged particle distributions, the results highlight clear differences between MC models and the measured distributions. Among the models considered EPOS reproduces the data the best, PYTHIA 8 - A2 and MONASH give reasonable descriptions of the data and QGSJET-II provides the worst description of the data. These discrepancies, together with those observed in the measurement of the fiducial inelastic cross-section, motivated a new ATLAS tune of the Pythia 8 event generator, the Pythia 8 - A3 tune. The current models in use for UE modelling typically describe this data to 5% accuracy, compared with data uncertainties of less than 1%. The model defects are particularly acute in the transition from generic soft inelastic interactions to secondary interactions in the presence of hard partonic scattering. Some improvement to MC tunes in light of these data will hence be of benefit to physics studies in LHC Run 2 and beyond.

References

[1] ATLAS Collaboration, New J. Phys. 13, 053033 (2011).
[2] CMS Collaboration, JHEP 1101, 079 (2011).
[3] CMS Collaboration, Phys. Rev. Lett. 105 (2010) 022002.
[4] CMS Collaboration, JHEP 1002 (2010) 041.
[5] ALICE Collaboration, Eur. Phys. J. C 68, 345?354 (2010).
[6] CDF Collaboration, Phys. Rev. D 79 (2009) 112005.
[7] CMS Collaboration, Phys. Lett. B 751 (2015) 143.
[8] ATLAS Collaboration, JINST 3 (2008) S08003.
[9] ATLAS Collaboration, Phys. Lett. B 758 (2016) 67.
[10] ATLAS Collaboration, Eur. Phys. J. C (2016) 76: 502.
[11] ATLAS Collaboration, Eur. Phys. J. C (2016) 76:403.
[12] CMS Collaboration, JINST 3 (2008) S08004.
[13] ALICE Collaboration, JINST 3 (2008) S08002.
[14] T. Sjstrand, S. Mrenna and P. Z. Skands, Comput. Phys. Commun. 178 (2008) 852.
[15] S. Porteboeuf, T. Pierog and K. Werner, arXiv:1006.2967 [hep-ph].
[16] S. Ostapchenko, Phys. Rev. D 83 (2011) 014018.
[17] ATLAS Collaboration, ATL-PHYS-PUB-2011-014 (2011), http://cds.cern.ch/record/1400677.
[18] A.D. Martin, W.J. Stirling, R.S. Thorne, G. Watt, Eur. Phys. J. C 63, 189 (2009).
[19] P. Skands, S. Carrazza, J. Rojo, Eur. Phys. J. C 74, 3024 (2014).
[20] NNPDF Collaboration, R.D. Ball et al., Nucl. Phys. B 867, 244 (2013).
[21] T. Pierog, Iu. Karpenko, J.M. Katzy, E. Yatsenko, K. Werner, Phys. Rev. C 92, 34906 (2015).
[22] S. Agostinelli et al., GEANT4 Collaboration, Nucl. Instrum. Methods A 506, 250 (2003).
[23] ATLAS Collaboration, Eur. Phys. J. C 70, 823?874 (2010).
[24] G. Piacquadio, K. Prokofiev, A. Wildauer, J. Phys. Conf. Ser. 119, 032033 (2008).
[25] ATLAS Collaboration, CERN-LHCC-2010-013. ATLAS-TDR-19 (2010), http://cdsweb.cern.ch/record/1291633.
[26] ATLAS Collaboration, ATL-PHYS-PUB-2015-018 (2015), http://cdsweb.cern.ch/record/2037683.
[27] ATLAS Collaboration, JINST 12 (2017) P12009.
[28] ATLAS Collaboration, ATL-PHYS-PUB-2015-051 (2015), https://cds.cern.ch/record/2110140.
[29] ATLAS Collaboration, JINST 7 (2012) P01013.
[30] ATLAS Collaboration, JINST 11 (2016) P11020.
[31] G. D'Agostini, Nucl. Instrum. Methods A 362, 487?498 (1995).
[32] ATLAS Collaboration, Phys. Rev. Lett. 117 (2016) 182002.
[33] ATLAS Collaboration, ATL-PHYS-PUB-2016-017, https://cds.cern.ch/record/2206965.
[34] ATLAS Collaboration,JHEP 03 (2017) 157.

IOP Conf. Series: Journal of Physics: Conf. Series **1070** (2018) 012023 doi:10.1088/1742-6596/1070/1/012023

PHENIX Results on Jet Modification with π^0-, Photon-, and Isolated Photon-Triggered Two Particle Correlations

Tyler Danley, for the PHENIX Collaboration
Ohio University, Athens, Ohio 45701
E-mail: td533712@ohio.edu

Abstract. PHENIX has measured π^0 and direct photon-triggered two-particle azimuthal correlations in a variety of collision systems ranging from $p+p$ to Au+Au at 200 GeV. In p+p, p+Au, and d+Au correlations, interesting new measurements related to angular correlation widths are made which may reveal 'cold' nuclear effects. In Au+Au collisions, we have developed new methods to obtain isolated photons by using an isolation cut like those used in the smaller systems which provide for a cleaner sample of direct photons but reveal new challenges in the high multiplicity Au+Au environment. We present first measurements of centrality-dependent isolated photon-hadron angular correlations that quantify the transition from low z_T $(= p_T^h/p_T^\gamma)$ enhancement to high z_T suppression relative to $p+p$ collisions.

1. Introduction

Direct photon-hadron correlations are used because the direct photons are color neutral, they originate from the initial hard scattering, they are produced alone, meaning isolated and without other fragmentation products. Being color neutral means that they undergo no strong interactions and therefore no energy loss as they traverse the QGP. This means that their flow coefficients are close to zero. The direct photon energy can be used a a proxy of the of the hadron jet energy because they are produced alone in the opposite direction of the jet. This allows the study of jet energy loss as they traverse the QGP. In the past, PHENIX has used two methods to obtain direct photons, a statistical subtraction method, for large systems, and an isolation cut method in small systems.

PHENIX has measured photon-hadron correlations in a variety of collision systems, $p+p$, p+Au, d+Au, and Au+Au. PHENIX consists of two central arms with azimuthal coverage π and $|\eta| < 0.35$ [1]. The photons and π^0's are measured in the electomagnetic calorimeter. The drift and pad chambers measure charged hadrons. Centrality is determined using beam-beam counters.

Most of the methods outline below, outside of the isolation cut method used in Au+Au, have been already described in PHENIX publications [2, 3, 4]. Two-particle correlation (2PC) analyses measure the azimuthal correlation function between photons and hadrons. This correlation function, $C(\Delta\phi_{AB})$, has two contributions, the contribution from the jet, the jet function, $J(\Delta\phi_{AB})$, and the background, $b_0 dN^{AB}_{comb}/d\Delta\phi_{AB}$, that is,

$$C(\Delta\phi_{AB}) = J(\Delta\phi_{AB}) + b_0 \frac{dN^{AB}_{comb}}{d\Delta\phi_{AB}}, \tag{1}$$

where A and B are the two correlated particles. The magnitude of the background, b_0, is usually found via zero yield at minimum or mean seeds mean partners [5]. The shape of the background, $dN^{AB}_{comb}/d\Delta\phi_{AB}$, comes from independent measurements of the single particle elliptic flow, that is,

$$\frac{dN^{AB}_{comb}}{d\Delta\phi_{AB}} \propto 1 + 2v_2^A v_2^B \cos(2\Delta\phi_{AB}). \tag{2}$$

In small systems, just a flat (in ϕ) background is subtracted–even though v_2 is found to be non-zero in p+Au and d+Au [6], the level of combinatoric background is so small that including v_n makes a negligible difference. The elliptic flow shape is subtracted from Au+Au data. Higher order flow harmonics are not included in the Au+Au subtraction because direct measurements are not available in all p_T bins used, and because the prompt photons have very small v_n, such that it is found that the effects of ignoring harmonics with $n > 2$ are negligible. Once the jet function is obtained for inclusive or isolated photon-hadron pairs, the statistical subtraction method can be applied to subtract the decay photon contribution from in order to find the direct photon-hadron yield. This is done using

$$Y_{direct} = \frac{R_\gamma Y_{inclusive} - Y_{decay}}{R_\gamma - 1}, \tag{3}$$

where Y_{direct} is the direct photon-hadron pair yield, $Y_{inclusive}$ is the inclusive photon-hadron pair yield, R_γ is the ratio of number of inclusive photons to number of decay photons, and Y_{decay} is the decay photon-hadron pair yield that is found by mapping the π^0 to decay photons based on the probability of the π^0 of a specific p_T to decay to two photons. When the isolation cut is applied, the statistical stubtraction is identical except $Y_{isolated} = Y_{inclusive}$ in Equation 3 and R_γ is modified to include the increased signal-to-background ratio for direct photons from the isolation cut. These methods are used in both small and, for the first time, large systems at RHIC.

2. Small Systems - $p+p$, p+Au, and p+Al Results

Measuring photon-hadron correlations in small systems aims to study 'cold' nuclear matter effects in the modification of jets and jet fragmentation. It has been predicted that in transverse momentum dependent (TMD) framework, QCD factorization breaks [7]. TMD evolution predicts that the the nonperturbative momentum widths increase with increasing hard scale interaction [8]. These momentum widths can be measured in PHENIX by measuring $p_{out} = p_T^{assoc} \sin\Delta\phi$, the component of the hadron transverse momentum perpendicular to the trigger direct photon or π^0 momentum vector, via direct photon-hadron and π^0-hadron correlations.

The p_{out} distributions on the left side of Figure 1 show a Gaussian at small p_{out} with power law tails at large p_{out}. The small p_{out} region is fit to a Gaussian and the widths are plotted as a function of trigger p_T shown on the right side of Figure 1. The away-side jet widths decrease with increasing hard scale (p_T^{trig}) for both $\pi^0 - h$ and $\gamma - h$ correlations. The lines show the slopes of the widths are different.

To check whether the trend of decreasing widths with increasing hard scale is ubiquitous across all collision species and energies, this analysis has been carried out on $p+p$, p+Al, and p+Au collisions at $\sqrt{s_{NN}} = 200$GeV. The left side of Figure 2 shows the direct photon-hadron

IOP Conf. Series: Journal of Physics: Conf. Series **1070** (2018) 012023 doi:10.1088/1742-6596/1070/1/012023

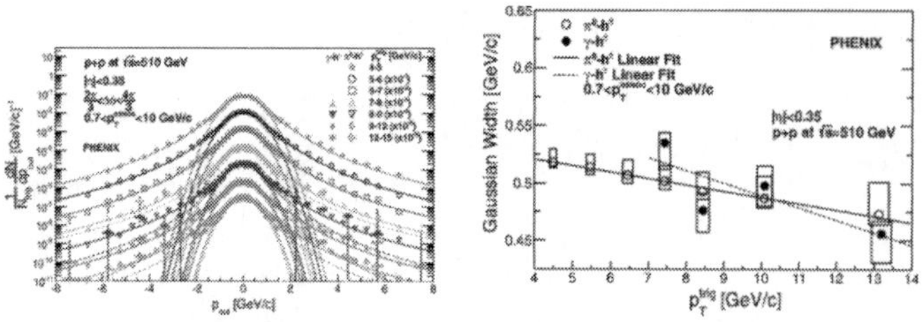

Figure 1. (Left panel) p_{out} distributions for π^0-hadron (open points) and direct photon-hadron correlations (closed points) in $p+p$ collisions at $\sqrt{s_{NN}} = 510\text{GeV}$. Each trigger p_T bin is a different color that has been offset for clarity. The lines are Gaussian fits at small p_{out} and power law fits at large p_{out}. (Right panel) Width of the low p_{out} Gaussian fit as a function of p_T^{trig} for π^0-hadron (open points) and direct photon-hadron correlations (closed points)[3].

p_{out} Gaussian widths as a function of trigger p_T for the $p+p$ collisions at $\sqrt{s_{NN}} = 510\text{GeV}$ shown on the right side of Figure 1 (red points) and $p+p$ collisions at $\sqrt{s_{NN}} = 200\text{GeV}$ (blue points). The decreasing trend is also shown in $p+p$ collisions at $\sqrt{s_{NN}} = 200\text{GeV}$ and the slopes of the widths seem similar. The right side of Figure 2 shows the direct photon-hadron p_{out} Gaussian widths as a function of trigger p_T for the $p+p$ (black points) and p+Au (red points) collisions at $\sqrt{s_{NN}} = 200\text{GeV}$. The widths between these two collision systems are consistent and the slopes are similar, though it is harder to discern because of the large error bars.

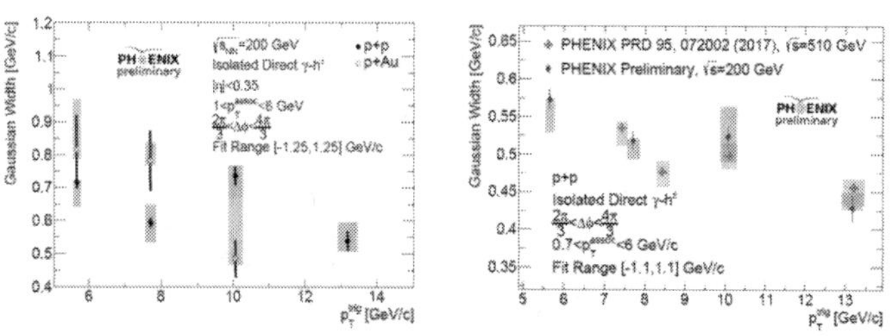

Figure 2. (Left panel) Direct photon-hadron p_{out} Gaussian width as a function of trigger p_T for the $p+p$ collisions at $\sqrt{s_{NN}} = 510\text{GeV}$ shown on the right side of Figure 1 (red points) and $p+p$ collisions at $\sqrt{s_{NN}} = 200\text{GeV}$ (blue points). (Right panel) Direct photon-hadron p_{out} Gaussian width as a function of p_T^{trig} for the $p+p$ (black points) and p+Au (red points) collisions at $\sqrt{s_{NN}} = 200\text{GeV}$.

The π^0-hadron correlations have more statistics so the error bars are smaller. Figure 3 shows the p_{out} Gaussian width for π^0-hadron correlations as a function of trigger p_T for various collision species and energies. The left and middle panel of this figure shows the Gaussian widths decrease in $p+p$, p+Au, and p+Al collisions but with different slopes. The right panel of Figure 3 shows the centrality dependence of the p_{out} Gaussian width in p+Au and p+Al collisions. There is a definite broadening from peripheral to central events and the effect is stronger in p+Au than p+Al.

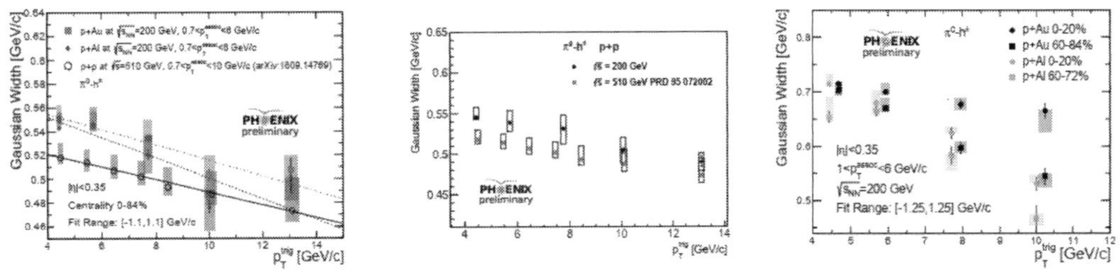

Figure 3. (Left panel) π^0-hadron p_{out} Gaussian width as a function of p_T^{trig} for the p+p collisions at $\sqrt{s_{NN}} = 510$GeV shown on the right side of Figure 1 (black points)cand p+Au (red points) and p+Al (blue points) collisions at $\sqrt{s_{NN}} = 200$GeV. (Middle panel) π^0-hadron p_{out} Gaussian width as a function of p_T^{trig} for the $p+p$ collisions at $\sqrt{s_{NN}} = 510$GeV shown on the right side of Figure 1(red points) and $p+p$ collisions at $\sqrt{s_{NN}} = 200$GeV (black points). (Right panel) π^0-hadron p_{out} Gaussian width as a function of p_T^{trig} for 0-20% central p+Au (black circles), 60-84% central p+Au (black squares), 0-20% central p+Al (green circles), and 60-72% central p+Al (green squares) collisions.

3. Small and Large System Comparison - Non Isolation Method d+Au and Au+Au Results

Photon-hadron correlations in large systems are used to measure jet energy loss in the QGP. This is done by measuring jet functions and comparing their away-side yields to the same measurement in $p+p$ collisions. Doing this is complicated by the presence of the sizable underlying event, for example in Au+Au collisions, the elliptic flow of the particles involved in the correlation cannot be neglected.

The upper panel of Figure 4 shows the integrated $\pi/2$ away-side yield in $p+p$ (blue points), d+Au (purple points), and Au+Au (black points) collisions as a function of $\xi = \ln(1/z_T)$ where $z_T = p_T^h/p_T^\gamma$. The ratio of the d+Au and Au+Au yield to the p+p yield, that is, the $I_{AA} = Y_{Au+Au}/Y_{p+p}$ and $I_{dA} = Y_{d+Au}/Y_{p+p}$ is shown in the lower panel of Figure 4. The I_{dA} shows that there is little or no fragmentation function modification in d+Au collisions. The I_{AA} result confirms the previous PHENIX result from Reference [9]. This also concludes that there is enhancement of the fragmentation function at low z_T, which is interpreted as recovered energy from the jet energy loss. This is confirmed with better statistical precision.

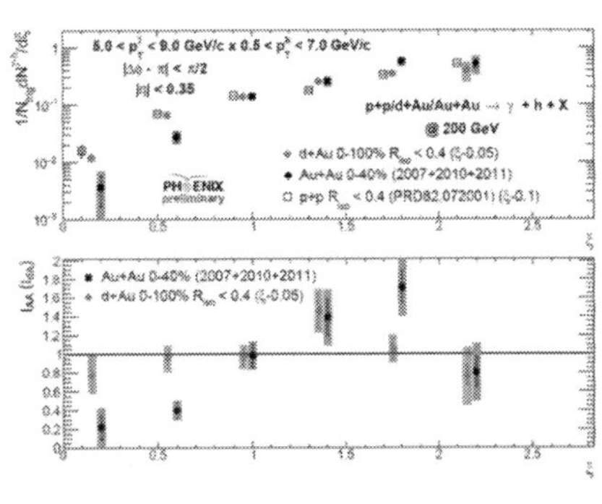

Figure 4. (Upper panel) Integrated $\pi/2$ away-side yield in $p+p$ (blue points), d+Au (purple points), and Au+Au (black points) collisions as a function of $\xi = \ln(1/z_T)$ where $z_T = p_T^h/p_T^\gamma$. (Lower panel) Au+Au and d+Au $\pi/2$ away-side I_{AA} as a function of ξ [10].

IOP Conf. Series: Journal of Physics: Conf. Series **1070** (2018) 012023 doi:10.1088/1742-6596/1070/1/012023

4. Centrality Dependence of Jet Energy Loss Effects: Isolation Method Au+Au Results

We have also developed a new isolation cut method in Au+Au. As usual for isolation methods, photons in each event are used to define a cone surrounding with a specific sized radius according to $R_{cone} = \sqrt{(\Delta\phi)^2 + (\Delta\eta)^2}$. The energy of all other particles inside the cone is then measured, and if it is smaller than the threshold energy, the photon is classified as isolated. If the energy is greater, the photon is not isolated and probably originated from decay products of jet fragments. In our method, the cone threshold energy is determined by $E_{TH} = aE_\gamma + b$, where E_γ is the central photon's total energy and a and b are parameters vary with centrality due to the magnitude of the underlying event.

Because we choose the parameters a and b as constant anywhere in a given event, the isolation cut preferentially accepts particles emitted "out-of-plane" with respect to the event plane as isolated since the underlying event modulates lower there. The converse is also true, it finds less isolated particles oriented "in-plane" relative to the event plane. This creates a bias when trying to measure the elliptic flow of the isolated particles. This bias can be parameterized by $1 + v_{2E}\cos\Delta\phi$ where the underlying event v_2, v_{2E}, is typically negative. This distribution can be derived by folding into the typical underlying event distribution to obtain

$$\frac{dN_{STSE}}{d\Delta\phi} = \mathcal{I}\Big(1 + 2\mathcal{J}\cos(2\Delta\phi) + \mathcal{K}\cos^2(2\Delta\phi) - 4\mathcal{L}\cos(4\Delta\phi)\Big), \qquad (4)$$

where $\mathcal{I}$ is a normalization factor, $\mathcal{J} = v_{2T} + v_{2E}\langle\cos(2\delta\Psi)\rangle$, $\mathcal{K} = v_{2T}v_{2E}$, and $\mathcal{L} \propto v_{2T}v_{2E}$. The $\langle\cos(2\delta\Psi)\rangle$ factor is the typical event plane resolution correction.

In Figure 5, we show the extracted isolated photon v_2 (taken from parameter $\mathcal{J}$) as a function of p_T for two centrality bins in Au+Au collisions, along with direct and inclusive photon v_2. One can see that the isolated photon v_2 (black points) is less than the inclusive photon v_2 (brown points) and consistent with the direct photon v_2 (red points). It is important to remember when comparing these v_2 that some decay photons survive the isolation cut so the black points are still a mixture of direct and decay photons, albeit with reduced decay contributions. The underlying event v_{2E} has not been corrected for here because the 2PC correlation functions also have this effect that must be subtracted, in Equation 2. This allows the net v_2 to even be slightly negative in some cases, because the true v_2 can be negligible for photons compared to the isolation-induced negative v_2.

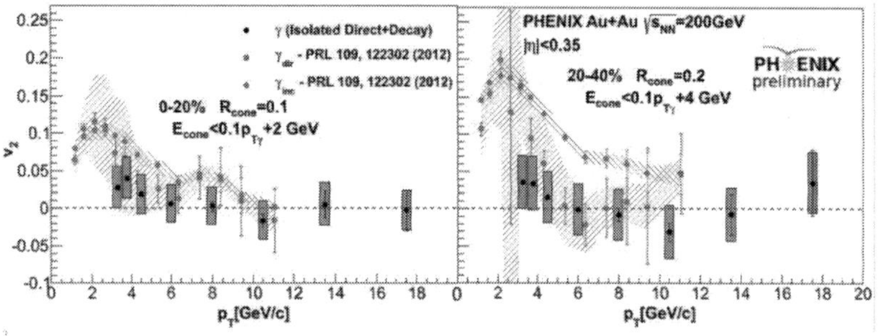

Figure 5. Isolated direct + decay photon v_2 ($v_{2T} + v_{2E}$) as a function of p_T for 0-20% (left) and 20-40% (middle) central events in Au+Au collisions. The boxes are systematic error bars. The red points are the direct photon v_2 and the brown points are inclusive photon v_2 for 0-20% and 20-40% central events that are taken from Reference [11].

An example isolated photon-hadron correlation function is shown in Figure 6. The black points are the correlation function, the black line is the background level, and the red lines show the error on the black line. Notice that the background shape is a negative cosine distribution for this momentum bin. This is due to the negative isolated photon v_2 discussed above. Once this background is subtracted from the correlation function, jet functions are obtained. These are shown in Figure 7 for four centrality bins in Au+Au collisions. The black points are $\gamma_{iso} - h$ pairs, the red points are $\gamma_{dec} - h$ pairs, and the blue points are $\gamma_{dir} - h$ pairs found using Equation 3.

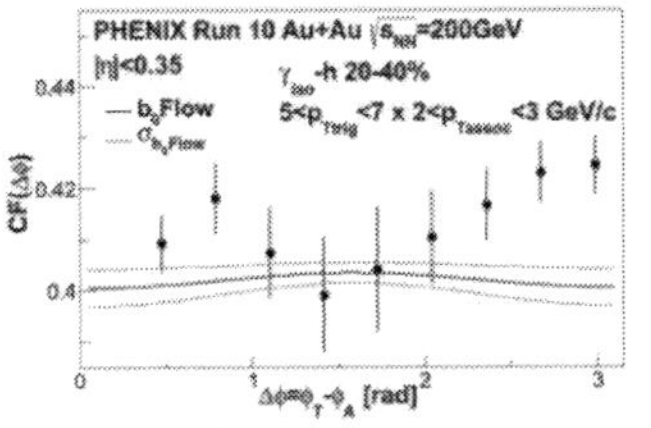

Figure 6. A Au+Au $\gamma_{iso} - h$ correlation function. The background shape is opposite of the expected shape because the isolated photon has negative v_2.

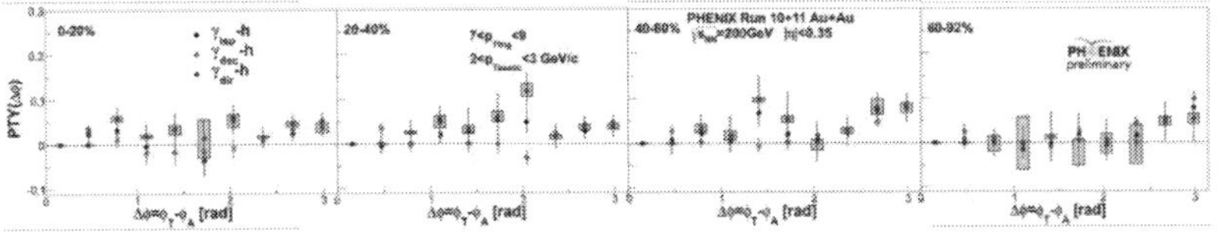

Figure 7. Au+Au $\gamma_{iso} - h$ (black points), $\gamma_{dec} - h$ (red points), and $\gamma_{dir} - h$(blue points) jet functions for $5 \leq p_T^{trigger} \leq 7$ and $3 \leq p_T^{associated} \leq 5$ for four centrality bins.

The direct photon-hadron jet functions can be integrated and compared to jet functions in $p+p$ collisions to find I_{AA} . Figure 8 shows the $\pi/2$ away-side region I_{AA} as a function of z_T for four centrality bins. Each trigger photon p_T bin is a different color. In each centrality bin, there is suppression of high z_T particles. By contrast there is no suppression of low z_T particles and in some cases even enhancement. This has been seen in other Au+Au correlation analyses [9, 10, 12]. To quantify this contrasting behavior, the low and high z_T regions are fit to constants (purple lines).

To see how the suppression varies with centrality, we plot the high z_T average I_{AA} as a function of N_{part} for the $\pi/3$ (Figure 9 left) and $\pi/5$ (Figure 9 right) away-side integration regions. The gray bands in these figures is the π^0 R_{AA} for π^0's with $p_T > 5$GeV/c. The high z_T I_{AA} approximately matches the π^0 R_{AA} which is expected because the energy loss geometry of the QGP is the same as the single inclusive jet geometry if other effects besides geometry are sub-dominant.

To see how both the high z_T suppression and low z_T non-suppression/enhancement vary with centrality, we plot the low and high z_T direct photon I_{AA} constant fit values as a function of N_{part} for three away-side integration regions, $\pi/2$ (Figure 10 left), $\pi/3$ (Figure 10 middle). and $\pi/5$ (Figure 10 right) away-side integration regions. Immediately obvious is that the low and and high z_T behaviors are different. There is high z_T suppression for all centrality bins while the low z_T points are not suppressed, so they are relatively flat with centrality. The low z_T behavior has previously been interpreted as being a signal of energy loss recovery by the medium. This is because the high z_T particle's energy loss enhances the production of low z_T particles. There is increasing low z_T enhancement for wider integration regions, the blue points increase right

IOP Conf. Series: Journal of Physics: Conf. Series **1070** (2018) 012023 doi:10.1088/1742-6596/1070/1/012023

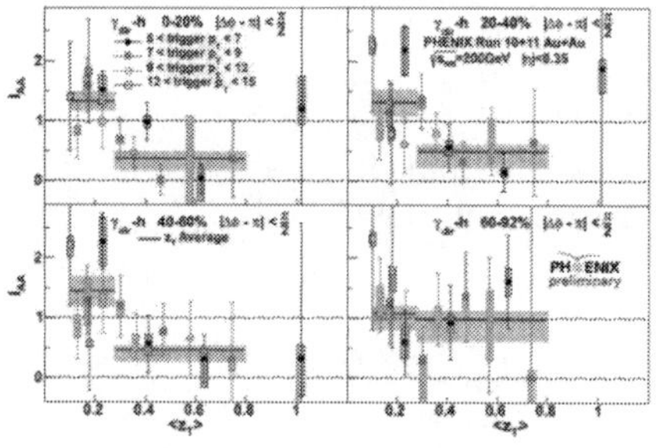

Figure 8. I_{AA} as a function of z_T for the $\pi/2$ away-side integration region for four centrality bins. Each photon p_T bin is a different color, $5 < p_T^{trigger} < 7$ GeV/c (black), $7 < p_T^{trigger} < 9$ GeV/c (red), $9 < p_T^{trigger} < 12$ GeV/c (green), and $12 < p_T^{trigger} < 15$ GeV/c (blue). The purple lines are constant fits to two z_T regions with the statistical error on the fit in light purple.

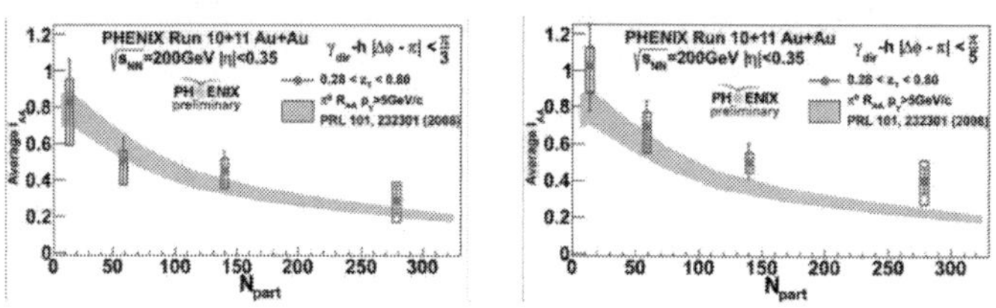

Figure 9. High z_T average I_{AA} as a function of N_{part} for the $\pi/3$ (left) and $\pi/5$ (right) away-side integration region $\pi/3$. The gray band is the π^0 R_{AA} for π^0 with $p_T > 5$GeV/c from [13].

to left. The isolation cut method allows for a more precise analysis of the semi-peripheral and peripheral centralities. This is the first measurement of the centrality dependence of the low z_T enhancement.

To judge true centrality dependence of enhancement however, we must account for overall reduction of jets due to suppression, because our normalization of the yields is made per trigger photon, and therefore includes the jet suppression factor–which is why the high z_T yields have $I_{AA} < 1$. For this reason, low z_T I_{AA} values in this observable, even if they do not exceed $I_{AA} = 1$ can reasonably described as "enhancement" as long as they are substantially higher than the overall suppression level measured by the high z_T I_{AA}. To quantify this, the ratio is made of the low z_T to high z_T average I_{AA} values, which we call the energy recovery factor (ERF) and is shown in the bottom panel of Figure 10. One can see that this ratio monotonically increasing towards central events for all away-side integration regions, qualitatively matching expectations of having low z_T energy recovery scaling inversely to the amount of energy loss/suppression.

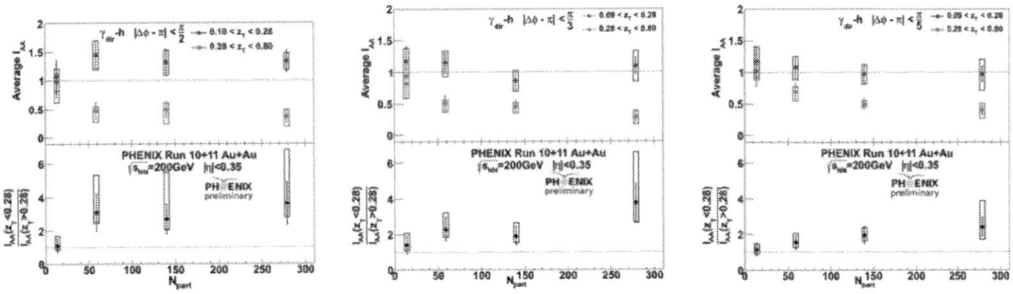

Figure 10. Average I_{AA} as a function of N_{part} for each away-side integration region $\pi/2$ (left), $\pi/3$ (middle), and $\pi/5$ (right). The red points in the top panels is the high z_T average and the blue points are the low z_T average. The bottom panels show the centrality dependence of ratio of high to low z_T average I_{AA}, the energy recovery factor.

5. Conclusion

PHENIX has measured nonpreturbative momentum widths and has seen a decrease with increasing hard scale. This has been measured in p+p at $\sqrt{s_{NN}}$ = 510GeV and $\sqrt{s_{NN}}$ = 200GeV, as well as p+Au, and p+Al collisions. This is opposite of TMD factorization expectations [3].

PHENIX has measured direct photon-hadron correlations in d+Au and Au+Au collisions. There is no modification in d+Au collisions within uncertainties when compared to $p + p$ collisions. The centrality integrated Au+Au result confirms the previously measured pattern of suppression to low z_T enhancement in Reference [9].

Finally, PHENIX has developed a new isolation cut method in Au+Au that allows for more precise analysis of the semi-peripheral and peripheral centralities. This method has lead to the first ever study of low z_T enhancement as a function of centrality. This enhancement, when properly accounting for the overall jet suppression level through the ERF ratio, is found qualitatively to be monotonic with centrality and scaling inversely to the jet-suppression centrality dependence. PHENIX still has high statistics datasets from RHIC Run-14 and Run-16 that need to be analyzed which should be promising for application of our new isolation method.

References

[1] Adcox K *et al.* 2003 *Nuclear Instruments and Methods in Physics Research Section A: Accelerators, Spectrometers, Detectors and Associated Equipment* **499** 469 – 479 ISSN 0168-9002 the Relativistic Heavy Ion Collider Project: RHIC and its Detectors
[2] Adare A *et al.* (PHENIX Collaboration) 2010 *Phys. Rev. D* **82**(7) 072001
[3] Adare A *et al.* (PHENIX Collaboration) 2017 *Phys. Rev. D* **95**(7) 072002
[4] Aidala C *et al.* (PHENIX) 2018 *Submitted to: Phys. Rev. D* (*Preprint* 1805.02450)
[5] Sickles A, McCumber M P and Adare A 2010 *Phys. Rev. C* **81**(1) 014908
[6] Aidala C *et al.* (PHENIX Collaboration) 2017 *Phys. Rev. C* **96**(6) 064905
[7] Rogers T C and Mulders P J 2010 *Phys. Rev. D* **81**(9) 094006
[8] Collins J 2012 (*Preprint* 1212.5974)
[9] Adare A *et al.* (PHENIX Collaboration) 2013 *Phys. Rev. Lett.* **111**(3) 032301
[10] Ge H 2017 *Nuclear and Particle Physics Proceedings* **289-290** 313 – 316 ISSN 2405-6014 8th International Conference on Hard and Electromagnetic Probes of High Energy Nuclear Collisions URL http://www.sciencedirect.com/science/article/pii/S2405601417302687
[11] Adare A *et al.* (PHENIX Collaboration) 2012 *Phys. Rev. Lett.* **109**(12) 122302
[12] Adare A *et al.* (PHENIX Collaboration) 2009 *Phys. Rev. C* **80**(2) 024908
[13] Adare A *et al.* (PHENIX Collaboration) 2008 *Phys. Rev. Lett.* **101**(23) 232301

IOP Conf. Series: Journal of Physics: Conf. Series **1070** (2018) 012024 doi:10.1088/1742-6596/1070/1/012024

sPHENIX - a new jet detector at RHIC

Christine Nattrass

Department of Physics and Astronomy, 1408 Circle Drive, Knoxville TN 37996-1200

E-mail: christine.nattrass@utk.edu

Abstract. sPHENIX is a new state-of-the-art detector optimized for measurements of jets and heavy flavor which will take advantage of increased luminosity available at the Relativistic Heavy Ion Collider (RHIC). It comprises vertex tracking and hadronic and electromagnetic calorimetry. This enables studies of jet structure, heavy flavor jets, open heavy flavor, and quarkonia, the pillars of the sPHENIX physics program. These measurements will complement measurements at the LHC and enable determination of the properties of the medium created in heavy ion collisions to high precision, including its temperature dependence. This talk will discuss the detector design and key measurements enabled by the new detector, as well as the status of the project itself.

1. Introduction

Measurements from the Relativistic Heavy Ion Collider (RHIC) and the Large Hadron Collider (LHC) have provided ample evidence for the production of a hot, dense Quark Gluon Plasma (QGP) [1–4], but key questions remain. Measurements of global observables and probes of soft physics are limited by systematic uncertainties and their precision is near what is possible with our current technology. By contrast, measurements of jets and heavy flavor production from RHIC are limited by both statistical and systematic uncertainties. Improvements in both accelerator and detector technology enable measurements of these observables with comparable precision to those achievable at the LHC. The sPHENIX detector will combine precision tracking and calorimetry capable of taking advantage of increased collision rates available at RHIC. This will enable constraints of the properties of the medium with uncertainties comparable to those at the LHC.

2. Detector

The sPHENIX detector, shown in figure 1, consists of a silicon tracking system, time projection chamber (TPC), electromagnetic calorimetor (EMCAL), and inner hadronic calorimeter (HCAL) in a 1.4 T magnetic field surrounded by an outer HCAL. The magnet was previously used in BaBar [5] and successful full field tests were completed in February 2018, demonstrating that the magnet performs to specifications. A 3-layer MAPS pixel sensor detector surrounded by four layers of silicon strips is proposed as the inner tracking system in order to provide the high precision distance of closest approach (DCA) resolution necessary for the identification of final state particles from heavy flavor decays. The TPC for the outer tracking system will be gateless with continuous readout in order to take advantage of high luminosities.

The calorimetry consists of an EMCAL, an inner HCAL inside the magnet, and an outer HCAL. The performance of this system has been demonstrated in a number of test beams [6].

IOP Conf. Series: Journal of Physics: Conf. Series **1070** (2018) 012024 doi:10.1088/1742-6596/1070/1/012024

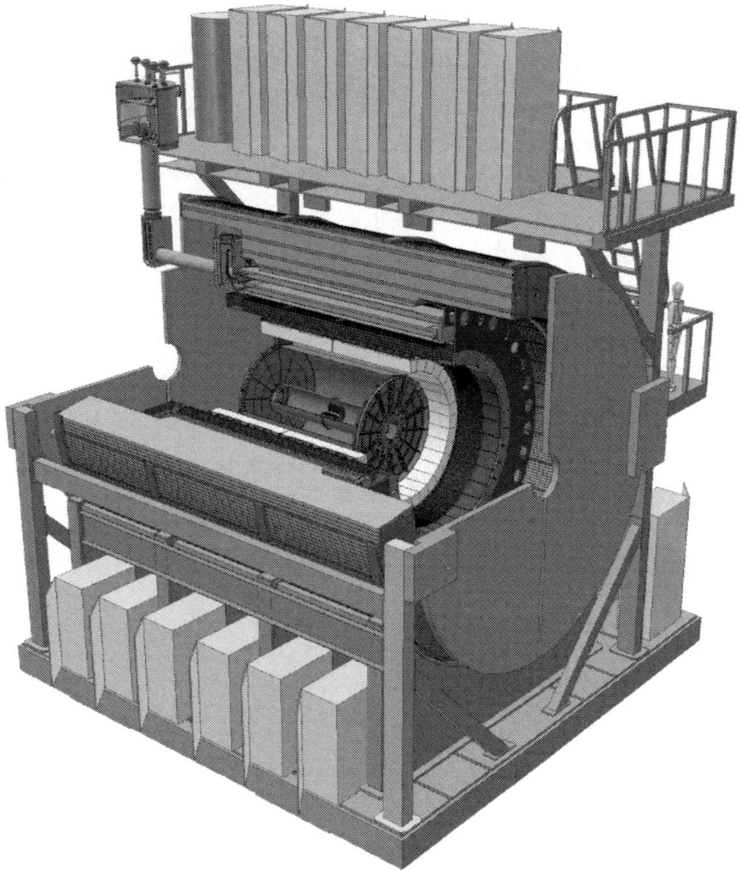

Figure 1. The sPHENIX detector consists of a silicon inner tracking system, a time projection chamber (TPC), an electromagnetic calorimeter (EMCAL) and inner hadronic calorimeter (HCAL) in a 1.4 T magnetic field surrounded by an outer HCAL.

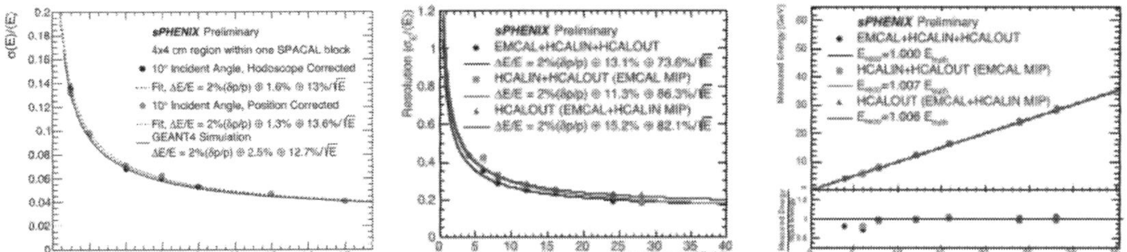

Figure 2. Energy resolution for electrons (left) and pions (middle) and the nonlinearity of the calorimeter response (right).

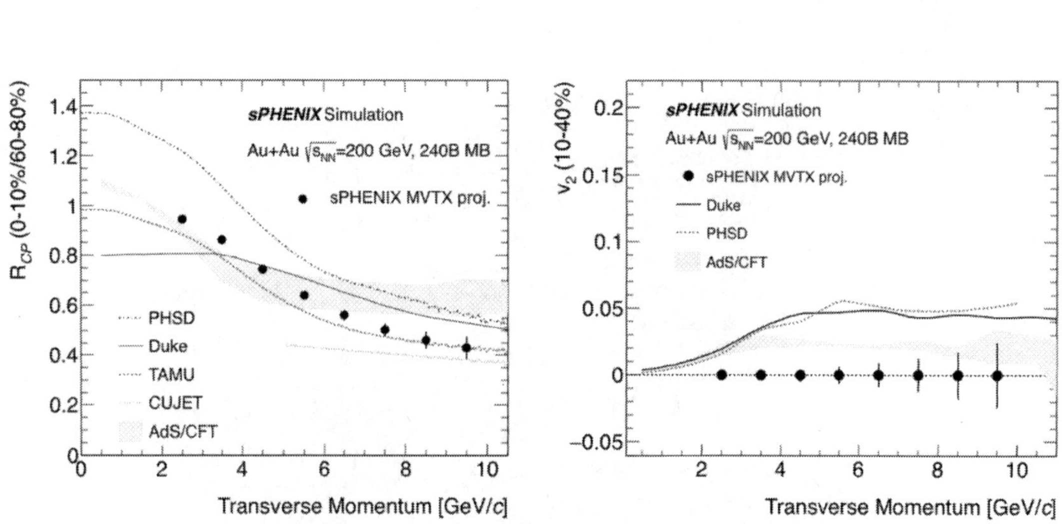

Figure 3. Projections for uncertainties on the nuclear modification factor R_{cp} and azimuthal asymmetry v_2 for non-prompt D mesons from B meson decay.

The energy resolution for electrons and pions and the nonlinearity are shown in figure 2. The combination of precision calorimetry and tracking enables identification of photons and electrons from decays of heavy flavor hadrons, including quarkonia, and measurements of jets. These detectors will enable measurements not possible with the previous generation of RHIC detectors.

3. Physics milestones

3.1. Heavy flavor

To date, measurements of heavy flavor production are limited by uncertainties and the limited ability to reconstruct final state hadrons. While D mesons have been reconstructed [7], no RHIC detectors enable measurements of reconstructed B mesons or have high enough statistics to measure them. In contrast, B measons have been measured at the LHC [8]. Figure 3 shows the projected uncertainties for measurements of the B meson nuclear modification factor R_{cp}, the ratio of B meson spectra in central to that in peripheral collisions, and azimuthal asymmetry v_2 for B mesons. These measurements will enable constraints of QGP properties at RHIC energies with precision comparable to that available at the LHC.

3.2. Quarkonia

Quarkonia measurements at RHIC to date [9,10] are also limited by their precision, in particular the ability to distinguish between the ground state and different excited states. Figure 4 shows the projected invariant mass peak for the $\Upsilon(1s)$, $\Upsilon(2s)$, and $\Upsilon(3s)$ (left) and uncertainties on the nuclear modification factor R_{AA},the ratio of B meson spectra in central collisions to that in p+p collisions, for the different Υ states, demonstrating that sPHENIX will be able to resolve the different quarkonia states and measure observables for quarkonia with precision comparable to that at the LHC.

3.3. Jets and jet structure

RHIC played a vital role in our understanding of partonic energy loss in the medium, providing critical evidence that hard partons lose energy in the medium and, as a result, final state jets are broader and softer in A+A collisions than in p+p and d+Au collisions [11]. By taking advantage

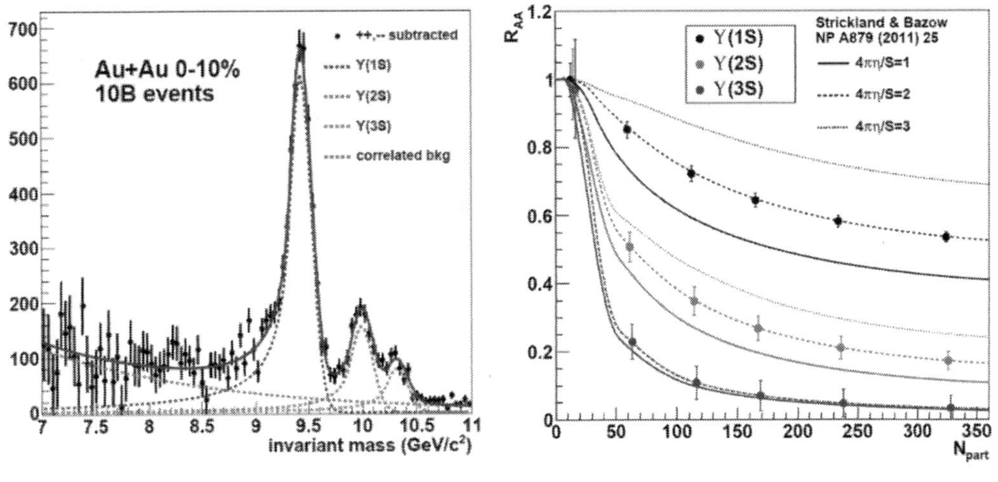

Figure 4. Projected invariant mass peak for the $\Upsilon(1s)$, $\Upsilon(2s)$, and $\Upsilon(3s)$ (left) and uncertainties on the nuclear modification factor R_{AA} for the different Υ states.

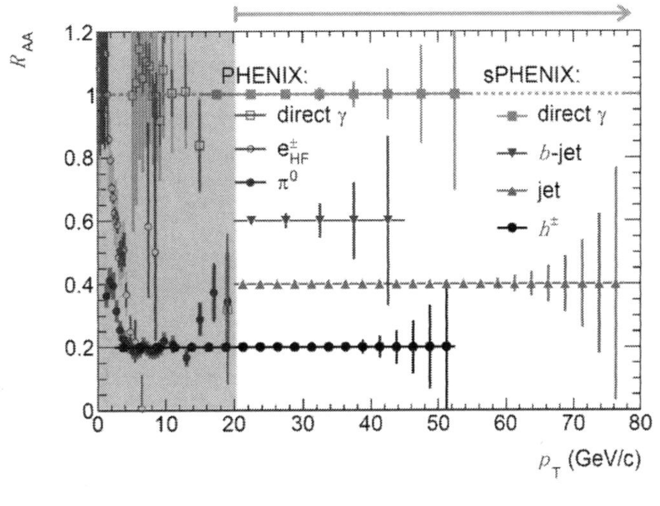

Figure 5. Projected uncertainties for single hadron and reconstructed inclusive jet and b-jet nuclear modification factors R_{AA} compared to measurements from PHENIX.

of higher rates, sPHENIX will extend the reach of single final state hadrons beyond that achieved at RHIC and enable high precision measurements of reconstructed jets. Figure 5 shows projected uncertainties for single hadron and reconstructed jet nuclear modification factors R_{AA} compared to measurements from PHENIX.

Reconstructed jets and the emerging study of jet structure are important for constraining the jet transport coefficients. Neither PHENIX nor STAR were optimized for reconstruction of jets in heavy ion collisions and to date there have only been three papers from heavy ion collisions at RHIC published on measurements of reconstructed jets [12–14]. Complete information on

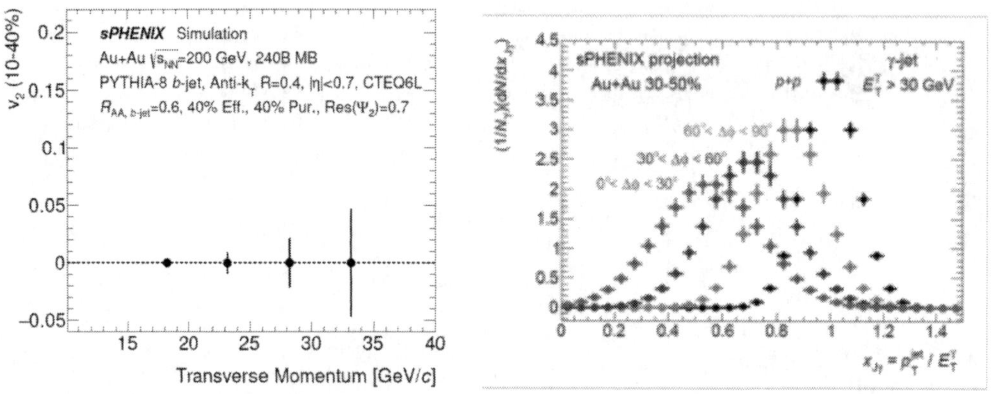

Figure 6. Projected uncertainties for the azimuthal asymmetry of reconstructed b-jets (left) and for the energy fraction of a jet in coincidence with a leading photon as a function of the angle relative to the reaction plane (right).

partonic energy loss comparable to that available at the LHC therefore requires a dedicated detector. sPHENIX is optimized to enable more differential measurements of jets. Figure 5 shows projected uncertainties for measurements of jets created from a bottom quark. Figure 2 shows the projected uncertainties for the azimuthal asymmetry of reconstructed jets and for the energy fraction of a jet in coincidence with a leading photon as a function of the angle relative to the reaction plane. The combination of these measurements will enable precision constraints of partonic energy loss in the medium.

4. Conclusions

Measurements at RHIC have provided critical information to establish the formation of the QGP and precision measurements of low momentum probes of the QGP enable some constraints of the properties of the medium. However, to date measurements at RHIC provide little differential information on jets and limited information on quarkonia states. sPHENIX is optimized to do these measurements, enabling determination of the properties of the QGP at RHIC energies with precision comparable to that available at the LHC.

References

[1] Adcox K *et al.* (PHENIX) 2005 *Nucl. Phys.* **A757** 184–283 (*Preprint* nucl-ex/0410003)
[2] Adams J *et al.* (STAR) 2005 *Nucl. Phys.* **A757** 102–183 (*Preprint* nucl-ex/0501009)
[3] Back B B *et al.* 2005 *Nucl. Phys.* **A757** 28–101 (*Preprint* nucl-ex/0410022)
[4] Arsene I *et al.* (BRAHMS) 2005 *Nucl. Phys.* **A757** 1–27 (*Preprint* nucl-ex/0410020)
[5] Fabbricatore P *et al.* 1996 *IEEE Trans. Magnetics* **32** 2210–2213
[6] Aidala C A *et al.* (sPHENIX) 2017 *Submitted to: IEEE Trans. Nucl. Sci.* (*Preprint* 1704.01461)
[7] Adamczyk L *et al.* (STAR) 2014 *Phys. Rev. Lett.* **113** 142301 (*Preprint* 1404.6185)
[8] Sirunyan A M *et al.* (CMS) 2017 *Phys. Rev. Lett.* **119** 152301 (*Preprint* 1705.04727)
[9] Adamczyk L *et al.* (STAR) 2014 *Phys. Lett.* **B735** 127–137 [Erratum: Phys. Lett.B743,537(2015)] (*Preprint* 1312.3675)
[10] Adare A *et al.* (PHENIX) 2016 *Phys. Rev.* **C93** 034903 (*Preprint* 1509.05380)
[11] Connors M, Nattrass C, Reed R and Salur S 2017 (*Preprint* 1705.01974)
[12] Adamczyk L *et al.* (STAR) 2014 *Phys. Rev. Lett.* **112** 122301 (*Preprint* 1302.6184)
[13] Adamczyk L *et al.* (STAR) 2017 *Phys. Rev. Lett.* **119** 062301 (*Preprint* 1609.03878)
[14] Adamczyk L *et al.* (STAR) 2017 *Phys. Rev.* **C96** 024905 (*Preprint* 1702.01108)

IOP Conf. Series: Journal of Physics: Conf. Series **1070** (2018) 012025 doi:10.1088/1742-6596/1070/1/012025

PHENIX flow results from the d+Au beam energy scan

D. McGlinchey (for the PHENIX collaboration)
Los Alamos National Laboratory, Los Alamos, NM 87545
E-mail: dmcglinchey@lanl.gov

Abstract. The appearance of flow signatures, typically associated with the formation of a Quark Gluon Plasma, in small collision systems was a surprise to many in the community. Experimental measurements of elliptic anisotropy, now in p/d/3He+Au collisions at RHIC and p+Pb collisions at the LHC, have been well described by both hydrodynamic models and models involving parton transport. To help investigate how these signals arise, RHIC delivered a beam energy scan of d+Au collision in 2016. PHENIX has measured the azimuthal anisotropy using event plane and multiparticle cumulant methods at the four collision energies and find nonzero signals at even the lowest collision energy of $\sqrt{s_{NN}}$ = 19.6 GeV. This presentation will discuss the results, comparisons to theoretical models, and their implications for our understanding of how these signals arise in small collision systems.

1. Introduction

Unexpected by many, azimuthal momentum anisotropies typically associated with quark gluon plasma (QGP) formation in A+A collisions have now been observed in $p/d/^3$He+Au collisions at RHIC as well as p+p and p+Pb collisions at the LHC. Recent measurements of v_2 and v_3 in central $p/d/^3$He+Au collisions [1], shown in Fig. 1, follow the same ordering as their respective ε_n's, indicating a geometric origin of the measured anisotropies. Further, hydrodynamic predictions, which include a QGP phase, provide a simultaneous and quantitative description of the data. If a small QGP droplet is formed in these small collision systems, an obvious question is what are the limitations on it's formation? How small of a droplet can be formed? How short lived can the droplet be and still translate initial spatial anisotropy into final state particle momentum anisotropy? In order to address these questions, Ref. [2] proposed a beam energy scan of d+Au. A beam energy scan would amount to varying the initial temperature, as well as lifetime, of a possible QGP droplet. In 2016, this proposal was carried out at RHIC, the results of which are discussed in these proceedings.

2. RHIC d+Au beam energy scan

The 2016 RHIC d+Au beam energy scan consisted of four collision energies at $\sqrt{s_{NN}}$ = 200, 62.4, 39, and 19.6 GeV. PHENIX collected data using a high multiplicity trigger in order to take nearly the full luminosity for central collisions. Details of the trigger conditions and event selection criteria can be found in Ref. [3]. The resulting event statistics are shown in Table 1 for central events, along with the mean number of participating nucleons ($\langle N_{\text{part}}\rangle$) and the average second order eccentricity ($\langle \varepsilon_2 \rangle$) as calculated from Monte Carlo Glauber.

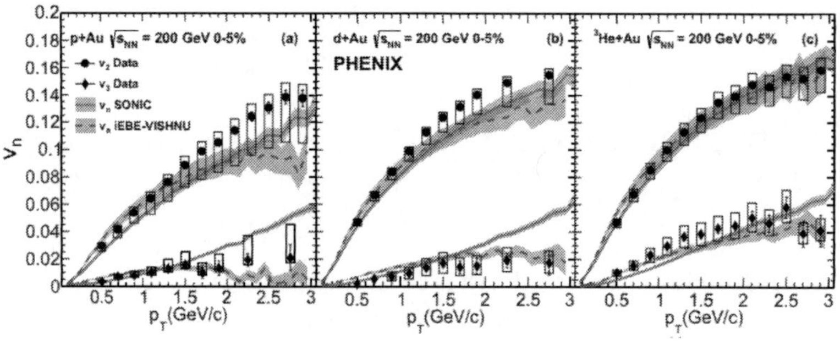

Figure 1. The measured v_2 and v_3 as a function of p_T in 0–5% central p+Au (a), d+Au (b), and ^{3}He+Au (c) collisions at $\sqrt{s_{NN}}$ = 200 GeV [1].

Table 1. Summary of the data analyzed by PHENIX from the 2016 d+Au beam energy scan.

	$\sqrt{s_{NN}}$ [GeV]	Centrality	$\langle N_{\text{part}}\rangle$	$\langle \varepsilon_2\rangle$	N_{events} [10^6]	$dN_{ch}/d\eta\|_{\eta=0}$
d+Au	200	(0–5%)	17.8±1.2	0.54±0.04	569	20.3±1.5
d+Au	62.4	(0–5%)	16.3±1.0	0.55±0.05	214	12.4±2.4
d+Au	39	(0–10%)	15.9±1.0	1.56±0.06	171	9.3±1.6
d+Au	19.6	(0–20%)	13.6±1.0	0.55±0.05	7	5.8±1.1

The charged particle multiplicity ($dN_{ch}/d\eta$) as a function of pseudorapidity (η) for each centrality and at each energy is shown in Fig. 2. The distributions show a clear asymmetry in the particle production for central events, with more particles being produced at $\eta < 0$ (backward or Au-going) compared to $\eta > 0$ (forward or d-going) due to the asymmetric collision system. This asymmetry decreases from central to peripheral events, with the most peripheral bin being symmetric, as is the case in p+p collisions. The charged particle production at midrapidity decreases by nearly a factor of 4 in central collisions between 200 and 19.6 GeV.

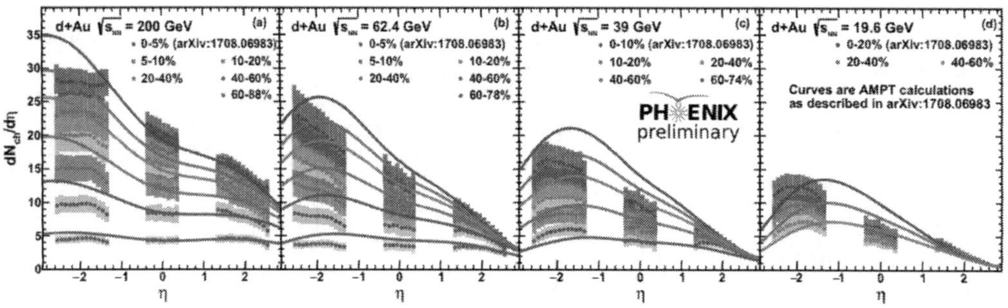

Figure 2. Charged particle multiplicity as a function of η and event centrality for d+Au collisions at $\sqrt{s_{NN}}$ = 200 (a), 62.4 (b), 39 (c), and 19.6 (d) GeV.

Also shown in Fig. 2 are calculations from A Multi-phase Transport (AMPT) model. The centrality bins in the AMPT results are determined similarly to the data, by selecting on the total number of charged particles in the region $-3.9 < \eta < -3.1$, which corresponds to acceptance of the PHENIX south beam beam counter (BBC). At 200 GeV, AMPT describes the data well at

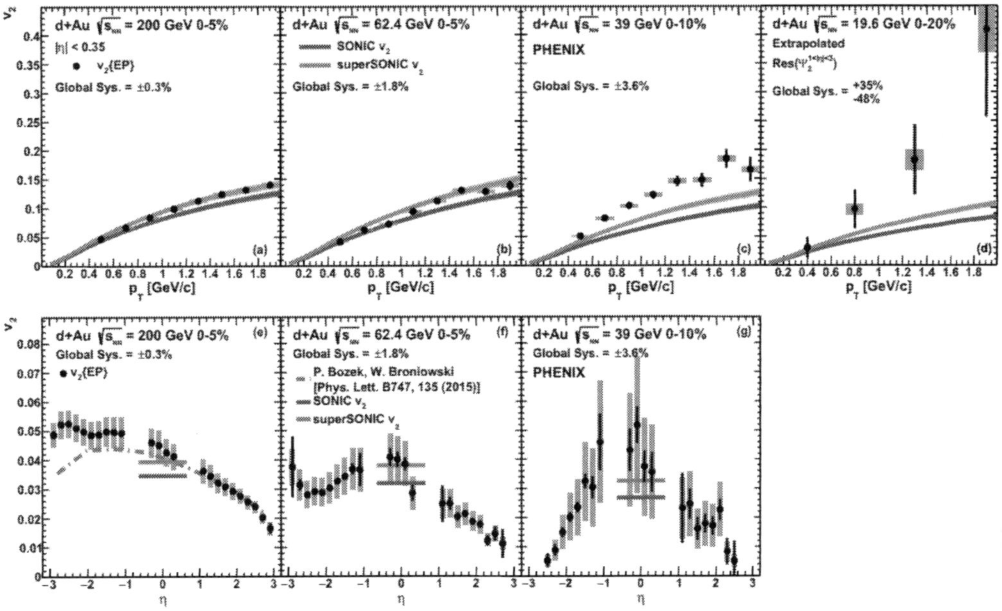

Figure 3. The $v_2\{\mathrm{EP}\}$ as a function of p_T (a-d) and η (e-g) for central d+Au collisions at $\sqrt{s_{NN}}$ = 200 (a, e), 62.4 (b, f), 39 (c, g), and 19.6 (d) GeV [3]. The solid orange (upper) and purple (lower) curves are predictions from superSONIC and SONIC, respectively [2]. The dashed [red] curve in panel (e) shows hydrodynamic predictions from Ref. [4].

mid and forward pseudorapidities for all centralities, but overpredicts the data at backward pseudorapidity. This may be due to a stronger bias from categorizing centrality based on multiplicity in the neighboring pseudorapidity region in AMPT than appears in the data. For the lower collision energies, AMPT overpredicts the $dN_{ch}/d\eta$ across pseudorapidity and centrality.

3. Establishing collectivity

Using the standard event plane (EP) method, the $v_2(p_T)$ of charged particles within $|\eta| < 0.35$ is measured for central d+Au collisions at all four energies. Additionally, the $v_2(\eta)$, integrated over p_T, of charged particles is measured for central d+Au collisions at $\sqrt{s_{NN}}$ = 200, 62.4, and 39 GeV. The results are shown in Fig. 3. A significant, nonzero, v_2 signal is measured at all four collision energies. When measuring $v_2(p_T)$, the EP is measured using the PHENIX forward silicon vertex detector (FVTX) in the Au-going direction which covers $-3.0 < \eta < -1.0$, yielding a pseudorapidity gap of $0.65 < \Delta\eta < 3.35$. For $v_2(\eta)$, the EP is measured in the Au-going BBC covering $-3.9 < \eta < -3.1$. In these results, no estimate or subtraction of nonflow contributions is performed. Based on comparing the v_2 measured using EP's in different regions, as well as alternative 2 particle correlation measurements (detailed in Ref. [3]), the nonflow contribution for 0–5% central d+Au at 200 GeV is expected to be small for $p_T < 3$. However, at the lowest energies it is expected that the nonflow contribution increases.

Alongside the EP method, the v_2 is also measured using multiparticle cumulants. Figure 4 shows the $v_2\{2\}$, with and without a minimum $\Delta\eta$ cut, $v_2\{4\}$, and $v_2\{6\}$ as a function of the number of tracks reconstructed in the north and south FVTX arms ($N_{\mathrm{tracks}}^{\mathrm{FVTX}}$) [5]. A positive $v_2\{2\}$, $v_2\{4\}$, and $v_2\{6\}$ is observed at 200 GeV. The positive $v_2\{4\}$ and $v_2\{6\}$ and the fact that

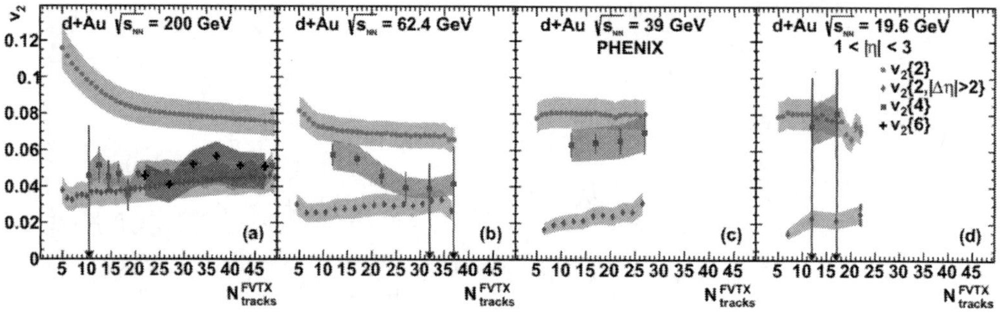

Figure 4. Multiparticle cumulants as a function of the number of offline tracks in d+Au collisions at $\sqrt{s_{NN}}$ = 200 (a), 62.4 (b), 39 (c), and 19.6 (d) GeV [5].

$v_2\{4\} \approx v_2\{6\}$ is a strong indication that the measured v_2 is truly collective, and the nonflow contributions are minimal. A positive $v_2\{4\}$ is also observed down to 19.6 GeV, again yielding evidence that the measured v_2 arises from a collective flow effect.

It is known that $v_2\{2\}$, without the requirement of a $\Delta\eta$ gap between pairs, is susceptible to a strong nonflow contribution. Typically, one then imposes a such a gap to help remove nonflow. Also shown in Fig. 4 is $v_2\{2, |\Delta\eta| > 2\}$, where a gap of $|\Delta\eta| > 2$ is imposed. Due to the pseudorapidity coverage of the PHENIX FVTX arms used in this analysis, this imposes a requirement that one track is measured in the Au-going FVTX while the other is measured in the d-going FVTX. Unfortunately, due to the asymmetric collision system, this imposes different weighting of the v_2 values at backward and forward pseudorapidities than in the $v_2\{2\}$ or $v_2\{4\}$ without an eta gap, making direct comparisons difficult. The $v_2\{2, |\Delta\eta| > 2\}$ is effectively a direct product of the v_2 in the d and Au-going directions, while the $v_2\{2\}$ and $v_2\{4\}$ are weighted by the respective multiplicity in each region, as shown in Fig. 2. This can explain why $v_2\{2, |\Delta\eta| > 2\}$ decreases with decreasing energy while $v_2\{4\}$ appears to increase with decreasing energy. These features are qualitatively reproduced in AMPT, which includes both η dependent multiplicity and v_2, as shown in Ref. [5]. These cumulant results strongly indicate that the v_2 arises from multiparticle correlations from 200 down to 19.6 GeV.

4. Hydrodynamic theory comparison

Predictions for $v_2(p_T)$ from the hydrodynamic SONIC and superSONIC models are shown compared to the measured data in Fig. 3. The main difference between the calculations being that superSONIC additionally includes pre-equilibrium flow. The calculations agree with the data well at 200 and 62.4, with superSONIC giving a showing a slightly better agreement with the data, perhaps indicating the importance of the pre-equilibrium flow. However, at 39 and 62.4 GeV, the calculations are significantly lower than the measured data. One likely explanation for the discrepancy is that the measurement includes contributions from both flow and nonflow. As discussed above, the results at 200 and 62.4 GeV are believed to be dominated by flow, particularly at low-p_T. However, it is expected that the nonflow contribution will increase with decreasing collision energy and with increasing p_T, which would cause an increase in the measured v_2 signal. It is worth noting that both SONIC and superSONIC are in good agreement with the p_T integrated v_2 at midrapidity, even at 39 GeV, as shown in Fig. 3(g).

5. Microscopic transport comparison

An alternative class of models which translates initial spatial anisotropy into final state particle momentum anisotropy include microscopic transport models. Once such model is AMPT. One

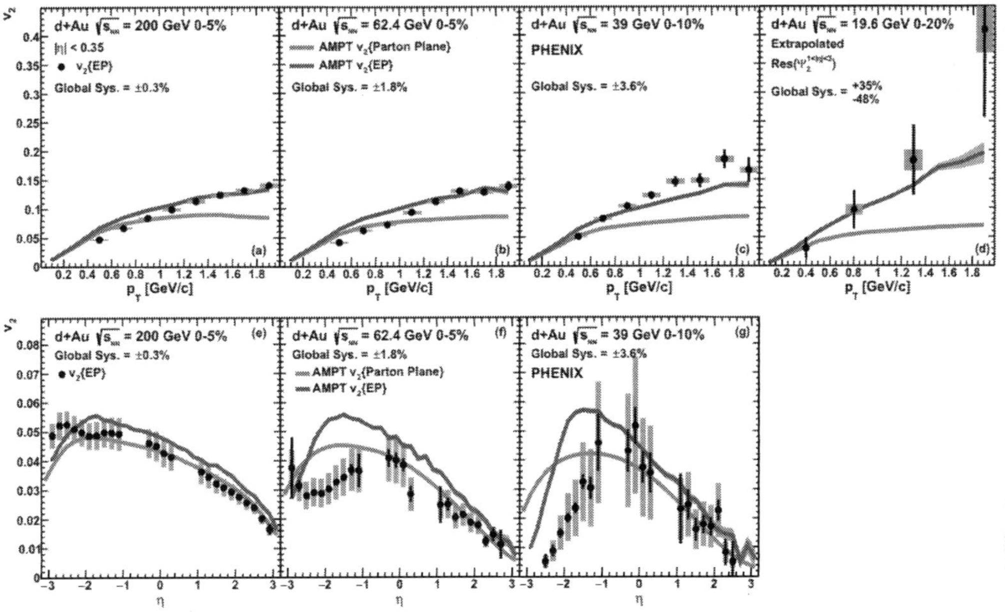

Figure 5. The $v_2\{\mathrm{EP}\}$ as a function of p_T (a-d) and η (e-g) for central d+Au collisions at $\sqrt{s_{NN}} = 200$ (a, e), 62.4 (b, f), 39 (c, g), and 19.6 (d) GeV [3]. The lower [green] curves show calculations from the AMPT model [6] where the v_2 is calculated relative to the parton plane. The upper [blue] curves show calculations from the AMPT model where the v_2 is calculated using the event-plane method as described in the Ref. [3].

advantage of AMPT is that it is a full Monte Carlo event generator which produces final state particles, as would be detected experimentally. Because of this the same analysis procedure used in data can be applied to the AMPT output, while simultaneously knowing the true initial conditions. This allowed us to produce two distinct "measures" of the v_2 within AMPT. The first calculates the v_2 of final state particles relative to the initial parton plane (PP). This correlates final state momentum with true initial geometry only, i.e. what we think of as flow. Additionally, for the same set of events, we can calculate the v_2 relative to the EP, as is done in real data. Since AMPT includes particle decays and mini-jets, that means that the calculated v_2 relative to the EP contains correlations from both flow and nonflow. Figure 5 shows the $v_2(p_T)$ and $v_2(\eta)$ compared to AMPT calculations using both the parton and event planes. By comparing the two AMPT curves, we can gain some intuition about the effect of nonflow as a function of p_T, η, collision energy, and event multiplicity. As expected, the nonflow contribution appears to grow with larger p_T, lower collision energy, and as the $\Delta\eta$ gap relative to the EP decreases. The AMPT EP calculations provide a much more consistent description of the data, particularly at high-p_T. An interesting phenomenon in $v_2(\eta)$ is also seen at 39 GeV, where the AMPT EP calculations show a lower value relative to the PP results for $\eta < -2$, indicating a more complicated coupling between flow and nonflow. For a more detailed discussion of the comparison to $v_2(\eta)$ data see Ref. [3].

Within AMPT, we can also disentangle how the various stages of the evolution contribute to the total v_2 signal. This can be done by turning off either or both of the partonic or hardonic interactions. The resulting calculations of $v_2(p_T)$ relative to the EP are shown in Fig. 6. For a full discussion including $v_2(\eta)$ calculations, as well as results relative to the parton plane,

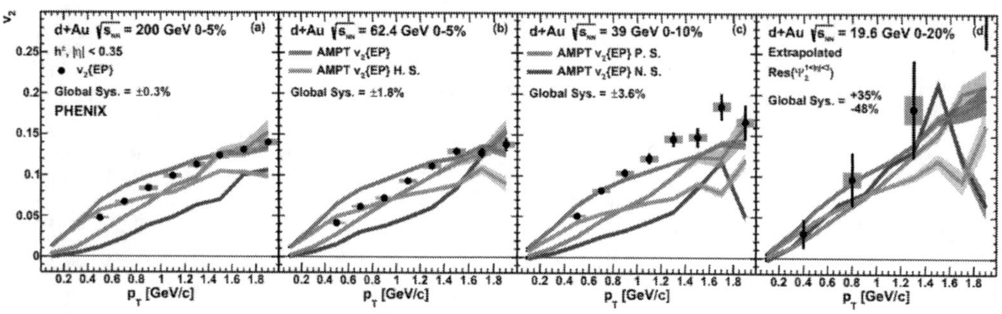

Figure 6. The value of v_2 as a function of p_T in central d+Au collisions at $\sqrt{s_{NN}} = 200$ (a), 62.4 (b), 39 (c), and 19.6 (d) GeV. The curves are calculations from AMPT under different conditions. The red curve is AMPT run with both partonic and hadronic scattering. The yellow curve is AMPT run with hadronic scattering only (H. S.). The cyan curve is AMPT run with partonic scattering only (P. S.). The purple curve is AMPT run with no scattering (N. S.). For all AMPT curves, the v_2 is calculated relative to the final state event plane.

see Ref. [3]. When both partonic and hadronic scattering are turned off, we find $v_2 = 0$ when calculated relative to the PP, indicating now true flow is present. However, when producing the same calculations relative to the EP, a nonzero signal is observed which increases strongly with increasing p_T and decreasing collision energy, showing the signal from nonflow alone. Once partonic scattering is turned on, a larger v_2 signal is observed which dominates at $p_T > 1$ GeV/c. When turning on hadronic scattering, but leaving partonic scattering turned off, the hadronic stage is isolated. Here we can see that the majority of the low-p_T signal in AMPT is driven by the hadronic scattering. A note of caution: the separate curves can not simply be added together to achieve the results with all scattering turned on, as turning off various pieces changes the number of particles produced, and hence the EP, as well as changing the amount of time spent in the various stages. However, it is still useful to gain some insight into the origins of v_2 within the microscopic transport approach as encoded within AMPT.

6. Summary

In summary, PHENIX has measured v_2 over a range of kinematics in d+Au collisions at $\sqrt{s_{NN}} = 200$, 62.4, 39, and 19.6 GeV using multiple techniques. These results provide further evidence that the measured v_2 is coupled to the initial geometry, and are consistent with hydrodynamic predictions. At $\sqrt{s_{NN}} = 200$ GeV, the measured v_2 appears to be dominated by contributions related to flow, while as the collision energy is lowered this contribution appears to be present, although it becomes less dominant.

References

[1] Aidala C *et al.* (PHENIX) 2018 (*Preprint* 1805.02973)
[2] Orjuela Koop J D, Belmont R, Yin P and Nagle J L 2016 *Phys. Rev. C* **93** 044910
[3] Aidala C *et al.* (PHENIX) 2017 *Phys. Rev.* **C96** 064905 (*Preprint* 1708.06983)
[4] Bozek P and Broniowski W 2014 *Phys. Lett. B* **739** 308
[5] Aidala C *et al.* (PHENIX) 2018 *Phys. Rev. Lett.* **120** 062302 (*Preprint* 1707.06108)
[6] Lin Z W, Ko C M, Li B A, Zhang B and Pal S 2005 *Phys. Rev. C* **72** 064901

IOP Conf. Series: Journal of Physics: Conf. Series **1070** (2018) 012026 doi:10.1088/1742-6596/1070/1/012026

Two- and three-pion Lévy femtoscopy with PHENIX

Máté Csanád (for the PHENIX Collaboration)

Eötvös Loránd University, Pázmány P. s. 1/a, H-1117 Budapest, Hungary

E-mail: csanad@elte.hu

Abstract. The last decades of high energy physics revealed, that in ultra-relativistic ion-ion collisions, a strongly interacting quark gluon plasma (sQGP) is created. Varying the collision energy allows for the investigation of the phase diagram of QCD matter. The nature of the quark-hadron transition can be studied via femtoscopy, as the investigation of momentum correlations in heavy ion reactions reveals the space-time structure of the hadron production of the sQGP. Going beyond the Gaussian assumption the shape of this source may be described by Lévy distributions. In this paper we report on recent femtoscopic measurements of PHENIX, utilizing Lévy sources.

1. Introduction

Pioneering interferometry discoveries were made by R. Hanbury Brown as he investigated correlations of intensity fluctuations with radio telescopes [1] at the Jordell Bank Observatory, where he was able to measure the angular diameter of two strong radiofrequency sources. Constructing two tabletop experiments [2,3], Hanbury Brown and R. Q. Twiss prove that these correlations are also observable in the optical spectrum. Subsequently, the angular diameter of Sirius was measured [4], utilizing this technique. These results can be regarded [5] as the extension of the work by A. A. Michelson and F. G. Pease [6], originally suggested by H. L. Fizeau and Michelson [5]. Brown and collaborators, building an optical interferometer [7] at the Narrabi Observatory, were able to measure the angular diameter of multiple stars [8, 9]. Nowadays, this technique can be applied to study the two dimensional structure of distant astrophysical objects [10].

These phenomena were interpreted in the language of quantum optics by R. J. Glauber [11–13], and observed in high energy physics independently by G. Goldhaber and collaborators in proton-antiproton collisions [14, 15]. These Hanbury Brown and Twiss (HBT) correlations have an important role in high energy physics, as they can be utilized to understand the femtometer scale space-time structure of the particle emitting source. Hence this field is often called femtoscopy [16]. The basic working principle of femtoscopy is, that if we define the probability density of particle creation at space-time point x and momentum p as $S(r, p)$ (a.k.a. the source), then the momentum correlation function, $C(q, K)$ (where q is the pair momentum difference and K is the average momentum of the pair) can be measured and (assuming thermal emission, lack of final state interactions, etc., see e.g. Ref. [17] for details) is related to the source as

$$C_2(q, K) = 1 + \left| \frac{\widetilde{S}(q, K)}{\widetilde{S}(0, K)} \right|^2, \tag{1}$$

where $\widetilde{S}(q,K) = \int S(x,K)e^{iqx}d^4x$ is the Fourier transformed of the source. It is worth noting that if we introduce the spatial correlation function as

$$D(x,K) = \int S(r + x/2, K)S(r - x/2, K)d^4r. \tag{2}$$

then the correlation function can be expressed as

$$C_2(q,K) = 1 + \frac{\widetilde{D}(q,K)}{\widetilde{D}(0,K)}, \tag{3}$$

where again $\widetilde{D}(q,K) = \int D(x,K)e^{iqx}d^4x$ is the Fourier transformed of the spatial correlation function. Note that in this case, no absolute value is taken, hence no sign or phase information is lost. Hence, momentum correlations measure the source, or more directly, the spatial correlation function.

The geometry of the source was often assumed to be Gaussian. However, the expansion of the medium and the corresponding increase of the mean free path suggests the onset of anomalous diffusion [18,19]. This leads to Lévy distributions, which in turn result in stretched exponential type of momentum correlation functions:

$$C_2(q,K) = 1 + \exp\left(-|qR(K)|^{\alpha(K)}\right), \tag{4}$$

where q is the momentum difference, R is the K dependent source size (or its length of homogeneity), and α is the Lévy-stability exponent, resulting from the anomalous diffusion, and probably also depending on K. The α parameter is interesting also because it may be connected to the critical exponent η, appearing in the power-law exponent of the spatial correlation function at the critical point [20]. Recently, Lévy sources have been utilized [17,21–24] to describe the particle emitting source. In present paper we report on the recent Lévy femtoscopy results of the PHENIX experiment.

Before we move on to the data analysis and the results, let us mention one important point. The above equations imply, that at zero relative momentum, the correlation function reaches a value of 2, i.e. $C_2(0,K) = 2$. Experimentally, the low-q region is usually inaccessible due to two-track resolution limitations. Hence one can only extrapolate to measure $C_2(0,K)$, the zero momentum intercept of the two-particle correlation function. It turns out, that this value is not 2, but has to be described by

$$\lim_{q\to 0} C_2(q,K) = 1 + \lambda(K), \tag{5}$$

where λ is the (K dependent) intercept parameter or correlation function strength. This observation can be easily understood in terms of the core-halo picture [25,26], where the source has two components: a hydrodynamically behaving fireball-type core, surrounded by a halo of the decay products of long lived resonances. In the core-halo picture, it turns out (as detailed e.g. in Ref. [17]) that

$$\lambda = f_c^2, \quad f_c = \frac{N_{\rm core}}{N_{\rm core} + N_{\rm halo}}, \tag{6}$$

where $N_{...}$ denotes the number of pions produced in a given component of the source. Hence λ carries indirect information on the number of long-lived resonances decaying to pions, with an important example being the η' meson, as discussed in detail in Refs. [17,27–29].

IOP Conf. Series: Journal of Physics: Conf. Series **1070** (2018) 012026 doi:10.1088/1742-6596/1070/1/012026

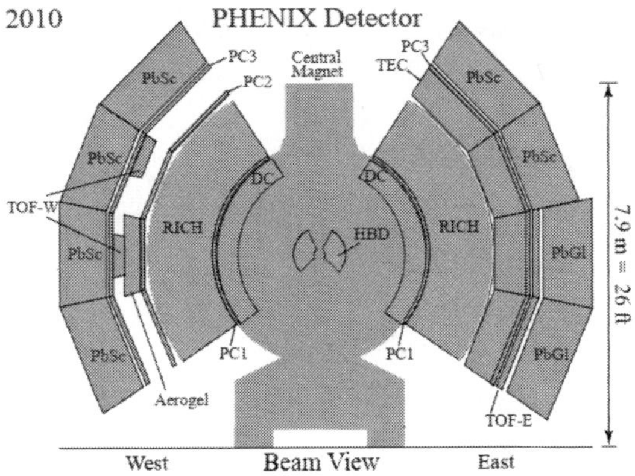

Figure 1. The 2010 PHENIX detector setup.

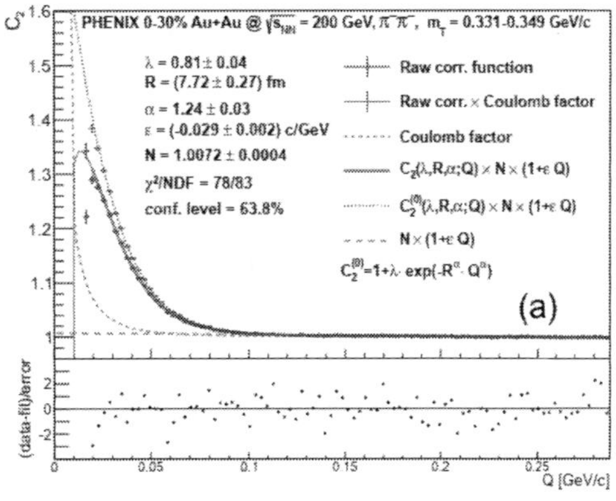

Figure 2. Example correlation function.

2. Two-pion femtoscopy in 200 GeV Au+Au collisions

The PHENIX experiment [30] recorded collisions of various nuclei, ranging from protons and deuterons to gold and uranium, at center of mass per nucleon energies ($\sqrt{s_{NN}}$) ranging from 7.7 to 200 GeV. This allows to study the QCD phase diagram. In this paper, we discuss the analysis of 0–30% $\sqrt{s_{NN}}$ = 200 GeV Au+Au collisions recorded in the 2010 data taking period. The setup of the experiment during this period is shown in Fig. 1. In our correlation measurements we applied particle identification via time of flight data from the PbSc and TOF detectors. In addition to track matching cuts, we applied pair cuts to reduce the effect of track merging and splitting, as detailed in Refs. [17, 21–24].

We measured pair momentum difference distributions in the longitudinally comoving system (LCMS) of each pair, for particles belonging to the same event, as a function of Q, the magnitude of the LCMS momentum difference three-vector. These distributions are usually called actual pair distributions, and are denoted by $A_2(Q, K)$. To investigate only effects which arise from the pair coming from the same event, we created the same number of mixed events (with the same multiplicity distribution). Taking a given actual event, we constructed a similar mixed events by selecting the same number of particles from a large pool of similar background events, making sure all particles are from different events. This resulted in the so-called background pair distribution, $B_2(Q, K)$. The correlation function was then calculated as

$$C_2(Q, K) = \frac{A_2(Q, K)}{B_2(Q, K)}, \tag{7}$$

separately for positive and negative charges. The q-dependence was investigated in 31 bins of pair transverse mass, related to average pair momentum K as

$$m_T = \sqrt{m^2 + K_T^2}, \tag{8}$$

where K_T is the pair average transverse momentum. The resulting 62 correlation functions were fitted assuming Lévy sources, as well as taking into account the Coulomb interaction, as detailed in Ref. [17]. An example fit is shown in Fig. 2. We then investigated the main fit parameters.

The fit parameters λ, R and α are shown in Fig. 3, as a function of pair m_T. We may observe that α is approximately constant (within systematic uncertainties), and takes an average value of 1.2, being far from the Gaussian assumption of $\alpha = 2$. We also observe, that despite being far

IOP Conf. Series: Journal of Physics: Conf. Series **1070** (2018) 012026 doi:10.1088/1742-6596/1070/1/012026

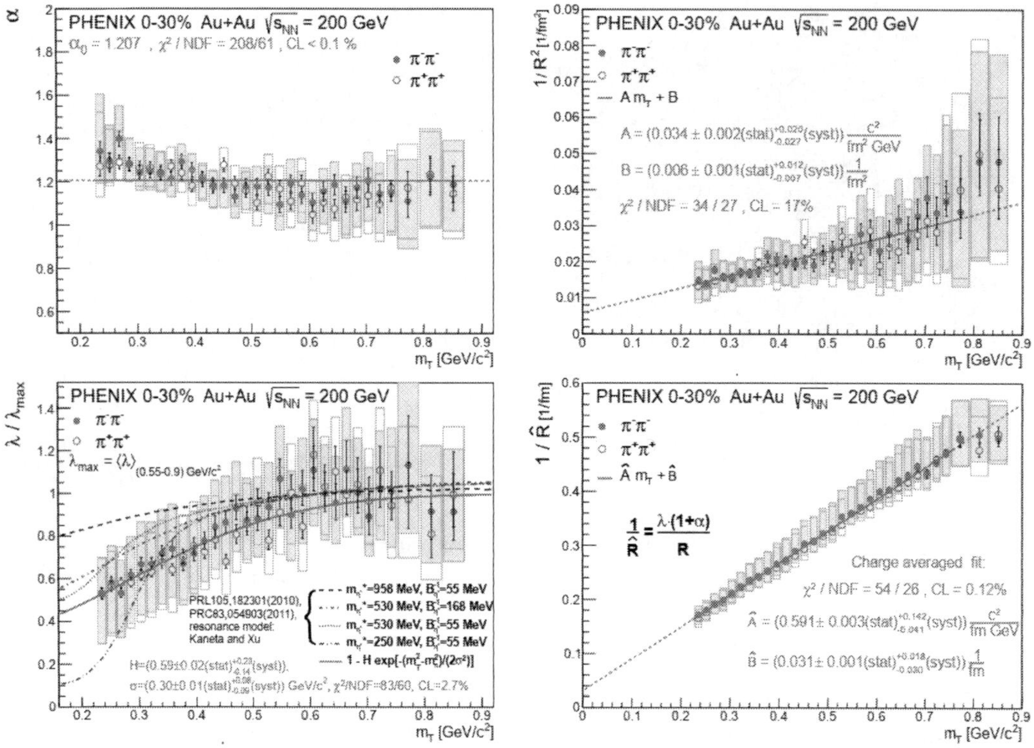

Figure 3. Lévy parameters α, R (shown as $1/R^2$) and λ (shown normalized as $\lambda/\lambda_{\max}$), as well as the new scaling parameter $\widehat{R}$ versus pair m_T show in the top left, top right, bottom left and bottom right panels, respectively. Statistical uncertainties are shown as bars, systematic ones with boxes.

from the hydrodynamic limit of Gaussian distributions, the hydro prediction of $1/R^2 \simeq Am_T+B$ still holds.

The measured correlation function strength λ is shown after a normalization by

$$\lambda_{\max} = \langle\lambda\rangle_{m_T\,=\,0.55-0.9\ \mathrm{GeV}/c^2}. \tag{9}$$

This clearly indicates a decrease at small m_T, which may be explained by the larger fraction of low momentum pions coming from resonance decays. Our measurement is in particular not incompatible with predictions [29] based on a reduced η' mass, see details in Ref. [17].

We also show, that a new, empirically found scaling parameter

$$\widehat{R} = \frac{R}{\lambda(1+\alpha)} \tag{10}$$

may be defined. This exhibits a clear linear scaling with m_T, and it also has much decreased statistical uncertainties.

While for the other parameters, approximate scalings may be observed (α is constant in m_T, $1/R^2$ is linear in m_T, $\lambda/\lambda_{\max}$ follows a unity minus a Gaussian type of m_T dependence), the scaling of $1/\widehat{R}$ is strikingly better than the others.

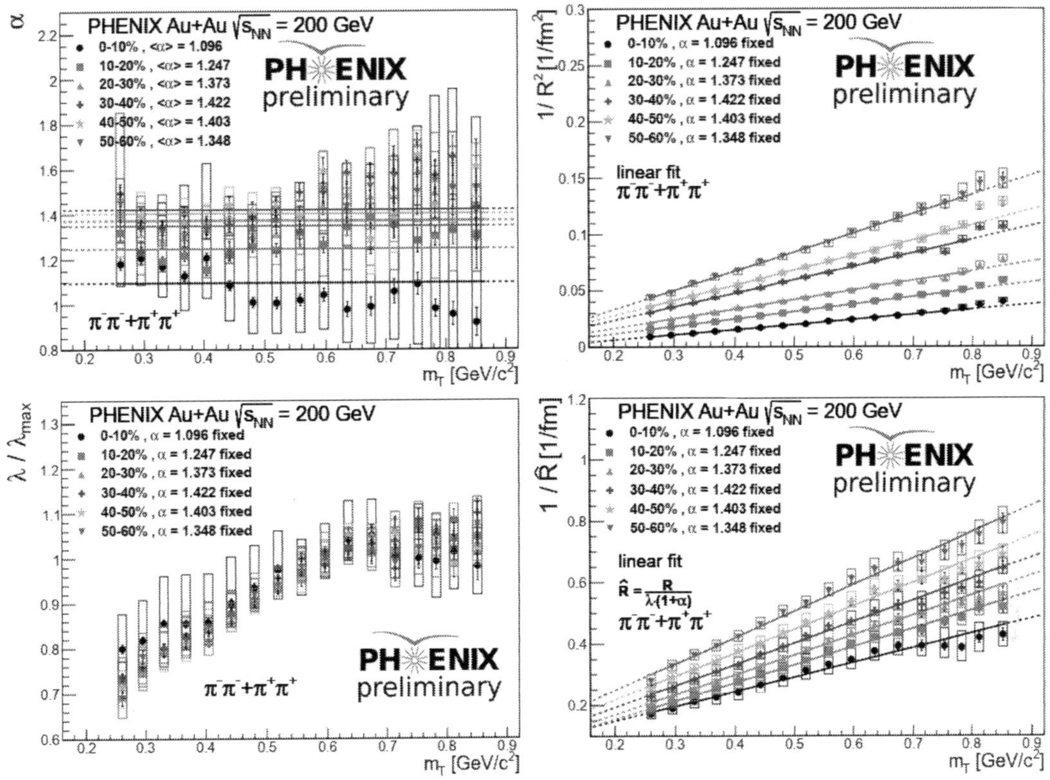

Figure 4. Lévy HBT fit parameters for various centralities in 200 GeV Au+Au collisions.

3. Centrality and collision energy dependence

We investigated the Lévy HBT correlation functions also at different collision energies, from $\sqrt{s_{NN}}$ = 39 to 200 GeV, and for various centralities. At 200 GeV, we defined six centrality classes from 0–10% to 50–60%, at 62 GeV four classes (from 0–10% to 30–40%) and two classes at 39 GeV (0–20% and 20–40%). After performing all the fits to all the measured correlation functions, we extracted the m_T-dependence of the fit parameters. For the 200 GeV analysis, we used 18 m_T bins, 8 bins at 62 GeV and 6 bins at 39 GeV. The extracted fit parameters (as a function of m_T) are shown in Figs. 4-5. At each energy, we observe an approximately m_T independent α value. To investigate the m_T dependence of the parameters, we performed fits with α fixed to its average value in the given centrality bin. This resulted in the clear trends shown in Fig. 4: in each centrality, $1/R^2$ and $1/\widehat{R}$ scale linearly with m_T, and a "hole" of similar magnitude is observed for $\lambda/\lambda_{\rm max}$. At the lower energies (shown in Fig. 5), we observe qualitatively similar trends, but statistics prevents us from drawing more definitive conclusions. In both case, the most clear trend is observable for the scaling variable $\widehat{R}$, which is apparently the less prone to statistical fluctuations of the correlation function and the fit.

4. Three-pion correlations

As mentioned above, the λ intercept parameter of the two-pion momentum correlations may be related to the fraction of pions coming from the decays of long lived resonances. If so, the possible enhancement of η' production may be connected to it. An important cross-check of this

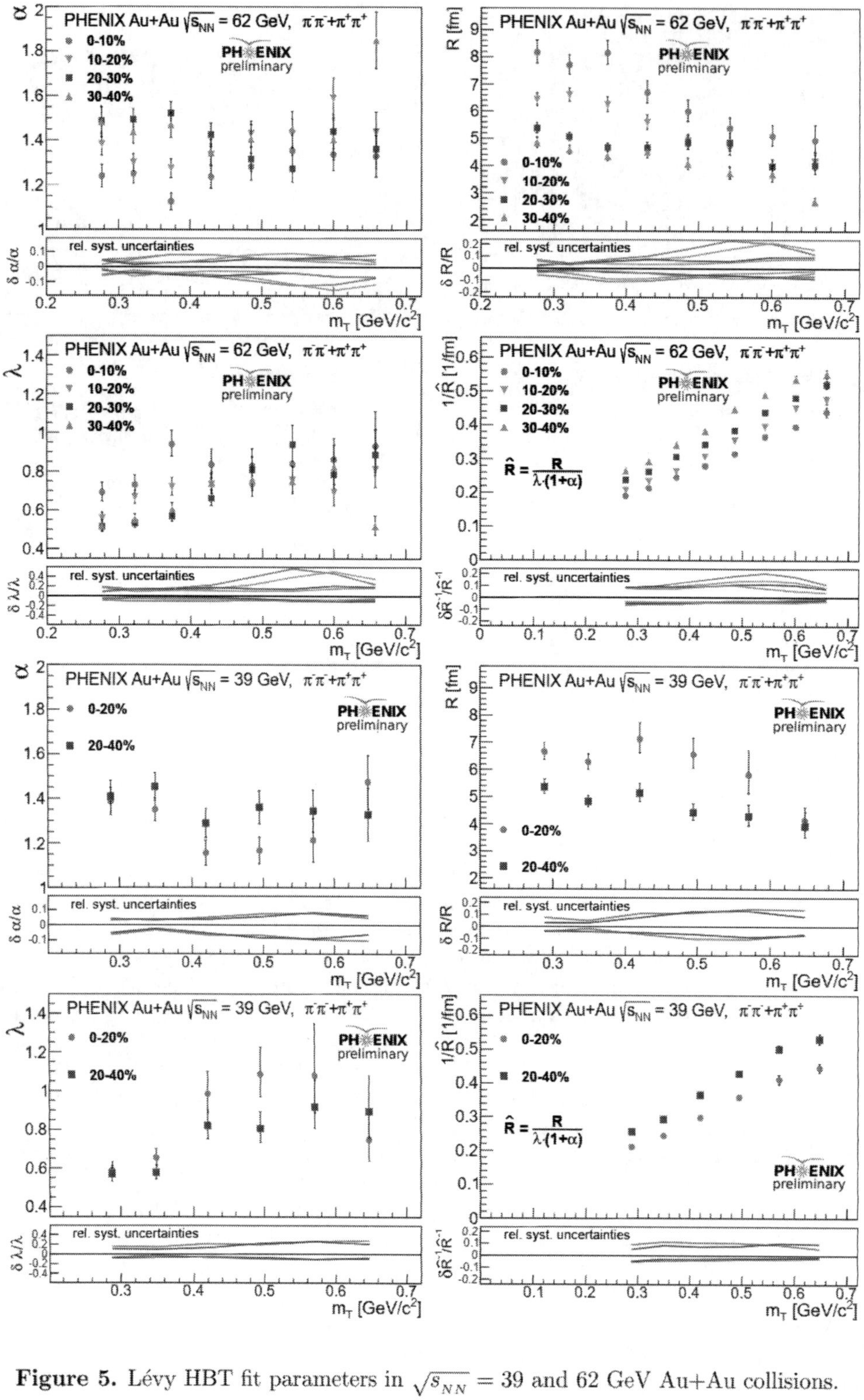

Figure 5. Lévy HBT fit parameters in $\sqrt{s_{NN}} = 39$ and 62 GeV Au+Au collisions.

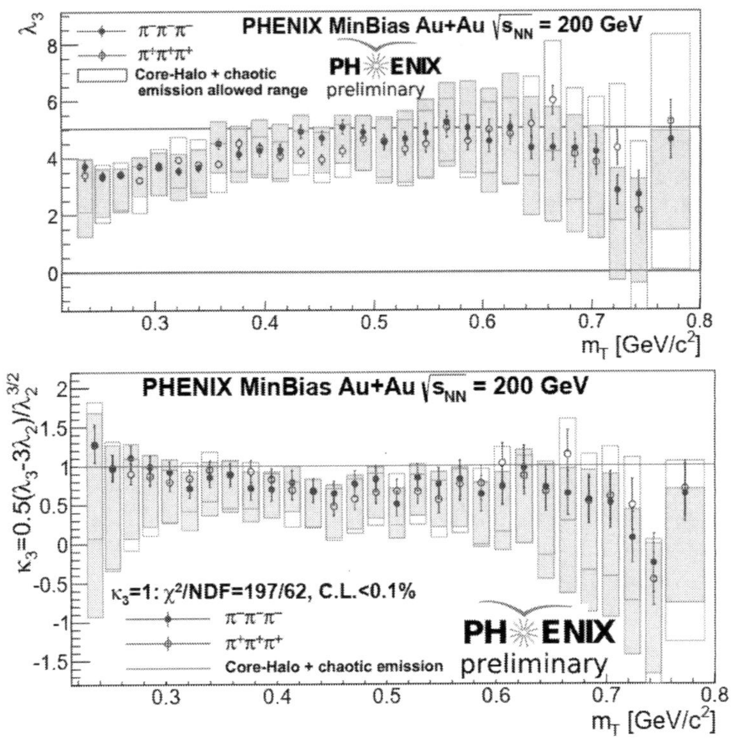

Figure 6. The three-particle correlation strength, λ_3, as function of m_T (top panel). Core-halo independent parameter κ_3 as function of m_T (bottom panel).

can be performed by the measurement of multi-pion correlations, since their zero momentum intercept is also related to the core/halo ratio. It turns out [31] that if we define two- and three-particle correlation intercepts (neglecting K dependence) as

$$\lambda_2 \equiv C_2(Q_{12} \to 0) - 1 \tag{11}$$

$$\lambda_3 \equiv C_3(Q_{12} \to 0, Q_{13} \to 0, Q_{23} \to 0) - 1 \tag{12}$$

then these can be expressed from the core fraction f_c as

$$\lambda_2 = f_c^2, \qquad \lambda_3 = 2f_c^3 + 3f_c^2. \tag{13}$$

It is immediately clear from this, that we may define

$$\kappa_3 = \frac{\lambda_3 - 3\lambda_2}{2\sqrt{\lambda_2}^3} \tag{14}$$

which is equal to one in case of the core-halo model. Hence if $\kappa_3 \neq 1$, that would point to pion production beyond the core-halo model, such as coherence, as also discussed in Ref. [24]. We measured the value of the above defined λ_3 and κ_3 parameters, as shown in Fig. 6. The preliminary uncertainties were calculated from simultaneous variation of two- and three-pion analysis settings. We observe that κ_3 is near unity, but the preliminary status of this analysis prevents us from making a more definitive conclusion.

IOP Conf. Series: Journal of Physics: Conf. Series **1070** (2018) 012026 doi:10.1088/1742-6596/1070/1/012026

5. Summary

In summary, we measured two- and three-particle correlation functions of identical pions at various collision energies and centralities, and fitted them with correlation functions calculated from Lévy sources. We found that this assumption yields a statistically acceptable description of all results. We extracted the m_T dependence of two-particle Lévy parameters α, R and λ. We found that the α parameter is far from 2, and has only a slight m_T dependence, but a more pronounced centrality dependence. We also found that $1/R^2$ is approximately linear with m_T, while λ decreases at low m_T, showing a "hole-like" structure, not incompatible with models utilizing a decreased η' mass. We also demonstrated the appearance of a scaling parameter $\widehat{R}$, the inverse of which shows a very clear linear scaling with m_T. Furthermore, we investigated if the interpretation of λ holds as being related to the core-halo ratio, by measuring three-particle correlations. The κ_3 parameter, which equals to unity in the core-halo model, was measured to be close to this conjectured value.

Acknowledgments

The author expresses gratitude for the excellent organization of the WWND 2018 conference. The author was supported by the Hungarian NKIFH grant No. FK-123842, by the János Bolyai Research Scholarship of the Hungarian Academy of Sciences and the ÚNKP-17-4 New National Excellence Program of the Hungarian Ministry of Human Capacities.

References

[1] Hanbury Brown R, Jennison R C and K D G M 1952 *Nature* **170** 1061–1063
[2] Brown R H and Twiss R 1956 *Nature* **177** 27–29
[3] Twiss R Q, Little A G and Hanbury Brown R 1957 *Nature* **180** pages 324–326
[4] Hanbury Brown R and Twiss R Q 1956 *Nature* **178** 1046–1048
[5] Townes C H 1999 *Astrophysical Journal* **525C** 148
[6] Michelson A A and Pease F G 1921 *Astrophysical Journal* **53** 249–259
[7] Brown R H, Davis J and Allen L R 1967 *Mon. Notices Royal Astron. Soc.* **137** 375–392
[8] Hanbury Brown R, Hazard C, Davis J and Allen L R 1964 *Nature* **201** pages 1111–1112
[9] Brown R H, Davis J, Allen L R and Rome J M 1967 *Mon. Notices Royal Astron. Soc.* **137** 393–417
[10] Dravins D, Jensen H, LeBohec S and Nuñez P D 2010 *Optical and Infrared Interferometry II* (*Proceedings of the SPIE* vol 7734) p 77340A (*Preprint* 1009.5815)
[11] Glauber R J 1963 *Phys. Rev. Lett.* **10** 84–86
[12] Glauber R J 2006 *Rev. Mod. Phys.* **78** 1267–1278
[13] Glauber R J 2006 *Nucl. Phys.* **A774** 3–13 (*Preprint* nucl-th/0604021)
[14] Goldhaber G, Fowler W B, Goldhaber S and Hoang T F 1959 *Phys. Rev. Lett.* **3** 181–183
[15] Goldhaber G, Goldhaber S, Lee W Y and Pais A 1960 *Phys. Rev.* **120** 300–312
[16] Lednicky R 2001 (*Preprint* nucl-th/0112011)
[17] Adare A *et al.* (PHENIX) 2017 (*Preprint* 1709.05649)
[18] Metzler R, Barkai E and Klafter J 1999 *Phys. Rev. Lett.* **82** 3563–3567
[19] Csanád M, Csörgő T and Nagy M 2007 *Braz. J. Phys.* **37** 1002–1013 (*Preprint* hep-ph/0702032)
[20] Csörgő T, Hegyi S and Zajc W A 2004 *Eur. Phys. J.* **C36** 67–78 (*Preprint* nucl-th/0310042)
[21] Csanád M (PHENIX) 2017 *Universe* **3** 85 (*Preprint* 1711.05575)
[22] Kincses D (PHENIX) 2018 *Universe* **4** 11 (*Preprint* 1711.06891)
[23] Lökös S (PHENIX) 2018 *Universe* **4** 31 (*Preprint* 1801.08827)
[24] Novák T (PHENIX) 2018 *Universe* **4** 57 (*Preprint* 1801.03544)
[25] Bolz J, Ornik U, Plumer M, Schlei B and Weiner R 1993 *Phys.Rev.* **D47** 3860–3870
[26] Csörgő T, Lörstad B and Zimányi J 1996 *Z. Phys.* **C71** 491–497 (*Preprint* hep-ph/9411307)
[27] Kapusta J I, Kharzeev D and McLerran L D 1996 *Phys. Rev.* **D53** 5028–5033 (*Preprint* hep-ph/9507343)
[28] Vance S E, Csörgő T and Kharzeev D 1998 *Phys. Rev. Lett.* **81** 2205–2208 (*Preprint* nucl-th/9802074)
[29] Csörgő T, Vértesi R and Sziklai J 2010 *Phys.Rev.Lett.* **105** 182301 (*Preprint* 0912.5526)
[30] Adcox K *et al.* (PHENIX Collaboration) 2005 *Nucl. Phys.* **A757** 184–283 (*Preprint* nucl-ex/0410003)
[31] Csörgő T 2002 *Heavy Ion Phys.* **15** 1–80 (*Preprint* hep-ph/0001233)

IOP Conf. Series: Journal of Physics: Conf. Series **1070** (2018) 012027 doi:10.1088/1742-6596/1070/1/012027

Factorization of two-particle distributions in AMPT simulations of Pb–Pb collisions at $\sqrt{s_{\mathrm{NN}}} = 5.02$ TeV

Christian Bourjau (for the ALICE Collaboration)
Niels Bohr Institute, University of Copenhagen
Blegdamsvej 17, 2100 Copenhagen, Denmark

Abstract. The flow ansatz states that the single-particle distribution of a given event can be described in terms of the complex flow coefficients V_n. Multi-particle distributions can therefore be expressed as products of these single-particle coefficients; a property commonly referred to as factorization. The amplitudes and phases of the coefficients fluctuate from event to event, possibly breaking the factorization assumption for event-sample averaged multi-particle distributions. Furthermore, non-flow effects such as di-jets may also break the factorization assumption. The factorization breaking with respect to pseudorapidity η provides insights into the fluctuations of the initial conditions of heavy ion collisions and can simultaneously be used to identify regions of the phase space which exhibit non-flow effects. These proceedings present a method to perform a factorization of the two-particle Fourier coefficients $V_{n\Delta}(\eta_a, \eta_b)$ which is largely independent of detector effects. AMPT model calculations of Pb–Pb collisions at $\sqrt{s_{\mathrm{NN}}} = 5.02$ TeV are used to identify the smallest $|\Delta\eta|$-gap necessary for the factorization assumption to hold. Furthermore, a possible $\Delta\eta$-dependent decorrelation effect in the simulated data is quantified using the empirical parameter F_2^η. The decorrelation effect observed in the AMPT calculations is compared to results by the CMS collaboration for Pb–Pb collisions at $\sqrt{s_{\mathrm{NN}}} = 2.76$ TeV.

1. Introduction

The Fourier coefficients of an event-sample averaged two-particle distribution are commonly described as

$$V_{n\Delta}(\eta_a, \eta_b) = \langle V_n(\eta_a) V_n^*(\eta_b) \rangle \,, \tag{1}$$

$$= \left\langle v_n(\eta_a) v_n(\eta_b) e^{in(\psi_n(\eta_a) - \psi_n(\eta_b))} \right\rangle , \tag{2}$$

where v_n are the flow coefficients and ψ_n are the symmetry planes. Either of these two quantities may fluctuate from event to even due to varying initial conditions, thereby breaking the factorization of the sample average even for simulations of ideal hydrodynamics [1]. By studying the factorization behavior of $V_{n\Delta}(\eta_a, \eta_b)$ one can therefore infer the properties of such fluctuations. Flow related analyses commonly assume that non-flow contributions decrease with an increasing η-separation of the particles. In order to minimize the impact of non-flow effects on the measurement a minimal longitudinal separation between particles, referred to as $|\Delta\eta|$-gap, is therefore often applied. Under the assumption that non-flow effects do not factorize identically to anisotropic flow it is possible to identify regions of the phase space where non-flow effects become negligible [2, 3].

Whether Eq. (2) may be written in a factorized form depends on the correlations between the four quantities $v_n(\eta_a)$, $v_n(\eta_b)$, $\psi_n(\eta_a)$, and $\psi_n(\eta_b)$. These proceedings focus on the effect of symmetry plane decorrelation effects. The phases at η_a and η_b are commonly assumed to be correlated with each other through a common symmetry plane angle Ψ_n, but may fluctuate from event to event. The fluctuations are equally likely to occur in either direction which ensures that $V_{n\Delta}(\eta_a, \eta_b)$ is a real quantity. The observed average is attenuated due to these fluctuations which are therefore also referred to as *decorrelation* effects. If the fluctuations at $\psi_n(\eta_a)$ and $\psi_n(\eta_b)$ in Eq. (2) exhibit a dependence on $\Delta\eta = \eta_a - \eta_b$ it may cause a factorization breaking of $V_{n\Delta}(\eta_a, \eta_b)$.

2. Observable definition

The results presented here are exclusively based on the Monte-Carlo (MC) simulations and would therefore not require considering detector effects on the observables. However, in order to present a generally applicable method, the analysis presented here is constructed around an observable which is largely independent of uncorrelated detector deficiencies.

At its core, this analysis is based on the single- and two-particle distributions averaged over many events. The single-particle distribution $\hat{\rho}_1$ is given by

$$\hat{\rho}_1(\eta, \varphi) = \left\langle \frac{d^2N}{d\eta d\varphi} \right\rangle \tag{3}$$

where N is the number of observed charged particles and φ is the azimuthal coordinate. In experimental measurements any azimuthal anisotropies in $\hat{\rho}_1$ are exclusively caused by detector effects.

The distribution of particle-pairs $\hat{\rho}_2$ is given by

$$\hat{\rho}_2(\eta_a, \eta_b, \varphi_a, \varphi_b) = \left\langle \frac{d^4N_{\text{pairs}}}{d\eta_a d\eta_d \varphi_a d\varphi_b} \right\rangle \tag{4}$$

where N_{pairs} denotes the number of particle pairs observed at $\eta_a, \eta_b, \varphi_a, \varphi_b$.

The two distributions $\hat{\rho}_1$ and $\hat{\rho}_2$ can be used to construct the *normalized two-particle density* r_2 which is largely independent of uncorrelated detector effects [4, 5].

$$r_2(\eta_a, \eta_b, \varphi_a, \varphi_b) = \frac{\hat{\rho}_2(\eta_a, \eta_b, \varphi_a, \varphi_b)}{\hat{\rho}_1(\eta_a, \varphi_a)\hat{\rho}_1(\eta_b, \varphi_b)} \tag{5}$$

The non-zero two-particle Fourier coefficients can then be computed by

$$V_{n\Delta}(\eta_a, \eta_b) = \left(\frac{1}{2\pi}\right)^2 \int_0^{2\pi} \int_0^{2\pi} r_2(\eta_a, \eta_b, \varphi_a, \varphi_b) e^{-in\varphi_a} e^{in\varphi_b} d\varphi_a d\varphi_b \tag{6}$$

3. Factorization

The functional form of $V_{n\Delta}(\eta_a, \eta_b)$ in the (η_a, η_b)-plane is assessed with two different models. Both models focused on the flow ansatz i.e., individual events can be described in terms of single particle distributions. Neither model attempts to describe any *non-flow* processes such as di-jets or weak decays.

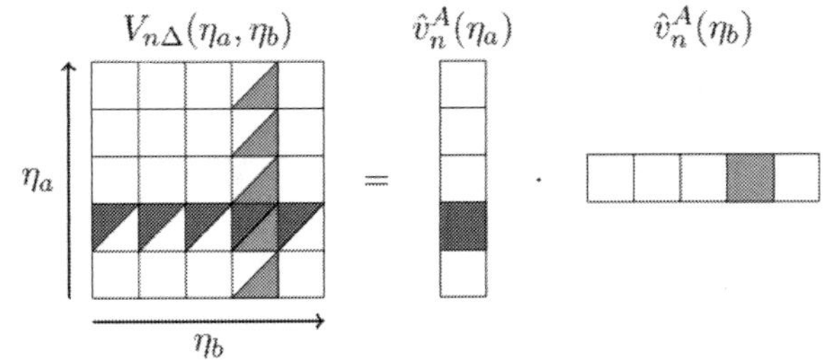

Figure 1: Schematic representation of Model A. Each element of $\hat{v}_n(\eta)$ affects several elements of $V_{n\Delta}(\eta_a, \eta_b)$.

3.1. *Purely factorizing model (Model A)*

This model assumes that the averaged two-particle coefficients may be described by a product of two identical functions $\hat{v}_n^A$

$$V_{n\Delta}(\eta_a, \eta_b) = \langle V_n(\eta_a) V_n^*(\eta_b) \rangle \tag{7}$$

$$= \left\langle v_n(\eta_a) v_n(\eta_b) e^{in(\psi_n(\eta_a) - \psi_n(\eta_b))} \right\rangle \tag{8}$$

$$= \hat{v}_n^A(\eta_a) \hat{v}_n^A(\eta_b) \tag{9}$$

If $\psi_n(\eta_a)$ is always equal to $\psi_n(\eta_b)$ within each event and if the fluctuations of v_n are uncorrelated along η, Eq. (9) holds and $\hat{v}_n^A(\eta)$ is the mean value of the event-by-event flow coefficients $v_n(\eta)$. The degree to which the measured $V_{n\Delta}(\eta_a, \eta_b)$ is compatible with Model A provides a limit to the size of factorization-breaking fluctuations of the flow coefficients, event-plane decorrelations and non-flow effects.

The flow coefficients $\hat{v}_n^A(\eta)$ are fit to the observed $V_{n\Delta}(\eta_a, \eta_b)$. The latter is computed as a histogram of finite bin size in η_a and η_b. Eq. (9) can thus be seen as a non-linear equation system

$$V_{n\Delta}(\eta_a^i, \eta_b^j) = \hat{v}_n^A(\eta_a^i) \hat{v}_n^A(\eta_b^j) \tag{10}$$

where i and j are the bin-indices along η_a and η_b respectively. Graphically, Eq. (10) can be represented as shown in Fig. 1 where $V_{n\Delta}(\eta_a, \eta_b)$ is a two-dimensional matrix and $v_n^A(\eta)$ a one-dimensional vector. Solving Eq. (10) for all points in $V_{n\Delta}(\eta_a, \eta_b)$ yields the "vector" $v_n^A(\eta)$ which best describes the observed data. Each value in $\hat{v}_n^A$ affects several points in the (η_a, η_b)-plane.

3.2. *Long-range decorrelating model (Model B)*

The second model presented here was suggested by the CMS collaboration, albeit based on a vastly different analysis method [6]. The model is given by

$$V_{n\Delta}(\eta_a, \eta_b) = \hat{v}_n^B(\eta_a) \hat{v}_n^B(\eta_b) e^{-F_n^\eta |\eta_a - \eta_b|} \tag{11}$$

where the parameter F_n^η is a measure for a $\Delta\eta = \eta_a - \eta_b$ dependent factorization breaking. Despite being empirical F_n^η provides insights into longitudinal fluctuations during the early stages of the collision [7, 8]. It should be noted that $\hat{v}_n^A(\eta) \neq \hat{v}_n^B(\eta)$ unless $F_n^\eta = 0$. Analogous to the previous model, the flow coefficients $\hat{v}_n^B(\eta)$ and the constant F_n^η are found by solving

$$V_{n\Delta}(\eta_a^i, \eta_b^j) = \hat{v}_n^B(\eta_a^i) \hat{v}_n^B(\eta_b^j) e^{-F_n^\eta |\eta_a^i - \eta_b^i|} \tag{12}$$

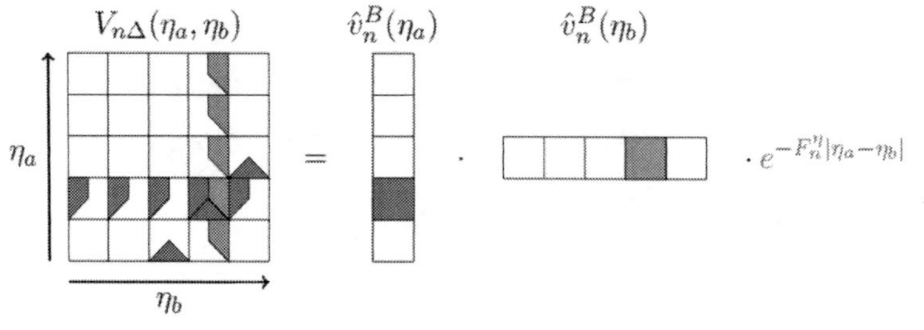

Figure 2: Schematic representation of Model B. Each point in $v_n(\eta)$ affects several elements in $V_{n\Delta}(\eta_a, \eta_b)$. The factor $e^{-F_n^\eta|\eta_a-\eta_b|}$ attenuates $V_{n\Delta}(\eta_a, \eta_b)$ along $|\Delta\eta|$.

The graphical representation of Eq. (12) is depicted in Fig. 2. The exponential factor causes an attenuation of $V_{n\Delta}(\eta_a, \eta_b)$ along $|\Delta\eta|$ and is constant along $\eta_a + \eta_b$.

Particle pairs with a small separation in $\Delta\eta$ are commonly excluded from flow analyses as this region of the phases space is known to exhibit large non-flow contributions. Furthermore, an experimentally measured $V_{n\Delta}(\eta_a, \eta_b)$ may exhibit acceptance gaps if various detector systems are combined in order to maximize the η coverage. Therefore, the procedure to numerically solve the equation systems in Eq. (10) and Eq. (12) needs to be able to be performed on arbitrary subsets of the (η_a, η_b)-plane. A minimization of a weighted sum of squares fulfills this requirement. The weighted sum S is given by

$$S = \sum_{i,j=1}^{N_a^{\mathrm{bin}}, N_b^{\mathrm{bin}}} \left(\frac{V_{n\Delta}(\eta_a^i, \eta_b^j) - M(\eta_a^i, \eta_b^j)}{\sigma_{n\Delta}(\eta_a^i, \eta_b^j)} \right)^2 \tag{13}$$

where N_a^{bin} (N_b^{bin}) represents the number of bins in the η_a (η_b) and M represent either Model A or Model B as defined in Eq. (10) or Eq. (12) respectively. The uncertainty associated with each point of $V_{n\Delta}(\eta_a^i, \eta_b^j)$ is given by $\sigma_{n\Delta}(\eta_a^i, \eta_b^j)$.

3.3. Factorization ratio

The agreement of $V_{n\Delta}(\eta_a, \eta_b)$ with either of the two models is assessed by means of a factorization ratio $f_n(\eta_a, \eta_b)$ defined by

$$f_n(\eta_a, \eta_b) = \frac{V_{n\Delta}(\eta_a, \eta_b)}{M(\eta_a, \eta_b)} \tag{14}$$

where $M(\eta_a, \eta_b)$ represents either model with the parameters fitted to the observed $V_{n\Delta}(\eta_a, \eta_b)$. Note that $f_n(\eta_a, \eta_b)$ can be computed for the entire (η_a, η_b)-plane even if S was only minimized for a subset of it.

4. Results

Figure 3 presents $V_{2\Delta}(\eta_a, \eta_b)$ obtained for collisions of 20–40% centrality for AMPT calculations of Pb–Pb collisions at $\sqrt{s_{\mathrm{NN}}} = 5.02$ TeV with string melting enabled. Every two-particle Fourier coefficient in the (η_a, η_b)-plane is computed independently with no a priori assumption about the event-by-event fluctuations.

Figure 4 (left) presents $\hat{v}_n^A(\eta)$ obtained from a fit to $V_{2\Delta}(\eta_a, \eta_b)$ without the requirement of a $|\Delta\eta|$-gap. Therefore, the factorization procedure included the short-range $\Delta\eta$ region of the

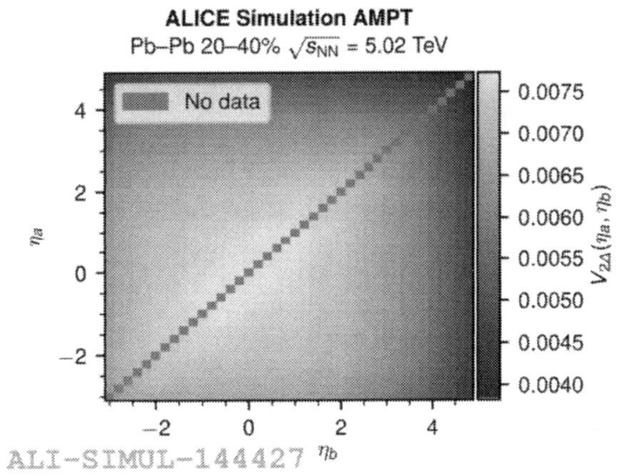

Figure 3: The two-particle Fourier coefficients $V_{2\Delta}(\eta_a, \eta_b)$ from AMPT calculations of Pb–Pb collisions at 5.02 TeV with a centrality of 20–40%. The measurement of the two-particle Fourier coefficients is conducted for every point in the (η_a, η_b)-plane independently.

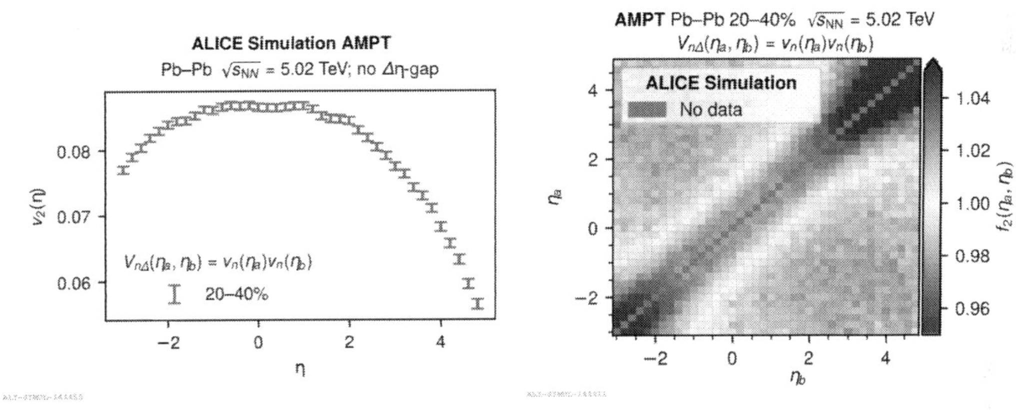

Figure 4: Factorization of $V_{2\Delta}(\eta_a, \eta_b)$ according to Model A without the use of a $|\Delta\eta|$-gap based on AMPT simulations. Left: The flow coefficients yielded by the minimization procedure. Right: Factorization ratio for the solution shown on the left. The found solution fails to describe the short-range and long-range pairs.

(η_a, η_b)-plane which exhibits non-negligible non-flow contributions. Figure 4 (right) shows the factorization ratio for the flow coefficients from the left panel. The short-range non-flow does not exhibit the same factorization behavior as the long-range regions. This caused the fitted solution to neither accurately describe the short-range nor the long range regions of the (η_a, η_b)-plane highlighting the need of a $|\Delta\eta|$-gap.

Figure 5 (left) presents the factorization ratio for Model A if only particle pairs with $|\Delta\eta| > 3$ are taken into account. By excluding the short-range pairs, good agreement to Model A is observed in the long-range region. However, the solution found for long-range pairs is not able to describe the short-range region of $V_{2\Delta}(\eta_a, \eta_b)$ further corroborating that non-flow effects in this region of the phase space do not factorize identically to the long-range anisotropic flow. In order to determine the minimal $|\Delta\eta|$-gap necessary to exclude pairs originating from non-flow processes from the fitting procedure, a projection of the factorization ratios for Model A onto

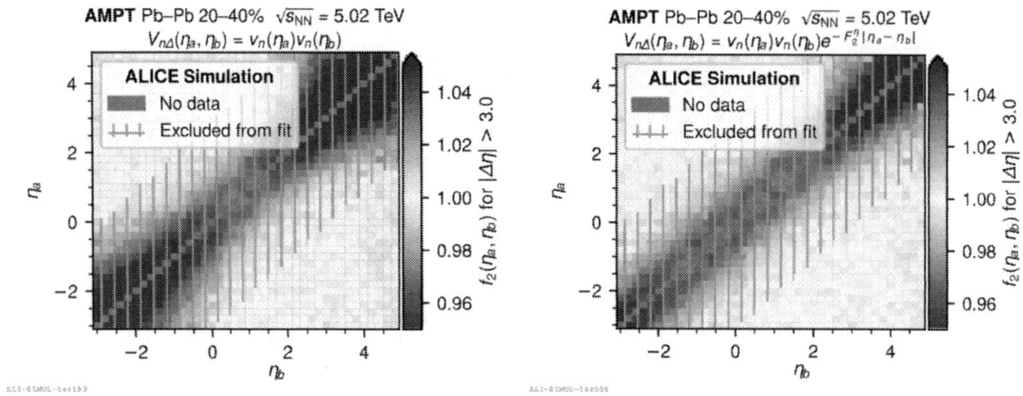

Figure 5: Factorization ratio for Model A (left) and Model B (right) with $|\Delta\eta| > 3$ in both cases. Either model provides good agreement for $|\Delta\eta| > 3$, while the short-range region is better described by Model B.

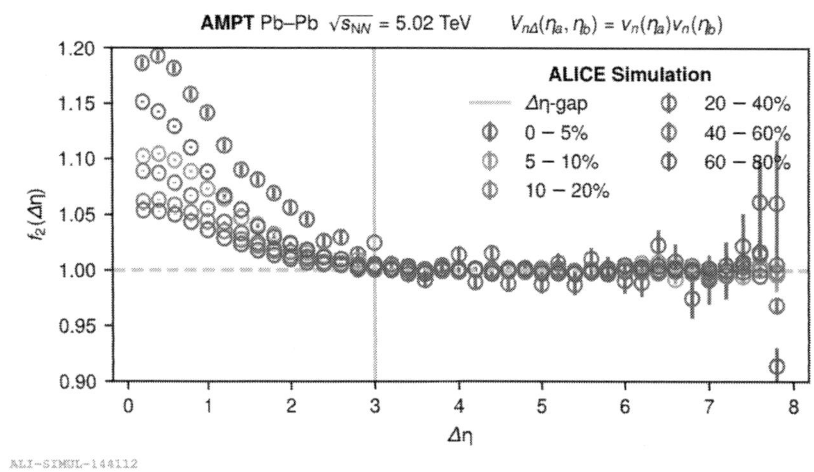

Figure 6: Projection of the factorization ratios of various centralities onto $\Delta\eta$. A $|\Delta\eta|$-gap of 3 was imposed during the factorization procedure with Model A. The shown uncertainties are statistical.

the $\Delta\eta$-axis is performed. The results for all analyzed centralities and for a $|\Delta\eta|$-gap of 3 are depicted in Fig. 6. A centrality dependence for the factorization breaking in the short-range region is observed. The deviation from unity is most pronounced for the most central events, decreases to a minimum for the 20–40% centrality class and increases for more peripheral events thereafter. This centrality dependence may originate from the centrality dependence of $\hat{v}_2^A(\eta)$.

Events of all centralities exhibit good agreement to Model A in the long-range region for $\Delta\eta > 3$. A further increase of the $|\Delta\eta|$-gap does not significantly alter the extracted flow coefficients. A decrease of the $|\Delta\eta|$-gap, includes short-range regions of the phase-space which are incompatible to the solution found in the long range region. Factorization procedures with a smaller $|\Delta\eta|$-gap therefore decrease the fit quality in the long-range region. Analyses which implicitly rely on the factorization assumption to hold should thus apply a minimal longitudinal

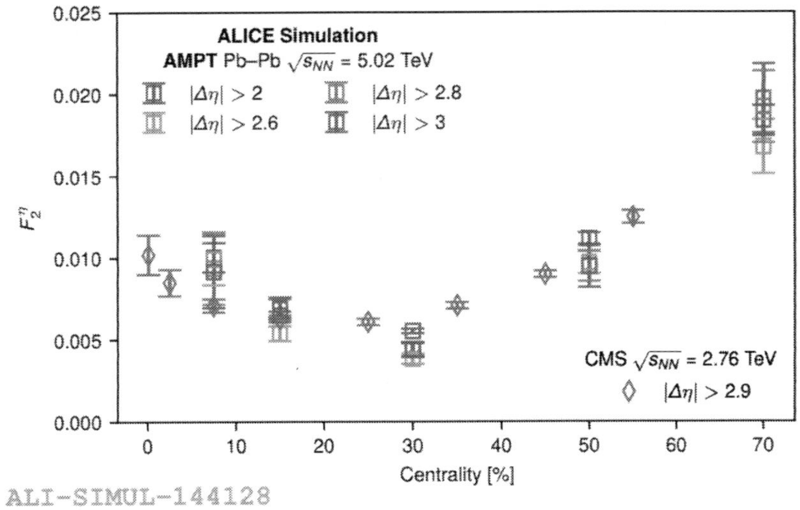

Figure 7: Empirical decorrelation parameter F_2^η for AMPT model calculations of Pb–Pb collisions at $\sqrt{s_{\mathrm{NN}}} = 5.02\,\mathrm{TeV}$. The simulation is compared to measurements by the CMS collaboration for $\sqrt{s_{\mathrm{NN}}} = 2.76\,\mathrm{TeV}$.

separation of $\Delta\eta_{\mathrm{min}} \approx 3$ for the kinematic region here studied. The method presented here can be used to improve the precision of similar previously published results which found that the factorization assumption holds to within 10% for Pb–Pb collisions at $\sqrt{s_{\mathrm{NN}}} = 2.76\,\mathrm{TeV}$ for $\Delta\eta > 2$ [9].

Performing the factorization using Model B allows for the measurement of the decorrelation parameter F_2^η. The decorrelation parameters computed with the method presented here is shown in Fig. 7 for various $|\Delta\eta|$-gaps and centralities. The data point for the 0–5% centrality bin is removed from the figure due to a poor statistical uncertainties. Results published by the CMS collaboration for Pb–Pb collisions at $\sqrt{s_{\mathrm{NN}}} = 2.76\,\mathrm{TeV}$ are included for comparison. The analysis used by the CMS collaboration corresponds to a $|\Delta\eta|$-gap of approximately 2.9 in this analysis. The centrality dependence observed for Pb–Pb collisions is reproduced by the AMPT simulations at 5.02 TeV supporting previous model comparisons [8] and studies of the energy dependence of F_2^η [10]. The centrality dependence of the decorrelation parameter is also found to reflect the centrality dependence of the short-range factorization breaking in Fig. 5. However, a quantification of possible non-flow contributions to the centrality dependence of F_2^η requires further research.

5. Summary

Depending on the nature of the event-by-event fluctuations, the two-particle Fourier coefficients $V_{n\Delta}(\eta_a, \eta_b)$ may assume different functional shapes. In these proceedings two distinct models were studied: The first model is a purely factorizing one as it is implicitly assumed in most flow analyses. The second model is an extension to the former and allows for a $\Delta\eta$ dependent attenuation of $V_{n\Delta}(\eta_a, \eta_b)$. AMPT calculations of Pb–Pb collisions at $\sqrt{s_{\mathrm{NN}}} = 5.02\,\mathrm{TeV}$ were used as the basis for the presented results. The first model was use to estimate the longitudinal extent of short-range non-flow correlations under the assumption that such effects are not well described by the factorized solution found from long-range particle pairs. For the studied kinematic region a minimal $\Delta\eta$-separation of $\Delta\eta_{\mathrm{min}} \approx 3$ is required for the factorization assumption to hold in the long-range region.

The second model was used to determine $\Delta\eta$-dependent decorrelation effects as they are to be expected from event-plane decorrelation effects. The empirical decorrelation parameter F_2^η is qualitatively compatible to measurements by the CMS collaboration at $\sqrt{s_{\rm NN}} = 2.76\,\text{TeV}$. This confirms previous studies suggesting that AMPT is able to reproduce the observed decorrelation effects as well as that these effects exhibit only a weak dependence on the center of mass energy [8, 10].

The method presented here offers a new way to investigate possible non-flow contributions to the observed decorrelation effects and will help to better understand the three dimensional initial conditions of heavy ion collisions in the future.

References

[1] Gardim F G, Grassi F, Luzum M and Ollitrault J Y 2013 *Physical Review C* **87** ISSN 1089-490X (*Preprint* 1211.0989) URL http://dx.doi.org/10.1103/PhysRevC.87.031901

[2] Kikoła D, Yi L, Esumi S, Wang F and Xie W 2012 *Physical Review C* **86** ISSN 1089-490X URL http://dx.doi.org/10.1103/PhysRevC.86.014901

[3] Alver B, Back B B, Baker M D, Ballintijn M, Barton D S, Betts R R, Bickley A A, Bindel R, Busza W, Carroll A and et al 2010 *Physical Review C* **81** ISSN 1089-490X URL http://dx.doi.org/10.1103/PhysRevC.81.034915

[4] Vechernin V 2015 *Nuclear Physics A* **939** 21 – 45 ISSN 0375-9474 URL http://www.sciencedirect.com/science/article/pii/S0375947415000731

[5] Ravan S, Pujahari P, Prasad S and Pruneau C A 2014 *Phys. Rev.* **C89** 024906 (*Preprint* 1311.3915)

[6] CMS Collaboration 2015 *Physical Review C* **92** ISSN 1089-490X (*Preprint* 1503.01692) URL http://dx.doi.org/10.1103/PhysRevC.92.034911

[7] Bożek P, Broniowski W and Moreira J 2011 *Physical Review C* **83** ISSN 1089-490X URL http://dx.doi.org/10.1103/PhysRevC.83.034911

[8] Pang L G, Petersen H, Qin G Y, Roy V and Wang X N 2016 *The European Physical Journal A* **52** ISSN 1434-601X URL http://dx.doi.org/10.1140/epja/i2016-16097-x

[9] Aad G, Abbott B, Abdallah J, Abdel Khalek S, Abdelalim A A, Abdesselam A, Abdinov O, Abi B, Abolins M, AbouZeid O S and et al 2012 *Physical Review C* **86** ISSN 1089-490X URL http://dx.doi.org/10.1103/PhysRevC.86.014907

[10] ATLAS Collaboration 2018 *The European Physical Journal C* **78** ISSN 1434-6052 (*Preprint* 1709.02301) URL http://dx.doi.org/10.1140/epjc/s10052-018-5605-7